MAKING ITS MARK

Proceedings of the Seventh ECMWF Workshop on the Use of Parallel Processors in Meteorology

MAKING ITS MARK

Reading, UK   November 2–6, 1996

Editors

Geerd-R Hoffmann
Norbert Kreitz

(European Centre for Medium-Range Weather Forecasts, UK)

World Scientific
Singapore • New Jersey • London • Hong Kong

Published by

World Scientific Publishing Co. Pte. Ltd.

P O Box 128, Farrer Road, Singapore 912805

USA office: Suite 1B, 1060 Main Street, River Edge, NJ 07661

UK office: 57 Shelton Street, Covent Garden, London WC2H 9HE

British Library Cataloguing-in-Publication Data
A catalogue record for this book is available from the British Library.

MAKING ITS MARK
Proceedings of the 7th ECMWF Workshop on the Use of Parallel Processors in Meteorology

ISBN 981-02-3350-7

Printed in Singapore.

PREFACE

Continuing the biennial workshop series on the "Use of Parallel Processors in Meteorology", the seventh was held at the European Centre for Medium-Range Weather Forecasts (ECMWF) on 2-6 December 1996. The title of the workshop "Making Its Mark" reflected the perception of its organizers that parallel processors have progressed sufficiently to support the production of operational weather forecasts. The selection of a distributed memory, vector parallel system by ECMWF in 1995 was an indication of the perceived maturity of parallel systems. The acquisition of highly parallel systems by a number of national meteorological services, for example DWD, FMI, UKMO, further supports this view.

The maximum performance of the parallel systems currently being used for operational weather prediction is of the order of 50 GFlops/s sustained. However, there are contracts already in place which will increase the sustained performance by up to a factor of 10 before the end of the millennium. To reach such performance levels, these systems will have to exceed a TFlop/s peak rate.

Having, to a large degree, solved the problems created by using distributed memory in these highly parallel systems for meteorology, the next challenge is to surpass a TFlop/s sustained performance. To reach this goal, major effort is still required to cope with load imbalances, I/O bandwidth requirements and, last but not least, with the creation of an operational environment for such systems, in particular to provide network connectivity and data archival support. It remains to be seen to what degree the meteorological community can benefit from the ASCI programme in the United States.

The programme of an eventual eighth workshop will therefore have to reflect the different emphasis of the problems ahead. It will have to focus more on systems aspects and less on programming methods.

Geerd-R. Hoffmann

CONTENTS

MPP: WHAT WENT WRONG? CAN IT BE MADE RIGHT?

PAUL N. SWARZTRAUBER
Scientific Computing Division
National Center for Atmospheric Research
Boulder, Colorado 80307, USA

ABSTRACT

In early 1988, NCAR benchmarked the Connection Machine 2 (CM2). A fully configured machine had 2048 processors and a peak performance of 32 Gflops. A finite difference version of a shallow-water model ran at 1.7 Gflops (5% of peak) and a spectral version ran at 1 Gflop (3% of peak). Later, a joint project to implement a working massively parallel processor (MPP) climate model was established with NCAR, Oak Ridge, Argonne, and Los Alamos. After considerable effort the model ran at about 7% of peak on the CM5 in 1995. At best, this performance was comparable to traditional architectures that achieved a much higher percentage of peak and required less code development effort. MPP had not lived up to expectations. The origins of the problem were quite evident. On the CM2, 185 cycles were required for a memory to memory transfer between adjacent processors compared with a processor performance of two floating point operations per cycle. The high cost of this interprocessor communication resulted in a significant degradation in sustained performance. Similar results have been reported by NCAR and other institutions for a number of machines that have come to the market. In this paper we present a combination of algorithms and hardware features that reduces processor to processor communication to a single cycle (on average) not only to an adjacent processor but to the destination processor and in parallel. If implemented, communication could be reduced by as much as two orders of magnitude and perhaps the original promise of MPP could be realized.

1. Introduction.

The two most often expressed concerns about massively parallel processors (MPP) are: (a) the high cost of interprocessor and related communication and (b) the extensive algorithmic and programming requirements. However, if the communication problem could be solved, the resulting performance increase would likely be worth the developmental effort. That is, if problem (a) was solved then problem (b) could also be solved. When vector processors were introduced in the early 70's they were considered difficult to program but the effort was considered worth the increased performance/cost. So the two problems of communication and development effort depend on a single problem; namely, the inter-processor and related MPP communication.

Although MPP communication has always been substantial, the hope was that parallel algorithms, data structures, and compilers would somehow reduce communication to acceptable levels. Although communication was reduced by these efforts, it nevertheless remained at levels that were considered excessive. For most large-scale scientific applications, current performance is a few percent of peak performance. This does not meet the expectations established by traditional shared-memory multiprocessors, which attain a much higher percentage of peak with less code conversion effort. This would not necessarily be a concern if the performance/cost ratio of MPP was superior; however, for many applications, the ratio for MPP is less than or equal to traditional shared-memory multiprocessors, particularly if the cost of code development is included.

In early 1988, Dick Sato and the author began experiments with a primitive weather/climate model on the Connection Machine 2 (CM2) built by Thinking Machines Company (TMC). Both a finite difference and spectral model of the shallow-water equations were implemented. The following results[1] were extrapolated to a fully configured CM2 with 2048 processors and a peak speed of 32 Gflops.

"A status report has been provided on a continuing project; namely, to determine the suitability of massively parallel processing for the computational needs of the atmospheric science community. As of this report, two data points have been obtained. The first is 1.7 Gflops for an elementary weather model in Cartesian coordinates using the finite difference method and the second is about 1 Gflop for the same model using the spectral method."

The spectral model used fast Fourier transform (FFT) software that was developed at TMC and optimized specifically for the CM2. Nevertheless the model ran at only about 3% of peak speed. John Richardson and Lennart Johnsson from TMC and Oliver McBryan from the University of Colorado also assisted in this effort. In spite of this extensive effort, the performance results did not meet the expectations of MPP at that time. Nevertheless, we were hopeful that these results could be improved and that future results would meet with expectations. This was **not** to be the case even though the effort was considerably expanded as described in the following paragraph. The origins of the problem were quite evident. On the CM2, 185 cycles were required for a memory to memory transfer between processors[2] compared with a processor performance of two floating point operations per cycle.

Following this work, a joint project to implement a working MPP weather/climate model was established with Oak Ridge, Argonne, and Los Alamos and supported by the Department of Energy (DOE). Initialy performance was further reduced by the additional complexity of the working model; however, new methods were developed that offset early losses so that sustained performance remained a few percent of peak with an average of about 7% on the Connection Machine 5 (CM5) in 1995. Although this was considered a significant achievement, the resulting performance/cost was comparable to existing traditional computers and NCAR, once again, delayed the selection of MPP as the primary general purpose computing resource for the atmospheric science community. The MPP climate model project was extensive and provided the knowledge base on which current multiprocessor models are developed. Indeed it is important to acknowledge TMC and their products, as well as the DOE, for providing the resources that enabled the community to develop an extensive repertoire of efficient multiprocessor methods for weather and climate modeling.

The sustained performance/cost ratios of the CM2 and CM5 were not unique. Similar results were obtained, at NCAR and other institutions, for all of the machines that were benchmarked. Although peak performance was impressive, communication would substantially degrade performance to the point that traditional architectures, with less peak performance, remained competitive because they were able to sustain a higher percentage of peak. Without a substantial gain in performance, scientists were understandably reluctant to expend the algorithmic and programming effort required to port their codes to MPP architectures.

This completes the answer to the first question in the title; namely, "MPP: What went wrong?" The main problem was, and continues to be, the high cost of communication which significantly degrades sustained performance, particularly on working applications such as weather/climate models. The rest of this paper is devoted to the second question in the title; namely, "can it [MPP] be made right?" More specifically: What is the best possible MPP performance? Will communication always dominate MPP performance? Is there a "best" MPP architecture? To answer these questions we start from scratch and: (a) develop optimum parallel communication algorithms and (b) list the architectural features that enable the algorithms to achieve optimum performance. Two cases are considered. First, the number of processors is assumed proportional to the number of latitudes. This is called a one-dimensional (1D) model distribution. Each processor contains one or more latitudinal lines. Second, the number of processors is assumed proportional to the number of grid points. This is called a two-dimensional (2D) model distribution. Each processor contains at most a few grid points.

In the next section we review the transpose method which is fundamental to the efficient parallel implementation of geophysical models. In section 3 we describe optimum parallel transpose algorithms for both packet and element based communication systems. We show that a matrix transposition can be performed at the average rate of two elements per cycle to every processor. This is over two orders of magnitude faster than the CM2 which required 185 cycles to transmit a single element between processors. Hardware features that support optimum algorithmic performance are listed in section 4. The results are summarized in section 5.

2. The transpose method.

In the '70s and thereafter, a considerable number of parallel computational algorithms were published. For example, parallel algorithms were presented for solving tridiagonal linear systems of equations and the fast Fourier transform (FFT). The cost of communication was essentially ignored in this literature. During the DOE project, it became quite evident that communication can dominate the cost of MPP computing. In this section we focus on the fundamental communication task used by weather/climate models and show that distributed computational algorithms are not as efficient as traditional serial algorithms when used in combination with the transpose method. The transpose method is broadly applicable to geophysical models and certain fundamental computations such as parallel matrix-vector multiplication and the FFT.

The spectral transform method for weather and climate models includes multiple FFTs in the longitudinal direction and multiple Legendre transforms in the latitudinal direction[3]. If all points on a fixed latitude are in the same processor, the Fourier transforms can be computed efficiently using single processor algorithms and software. On the other hand, if the arrays are distributed in longitude, but not latitude, the Legendre transforms can be computed efficiently in each processor.

Consider the following two methods. First, a fixed or static distribution could be selected. For example, the arrays could be distributed in longitude but not latitude. The Legendre transforms could then be performed in-processor without interprocessor communication. However, the Fourier transforms would then require a distributed or parallel FFT algorithm[4]. If the arrays are distributed in latitude but not longitude a distributed Legendre transform is required. Alternately, the arrays could be dynamically redistributed (transposed) between the computational modules, and both the Legendre and Fourier transforms could be performed in-processor. This approach is called the transpose method.

There are several reasons why the transpose method is preferred even though it requires the global redistribution of the array. First, following the development of efficient transpose software, it is relatively straightforward to port a uniprocessor code to a multiprocessor. Second, the computations can be performed in-processor using sequential algorithms that usually require less computation than their parallel counterparts. Third, the communication requirements of the transposition are less than the communication requirements of the distributed algorithm. For example, assume an $N \times N$ array is distributed columnwise on $P = N$ processors but we wish to Fourier transform the rows. If we transpose the array then each processor contains a complete row and the FFT can be performed using efficient uniprocessor software. It will be shown that the communication requirements for this approach are $O(N)$ for the transposition. On the other hand, if the rows are transformed in-place, with elements of the row in different processors, then the communication requirements for each row are $O(\log N)$ and therefore $O(N \log N)$ for N rows. Where possible, it is preferable to move data into a processor with a communication requirement proportional to the amount of data rather than implement a distributed algorithm with communication requirements that are proportional to the amount of data times the parallel communication complexity of the distributed algorithm.

The transpose method is applicable to a number of fundamental problems including parallel matrix-vector multiplication. Let an $N \times N$ matrix be distributed row-wise across $N = P$ processors and let the vector be distributed element-wise. The single element in each processor is replicated N times and the resulting $N \times N$ array is transposed so that each processor contains a copy of the vector. The components of the matrix-vector product are then computed simultaneously as the inner product of a matrix row and the vector. Like the original vector, the resulting matrix-vector product is distributed elementwise across the processors. The inner products require $O(N)$ computational time and in section 4 we show that the transposition requires $O(N)$ communication time.

The transpose method is also applicable to the FFT of a long sequence. For example, let a sequence with length N^2 be uniformly distributed over $N = P$ processors. An ordered-to-ordered FFT can be performed with three transpositions and an unordered-to-unordered FFT can be performed with a single transposition[4]. A interesting aspect of this algorithm is $O(N)$ communication which is less than $O(N \log N)$ computation. Although the FFT requires global communication, for sufficiently large N, communication becomes negligible. In the next section we describe an efficient $O(N)$ parallel algorithm for transposition and list several extensions.

3. A transpose algorithm for the hypercube

In this section we first describe a transposition algorithm for a hypercube with an element based communication system like the CM2. The time required to transmit l elements is $l\beta$ where β is the time required to transmit a single element. The packet based version is outlined at the end of the section. A brief description of the algorithm will be given because the details can be found elsewhere[5]. The algorithm will be illustrated by transposing an 8×8 array distributed columnwise across $p=8$ processors. It consists of two phases. First, the elements in each processor (column) are reordered. Second, the elements are transmitted according to a schedule that is the same for all processors. The order phase for the elements in processor $p=5$ is given in Table 1. The choice $p=5$ is arbitrary. A table can be constructed in a similar manner for the other processors.

Table 1
The order phase for the elements in processor $p=5$
for an 8×8 transposition

element $x_{i,2}$	destination processor p_d	relative address $p \oplus p_d$	de Bruijn ordered rel. add.	de Bruijn ordered elements
$x_{0,5}$	000	101	001	$x_{4,5}$
$x_{1,5}$	001	100	011	$x_{6,5}$
$x_{2,5}$	010	111	111	$x_{2,5}$
$x_{3,5}$	011	110	110	$x_{3,5}$
$x_{4,5}$	100	001	101	$x_{0,5}$
$x_{5,5}$	101	000	010	$x_{7,5}$
$x_{6,5}$	110	011	100	$x_{1,5}$
$x_{7,5}$	111	010	000	$x_{5,5}$

Column 1 contains the fifth column of the array. Column 2 contains the binary representation of the destination processor p_d of the corresponding element in column 1.

Column 3 contains the bit-wise exclusive-or of the origination processor $p = 101$ with the destination processor p_d. These bits determine the channels that must be traversed by the corresponding element in the first column to reach its destination processor. For example, element $x_{6,5}$ must be transmitted on channels 0 and 1 to reach its destination processor 6. Channel numbers correspond to bit numbers with the least significant being bit number 0. Each bit in the fourth column is equal to the one above and to it's right and therefore all bits are determined by the least significant (rightmost) column of bits; namely 11101000. This particular sequence of bits is called a **de Bruijn sequence**[6] because all of the integers 0 through 7 occur in column 4. Therefore column 4 is a reordering of the integers in column 3. Column 5 is the same reorder of column 1, and therefore the bits in column 4 determine the channels that must be traversed by the corresponding element in column 5. The transmissions specified by column 4 are implemented by the following communication cycles.

Cycle 1: Transmit elements in locations 0, 1, and 2 on channels 0, 1, and 2 respectively.
Cycle 2: Transmit elements in locations 1, 2, and 3 on channels 0, 1, and 2 respectively.
Cycle 3: Transmit elements in locations 2, 3, and 4 on channels 0, 1, and 2 respectively.
Cycle 4: Transmit elements in locations 4, 5, and 6 on channels 0, 1, and 2 respectively.

Column 4 is the same for all processors and therefore the routing information for the new elements is the same as the elements they replaced. This implies that column four remains unchanged from processor to processor **and** from cycle to cycle. Therefore it is not necessary to transmit routing information and the communication cycles listed above are the same for all the processors. It can be verified that each element in column 5 is transmitted on the channels specified by the 1-bits in column 4.

The contents of processor 5, following each communication cycle, are listed in Table 2. The first column contains the 5th column of the array and last column contains the 5th row, both are in de Bruijn order. The elements received on each cycle are stored into the locations vacated by the transmitted elements. For example, following the first cycle the elements $x_{5,4}$, $x_{4,7}$, and $x_{6,1}$ occupy locations 0, 1, 2 respectively.

Table 2

Contents of processor $p = 5$ for each element based communication cycle

0	1	2	3	4
$x_{4,5}$	$x_{5,4}$	$x_{5,4}$	$x_{5,4}$	$x_{5,4}$
$x_{6,5}$	$x_{4,7}$	$x_{5,6}$	$x_{5,6}$	$x_{5,6}$
$x_{2,5}$	$x_{6,1}$	$x_{4,3}$	$x_{5,2}$	$x_{5,2}$
$x_{3,5}$	$x_{3,5}$	$x_{7,1}$	$x_{5,3}$	$x_{5,3}$
$x_{0,5}$	$x_{0,5}$	$x_{0,5}$	$x_{4,1}$	$x_{5,0}$
$x_{7,5}$	$x_{7,5}$	$x_{7,5}$	$x_{7,5}$	$x_{5,7}$
$x_{1,5}$	$x_{1,5}$	$x_{1,5}$	$x_{1,5}$	$x_{5,1}$
$x_{5,5}$	$x_{5,5}$	$x_{5,5}$	$x_{5,5}$	$x_{5,5}$

The de Bruijn orderings can be performed in-processor using indirect addressing. Only a single initial ordering is required because the elements can remain de Bruijn ordered throughout the weather/climate model computations. In general, an $N \times N$ transposition can be performed in $N/2$ cycles; that is, the array can be transposed in time $\beta N/2$. The optimality of this result can be established from column 4 in Table 1 which is the same for all processors. Therefore the total number of transmissions is equal to the number of 1-bits in column 4 times the number of processors which for general N is $(N^2 \log_2 N)/2$. Since there are $N \log_2 N$ hypercube channels, a minimum of $N/2$ communication cycles is required. The significance of this result becomes evident with the observation that a transposition can be completed in half the time it would take to transmit a column of the array on a single channel. This is also half the time required by a crossbar interconnect because each processor communicates with the crossbar through a single channel.

Now consider the algorithm for packet based communication systems. The time required to transmit l elements in a single packet is $\alpha + \gamma l$ where alpha is the latency and γ^{-1} is the channel bandwidth. From columns 4 and 5 in Table 1 it is evident that $x_{0,5}$ could also be transmitted with $x_{4,5}$ on the first cycle. Indeed, on the first cycle, the packets $(x_{4,5}, x_{0,5})$, $(x_{6,5}, x_{7,5})$, and $(x_{2,5}, x_{1,5})$ could be transmitted simultaneously on wires 0, 1, and 2 respectively. The addresses of elements in a given packet must differ by at least $\log_2 P$ because each element must be transmitted sequentially on the channels specified by its relative address in column 4, Table 1. The total number of transmitted elements is the same as the element based algorithm or

$$\frac{N}{2} = \sum_{l=1}^{\log_2 N} \delta_l \tag{2.1}$$

where δ_l is the length of the packets on the lth communication cycle. Therefore the average packet length is $N/(2\log_2 N)$ and the total time is

$$T = \sum_{l=1}^{\log_2 N} (\alpha + \delta_l \gamma) = \alpha \log_2 N + \gamma N/2 . \tag{2.2}$$

The communication cycles for transposing a 16 × 16 array are listed in Table 3. Its columns correspond to the contents of processor 10 (chosen at random) following each cycle. The transmissions are based on the de Bruijn sequence 1111011001010000. The packet lengths vary from cycle to cycle because the pattern of zeros and ones in the de Bruijn sequence is irregular. The packet lengths δ_l for $l = 1, 2, 3,$ and 4 are 1, 3, 2, and 2 respectively. Column 1 contains the 10th column of the array and column 5 contains the 10th row. Both columns are de Bruijn ordered.

Table 3

Contents of processor $p = 10$ for each packet based communication cycle

0	1	2	3	4
$x_{11,10}$	$x_{10,11}$	$x_{10,11}$	$x_{10,11}$	$x_{10,11}$
$x_{9,10}$	$x_{11,8}$	$x_{10,9}$	$x_{10,9}$	$x_{10,9}$
$x_{13,10}$	$x_{9,14}$	$x_{11,12}$	$x_{10,13}$	$x_{10,13}$
$x_{5,10}$	$x_{13,2}$	$x_{9,6}$	$x_{11,4}$	$x_{10,5}$
$x_{4,10}$	$x_{4,10}$	$x_{12,2}$	$x_{8,6}$	$x_{10,4}$
$x_{7,10}$	$x_{7,10}$	$x_{6,11}$	$x_{14,3}$	$x_{10,7}$
$x_{1,10}$	$x_{1,10}$	$x_{3,8}$	$x_{2,9}$	$x_{10,1}$
$x_{12,10}$	$x_{12,10}$	$x_{8,14}$	$x_{10,12}$	$x_{10,12}$
$x_{6,10}$	$x_{6,10}$	$x_{14,2}$	$x_{10,6}$	$x_{10,6}$
$x_{3,10}$	$x_{3,10}$	$x_{2,11}$	$x_{10,3}$	$x_{10,3}$
$x_{8,10}$	$x_{8,10}$	$x_{10,8}$	$x_{10,8}$	$x_{10,8}$
$x_{15,10}$	$x_{15,10}$	$x_{11,14}$	$x_{11,14}$	$x_{10,15}$
$x_{0,10}$	$x_{0,10}$	$x_{8,2}$	$x_{8,2}$	$x_{10,0}$
$x_{14,10}$	$x_{14,10}$	$x_{14,10}$	$x_{14,10}$	$x_{10,14}$
$x_{2,10}$	$x_{2,10}$	$x_{2,10}$	$x_{2,10}$	$x_{10,2}$
$x_{10,10}$	$x_{10,10}$	$x_{10,10}$	$x_{10,10}$	$x_{10,10}$

So far we have discussed only the transposition of a square matrix that is distributed row or columnwise. The implementation of a working climate model is facilitated by the extensions listed below but developed elsewhere[5].

1. Index reversal, i.e. given a distributed vector $x(i)$ then distribute $y(i) = x(N-i-1)$.

2. Column permutations and the distribution of a translated vector.

3. Generalized transpositions bit reverse, shuffles, rectangular arrays and the case $P \neq N$.

4. Arrays with other than power-of-two dimensions.

5. The general one-to-one communication problem using repeated transpositions.

6. Generalization to mesh connected architectures.

4. The optimum architecture

Communication performance has two components; namely, algorithmic and architectural. Optimal algorithms for transposition and related communication tasks were discussed in section 3. In this section we first list, then discuss, the architectural features that provide optimum performance for the these algorithms.

1. Hypercube interconnection network

2. Element based communication system

3. Bi-directional network communication

4. Multichannel network communication

5. Programmable network

6. Collation of memory and network communications

The algorithms in section 3 were designed for the hypercube. However, mesh connected multiprocessors may offer comparable performance if the channels are sufficiently broad relative to the hypercube channels. If β^{-1} is the channel bandwidth then feature 2 implies the time required to transmit l words or elements over a single channel is $l\beta$ for all l. The time for a element based transposition is $\beta N/2$ compared with $\alpha \log_2 N + \gamma N/2$ for a packet based system. If the bandwidth of the the two systems differ, i.e. $\gamma \neq \beta$, the choice is not clear and a low-latency, high-bandwidth packet based system might have satisfactory performance. Nevertheless an element based system with comparable bandwidth is preferred because it provides the greatest degree of algorithmic and programmatic flexibility.

Feature 3 simply doubles the communication bandwidth, Feature 4 implies that all hypercube channels can be active simultaneously. That is, the full bandwidth of the network can be realized. Feature 5 implies that the network is programmable. That is, the physical channels traversed by each element can be specified on each communication cycle. The CM2 offered this important feature. However, elements were stored and subsequently retrieved from each processor along the path from origination to destination processor, resulting in a significant performance degradation. Feature 6 provides a programmable pass-through capability so that elements arriving at a processor communication interface switch Fig. 1 can be transmitted back onto the network on the next cycle without being stored to memory. In addition, as required by the algorithms, the interface switch collates memory and network traffic. Feature 5 corresponds to programming the interface switch which is required for optimum communication performance.

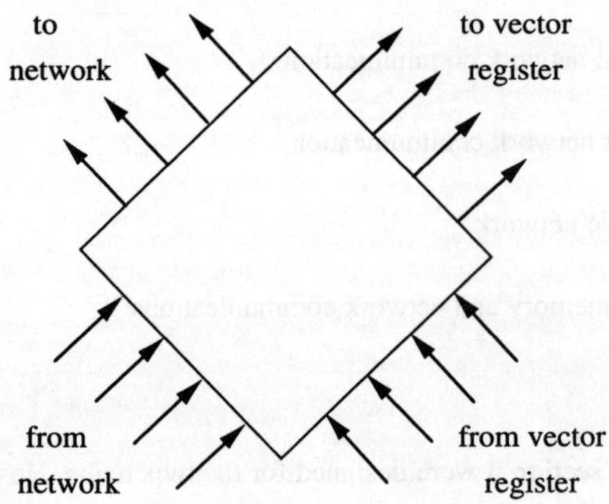

Figure 1. The processor interface communication unit consists of a programmable $2\log_2 P \times 2\log_2 P$ crossbar switch shown for the case $P = 16$.

Hardware features 1 through 6 provide the architectural functionality that enables the transposition algorithms to realize peak performance of two elements per cycle (on average) to every processor. Column permutation or index reversal can be implemented at the rate of one element per cycle to every processor. This level of performance prompts the thought that perhaps the multiprocessor can be vectorized. That is, perhaps the communication tasks induced by vector index syntax can also be performed in a time proportional to the length of the array in each processor. For example, an index reversal is equivalent to the index syntax $x(i)=x(N-i-1)$ which can be implemented at the rate of one element per cycle to every processor. This leads to the **vector parallel paradigm**[7] in which all arrays are uniformly distributed across the processors. An array $x(j)$, with N elements, is distributed with N/P elements per processor. New arrays, defined by index syntax, are also distributed uniformly across the processors at an average rate of one element per cycle to every processor. Computations then proceed in-processor using the parallelism that is implicit in traditional vector codes. Programs written for a single vector processor will port without modification to a multiprocessor with the vector parallel paradigm. The vector parallel paradigm brings to the multiprocessor what vectorization brought to the uniprocessor.

5. Summary.

The optimum MPP performance of the harmonic transform and hence climate/weather model dynamics has been determined. If N is both the number of processors and the number latitudes (1D model distribution), communication time is $\alpha \log_2 N + \gamma N/2$ and computation time is $2\eta N^2$ where α is the packet latency, γ^{-1} is the bandwidth of a single channel, and η is the time for a single floating point operation. If each processor contains a single point (2D model distribution) communication time remains unchanged but computation time is reduced to $2\eta N$. If an element based communication system is used then the time for communication is $\beta N/2$ where β is the time required to transmit a single element.

These results demonstrate that the spectral method for climate modeling is able to take full advantage of a suitably configured MPP. A single time step and level of model dynamics requires $O(N^3)$ time on a uniprocessor. Here it has been shown that the model requires $O(N^2)$ time using $P=N$ processors, and $O(N)$ time using $P=N^2$ processors. Furthermore, for the 1D model distribution and sufficiently large $P=N$, communication is negligible relative to computation. For the 2D model distribution, it can be shown that the asymptotic efficiency is $\eta/(\gamma/4+\eta)$ which is bounded away from zero.

These results differ significantly from those obtained in practice because the required architectural features are not currently available in the marketplace. The high cost of communication reduces sustained performance of MPP to a few percentage of peak which, at best, results in a performance/cost that is comparable with traditional architectures that attain a higher percentage of peak and require less development effort. Here, a two-component solution to the communication problem is proposed. The first component is parallel communication algorithms that provide optimum performance for a wide class of problems including weather and climate modeling. The second component is hardware features that are capable of realizing the optimum performance offered by the algorithms.

If these hardware features had been available on the CM2, communication could have been reduced by as much as two orders of magnitude. One can conservatively estimate a sustained performance of up to 25% of peak or 8 Gflops on a fully configured 2048 processor CM2. This performance would have been highly competitive in 1988 and well worth the developmental effort required to port uniprocessor codes. The history of MPP would have been quite different.

Interprocessor and related communication dominate MPP computing time and also present a growing problem for traditional shared-memory multiprocessors. In this paper, parallel communication algorithms and hardware features that support optimum performance are the suggested tools for solving the communication problem. More specifically, we suggest communication algorithms and the capability to program them as the tools required to most effectively speed communication, just as these tools have always been used to speed computation on a uniprocessor. A potential 4-fold increase in sustained performance is conservatively estimated which, if implemented could very well fulfill the original promise of MPP.

REFERENCES

1. P. N. Swarztrauber and R.K. Sato, "Solving the shallow water equations on the Cray X-MP/48 and the Connection Machine 2," in *The Dawn of Massively Parallel Processing in Meteorology,* G.-R. Hoffmann and D. K. Maretis, eds., Topics in Atmospheric and Oceanic Sciences, Springer-Verlag, 1989.

2. S. L. Johnsson and C.-T. Ho, "Generalized shuffle permutations on Boolean cubes," *J. Parallel and Distrib. Comput.,* **16**(1992), pp. 1-14.

3. P. N. Swarztrauber, "Spectral transform methods for solving the shallow-water equations on the sphere," *Mon. Wea. Rev.,* **124**(1996), pp. 730-744.

4. P. N. Swarztrauber, "Multiprocessor FFTs," *Parallel Comput.,* **5**(1987), 197-210.

5. P. N. Swarztrauber, "Transposing the k-ary n-cube," *J. Parallel Distrib. Comput.*, submitted for publication.

6. S. B. Maurer and A. Ralston, *Discrete Algorithmic Mathematics*, Addison Wesley, New York, 1991.

7. P. N. Swarztrauber, "The vector parallel paradigm," *Proceedings of the International Conference: Advanced Mathematics, Computations & Applications*, A. S. Alekseev & N. S. Bakhvalov, eds., Novosibirsk, June 20-24, 1995, pp. 675-687. Table of contents: http://www.comcen.nsk.su/amca95.html

ECMWF Operational Forecasting on a Distributed Memory Platform: General Overview

D.Dent
European Centre for Medium-range Weather Forecasts
Shinfield Park, Reading, RG2 9AX, U.K.
e-mail : david.dent@ecmwf.int

Abstract

Since September 1996, the ECMWF operational forecasts have been generated on a distributed memory platform. This has necessitated substantial migration effort. We give an overview of the most significant components, particularly from the point of view of efficient use of distributed memory.

1 Introduction

The analysis and forecast model of ECMWF have been running operationally for many years on multi-processor shared memory computer systems (PVP). During 1996, the analysis scheme was changed from one based on optimal interpolation (OI) to a 3 dimensional variational scheme [1]. This scheme is a part of the Integrated Forecasting System (IFS) which has been developed jointly by scientists from Meteo-France and ECMWF. It contains:

1. a spectral, semi-Lagrangian atmospheric model
2. 3D and 4D variational analysis schemes
3. an ensemble prediction system
4. interpolation and post-processing software

This overview forms an introduction to subsequent papers in this workshop, namely:

1. Parallelization of Observation Processing
2. Analysis
3. Forecast

Specifically outside of the scope of IFS is the extensive observation handling software needed in order to support an operational data assimilation. Historically, this was contained inside OI and implemented assuming a shared memory platform. Hence, extensive development had to be carried out in this area. Part of this work is discussed in [2].

Some aspects of the 3D variational analysis were also originally developed on a shared memory platform and therefore also required development. These are described in [1].

Developments of the atmospheric model for a distributed memory platform have been described extensively during the previous ECMWF parallel processing workshop [3]. On this occasion, we restrict the model discussion to some recent developments and investigations, and also include some performance figures for the current ECMWF production platform. [4]

2 Code developments

Historically, the distributed memory developments of IFS have progressed through a number of stages:

1. Creation of a portable/serial atmospheric model, based on Fortran77. Removal of memory management features meant the loss of resolution independence.
2. Creation of a portable/parallel atmospheric model for benchmarking purposes, based on Fortran77 + PARMACS message passing.
3. Unification of the production version and the benchmark version since the maintenance cost for two versions is unsustainable. This unified code is based on Fortran90 and MPI message passing. However, we localise the message passing to a small set of 'support' routines so that any flavour of message passing can easily be accommodated. Only a small subset of MPI is used.
4. Retention of MACROTASKING (for PVP platforms) in order that the shared memory parallel version remains available (CRAY C90 is the production platform in use at Meteo-France).

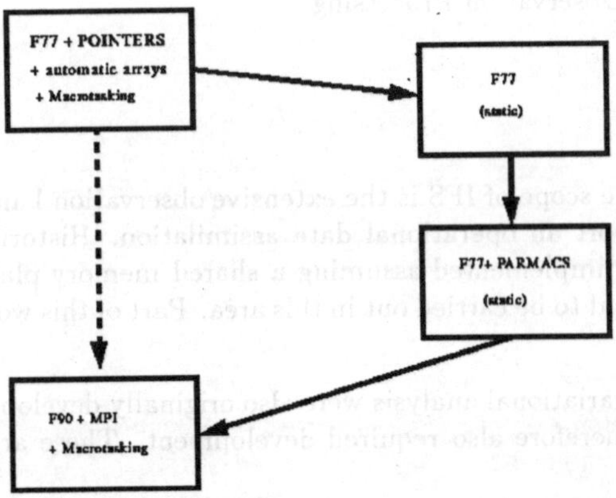

Figure 1: Code development history.

5. The choice between 'serial', 'CRAYPVP' and 'message passing' is made at run time by namelist switches.

6. The distributed memory version has no I/O scheme, although one is retained in the PVP version. This makes the parallel version very expensive to run on PVP in terms of primary memory requirements.

The history of development is illustrated schematically in figure 1 where the dotted line on the left side represents substantial meteorological developments carried out over several years on the operational shared memory version.

3 Analysis Cycling

In order to understand the relationship between analysis and forecast, it is useful to remind ourselves of the time dependent cycling nature of the 3 dimensional data assimilation. Figure 2 illustrates the mechanism which is based on 6 hourly collections of observations. The analysis is fed a grouping of observations, timed within a window of plus or minus 3 hours from the nominal analysis time. The analysis is also supplied with a 'first-guess' forecast which itself commences from the previous analysed state 6 hours earlier. Note that this is exactly the same mechanism as employed when the analysis process utilised OI techniques. The only difference is that the box labelled "ANALYSIS' now consists of a 3D variational analysis.

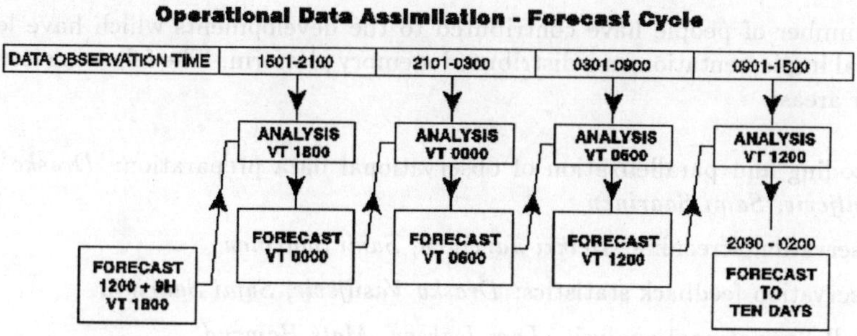

Figure 2: Data assimilation cycling

4 Run Times

From the point of view of computer resources, it is useful to have a feel for the relative expense of the various components of the operational suite. Table 1 shows wall clock run times extracted from the production suite running on the new ECMWF distributed memory platform (Fujitsu VPP700). Note that these are timings from the very first production implementation on VPP700. Many improvements are planned to reduce the wall clock times. Some of these are discussed in the subsequent papers.

Table 1: Wall clock run times from production suite on VPP700.

Component	Number of PEs used	Time (minutes)
observation preparation	1	6
observation preparation	6	6
3D VAR	6	38
6 hour forecast	10	8
10 day forecast	10	106

Acknowledgements

A large number of people have contributed to the developments which have led to operational implementation on a distributed memory platform. The following itemises the major areas:

Recoding and parallelization of observational data preparation: *Drasko Vasiljevic, Sami Saarinen*

Observation screening: *Heikki Jarvinen, Sami Saarinen*

Observation feedback statistics: *Drasko Vasiljevic, Sami Saarinin*

Parallel variational analysis: *Lars Isaksen, Mats Hamrud*

Forecast model parallel improvements: *George Mozdzynski, Fritz Wollenweber, Mats Hamrud*

Operational implementation: *Otto Pesonen*

References

[1] L.Isaksen & M.Hamrud : *ECMWF Operational Forecasting on a Distributed Memory Platform : Analysis*, in Making its Mark, Proceeding of the 7th ECMWF Workshop on the Use of Parallel Processors In Meteorology, December 2-6, 1996.

[2] S.Saarinen, D.Vasiljevic & H.Järvinen : *Parallelization of Observation Processing*, in Making its Mark, Proceeding of the 7th ECMWF Workshop on the Use of Parallel Processors In Meteorology, December 2-6, 1996.

[3] Barros, S. R. M., Dent, D., Isaksen, L., Robinson, G., Mozdzynski, G. and Wollenweber, F. : *The IFS Model: A parallel production weather code*, Parallel Computing **21** (1995), 1621-1638

[4] D.Dent & G.Mozdzynski : *ECMWF Operational Forecasting on a Distributed Memory Platform : Forecast Model*, in Making its Mark, Proceeding of the 7th ECMWF Workshop on the Use of Parallel Processors In Meteorology, December 2-6, 1996.

ECMWF Operational Forecasting on a Distributed Memory Platform : Analysis

Lars Isaksen and Mats Hamrud
European Centre for Medium-Range Weather Forecasts
Shinfield Park, Reading, Berkshire RG2 9AX, U.K.
e-mails: `firstname.lastname@ecmwf.int`

Abstract

Since September 1996, the ECMWF operational forecasts have been generated on a distributed memory platform. A significant component of the system is the 3 dimensional variational analysis scheme which involves use of both the forward and adjoint forecast models. New aspects like calculation of cost functions, calculations related to observations, minimisation of cost functions, and treatment of trajectories for adjoint calculations have to be considered in the distributed memory implementation. This paper focuses on design aspects and efficiency issues such as the message-passing implementation, load imbalance in the calculations related to observations, and minimisation calculations.

We include performance data from the current production on the FUJITSU VPP700.

1 Introduction

An important part of an operational meteorological centre is the analysis code. ECMWF's analysis calculates a good estimate of the state of the atmosphere four times daily. A so-called first guess, which is a six hour forecast valid at the analysis time, is updated by use of observations. Since January 1996 ECMWF has used the 3 dimensional variational technique to absorb information available from observations. The variational method makes use of the adjoint forecast model in order to minimise a cost function that describes the difference between first guess and observations. The cost function also constrains the modifications to deliver a balanced meteorologically sensible result.

This paper will describe how a parallel distributed memory version of the code has been implemented on ECMWF's Fujitsu VPP700. We have to a large degree made an implementation of the design plans described by Isaksen and Barros in [5] (below referred to as LISB).

In LISB, 4 dimensional variational analysis (4D-Var) was the main topic. Even though we technically have implemented 4D-Var for distributed memory machines, we prefer to present the design and performance data from ECMWF's operational 3D-Var implementation. From a technical point of view 3D-Var and 4D-Var are very similar. In 4D-var additional forward tangent linear and adjoint model integrations in time are done.

Most of the design considerations and related matters have already been described in LISB, so they will not be repeated in this paper. LISB covers our general parallelisation strategies and explains 4D-Var, including use of spectral transforms, minimisation, problems related to uneven distribution of observations, reproducibility of results, and the cost function components J_O, J_B, and J_C.

2 General dataflow

The main philosophy in the distributed memory implementation is to use what we will call a 'generalized transposition approach'. This involves reordering of data via message passing among a number of computational spaces, and keeping the computational phases as they are in the serial implementation. Within each computational phase data must be distributed in a way to ensure optimum load balance. 3D-Var has the following computational spaces :

- Grid Point Space
- Fourier Space
- Legendre Space
- Spectral Vertical Column Space
- Spectral Vertical Hough Space
- Observation Space
- Control Variable Space

In LISB sections 4, 5, 6, and 7 respectively the structure and data flow of J_O, J_B, J_C, and minimisation in the distributed memory implementation are explained. The only part that differs in the actual implementation is the calculation of J_O, so this will be described in more detail in this section.

The dataflow of an iteration in ECMWF's 3D-Var implementation is shown on figures 1 to 5. In these figures rectangles describe computations, octagons describe data exchanges, and rounded boxes describe 'spaces'. The complexity involved and the many transforms required between spectral and grid point space in order to do the scaling of variables for the minimisation and for the J_B and J_C calculations, becomes clear from these pictures. There are scientific reasons behind the choice of doing some scalings in spectral space and some in grid point space, which must be obeyed in a distributed memory implementation, despite the additional communication costs.

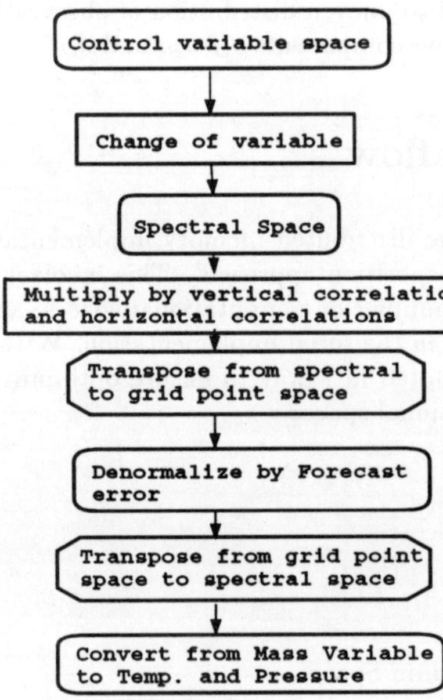

Figure 1: Dataflow in 3D-Var — part 1

The description starts in 'control variable space' where the minimisation is performed. The minimisation is done serially on each PE (Processing Element) because the minimisation code [4] is only available in a serial version and because these calculations take up a very small fraction of an iteration. By letting all PE's do the minimisation, communication of the results is only required for the inverse canonical injection from spectral space to control variable space (see figure 2).

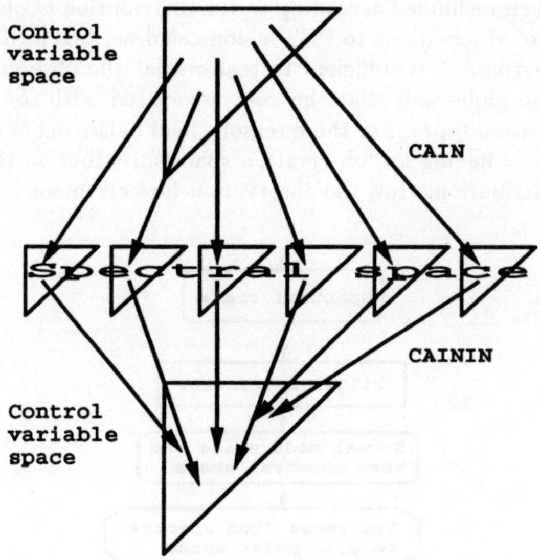

Figure 2: Transformations between Control Variable Space and Spectral Space. CAIN is the canonical injection, CAININ the inverse canonical injection

With the data back in spectral space, fields are multiplied by vertical and horizontal correlation matrices. Then a normalisation by forecast error is performed in grid point space. This requires a standard inverse transform from spectral to grid point space. Next step is a transformation from the mass variable, P, to the model variables, temperature and surface pressure. This is done in spectral column space, so this requires a direct transform and possibly an additional communication step to bring together all vertical levels. To remove noise a normal mode filtering is performed, so a transform from spectral to Hough space is required. This involves message passing if vertical levels are distributed (see [2] or [1] for details). Then a transform back to spectral space is performed.

We are now ready to do the J_O cost function calculations. The gradient of J_O with respect to the model variables is needed for the adjoint calculations. These calculations are done in grid point space, so first we transform from spectral to grid point space. Here the first step is to interpolate model field variables to observation point locations (the so-called observation equivalents). Contrary to the suggestion in LISB each PE is responsible for the interpolation for all the observations that are inside the PE's grid point area. This is convenient because the model field values needed are available on the PE, after halo regions have been filled to cater for observations near the border. In this way only one grid point distribution is required, whereas we would have needed two in a LISB based implementation. The calculated observation equivalents are then reshuffled according to the distribution of observations to PE's. The distribution of observations to PE's is done as described in LISB (see figure 1 in LISB). For our purpose it is sufficient to realise that the observations are unevenly distributed on the globe and that the cost associated with J_O calculations varies among the observation types. For these reasons load balancing of J_O calculations can only be achieved by having an 'observation space' distribution that will differ from the grid point distribution. How the observation files are made is explained in [6].

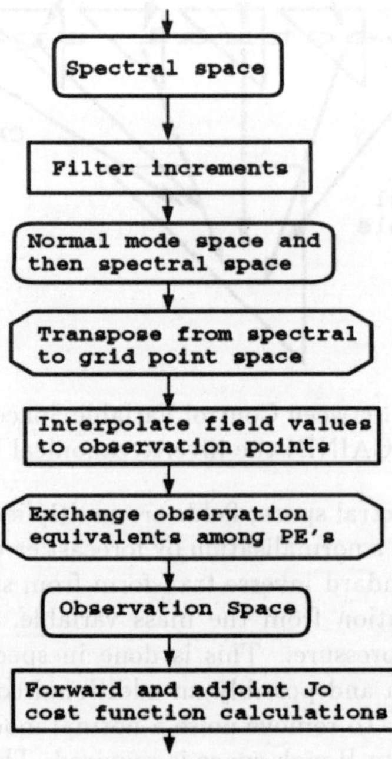

Figure 3: Dataflow in 3D-Var – part 2

After the reshuffling of observation equivalents we are now ready to do the J_O calculations. For some observation types like TOVS radiances the interpolated model values have to be transferred to real observation equivalents by the use of a radiative transfer function; this is done in 'observation space'. The J_O cost function contribution is calculated as the normalised difference between observation equivalents and observed quantities. We also have to calculate the gradient of J_O with respect to the model variables, and this is done just after the forward J_O calculations, still in 'observation space'. All these 'observation space' calculations are done independently on each PE and do not require synchronization within a PE. The only concern is to try to obtain the same amount of 'observation space' work on each PE via intelligent distribution of observations to PE's. Here it is necessary to consider the different costs associated with different observation types. The implementation (see [6]) allows us to associate weights with each observation type, based on statistical knowledge of the processing cost. The importance of this will be discussed in the result section.

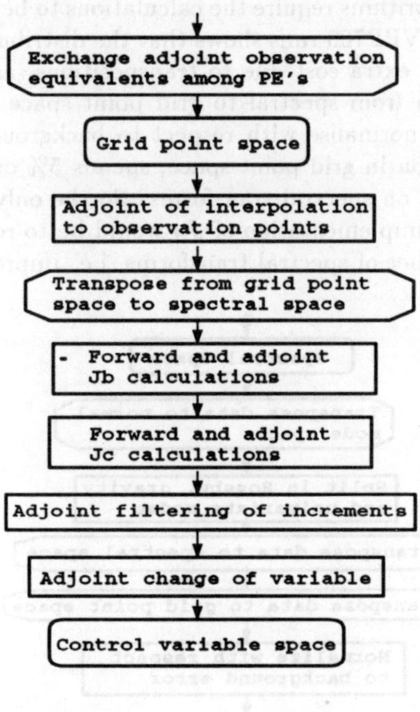

Figure 4: Dataflow in 3D-Var – part 3

The adjoint calculation of observation equivalents is performed in observation space. A certain grid point will be influenced by contributions from all the observations within the 'interpolation vicinity'. Some of these contributions may be calculated on surrounding PE's, so a communication step is performed to exchange adjoint observation equivalents. After the exchange, care has to be taken when contributions to a grid point are added up : a unique observation number makes sure it is done in a reproducible manner. We are now back in grid point space and perform the adjoint of interpolation to observation points. This is done by each PE for all the observations that reside inside its grid point area. As for the forward integration this is convenient because the interpolation framework in the IFS is designed to take place in grid point space. As we will show and discuss in the result section this will cause some static load imbalance.

After J_O, data is transformed to spectral space ready to perform J_B and J_C calculations. The handling of these two cost functions follows closely the description in LISB. Here we will only describe the flow of J_B, which is shown on figure 5. It is seen that a large number of transpositions are required, especially when the model levels are distributed among processors. It is impossible to avoid these transpositions because the scientific algorithms require the calculations to be done in a certain order. Results from our Fujitsu VPP700 runs shows that the distributed memory implementation adds only a small extra cost due to transpositions, and the main 'overhead' is the spectral transform from spectral to grid point space and vice versa. As an example the J_B step to normalise with respect to background error, which admittedly is a simple operation in grid point space, spends 5% on calculations, 10 % on transpositions, and 85 % on spectral transforms. So the only way really to improve the distributed memory implementation of J_B would be to rearrange calculations in order to reduce the number of spectral transforms, i.e. improve the efficiency of the serial code design.

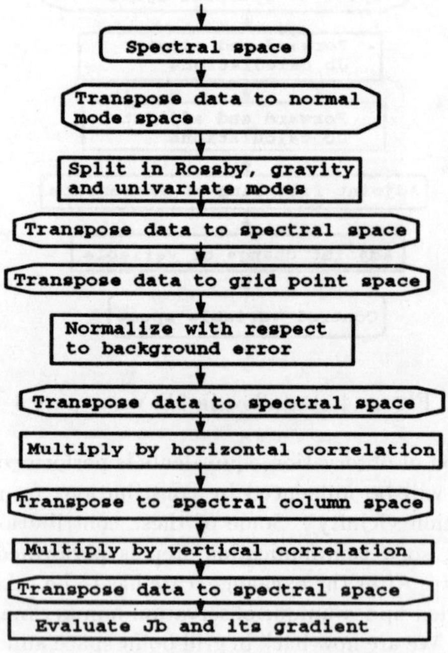

Figure 5: Dataflow in J_B calculations

After J_B and J_C follows adjoint filtering of increments and adjoint change of variable which technically are very similar to the forward calculations. The cost function contributions are scattered from all PE's and gathered on all PE's, so all end up having the same global reproducible cost function. All PE's are now ready to perform the minimisation and start on the next iteration. The operational 3D-Var performs 70 iterations.

3 Results

We will show results from an operational style run on ECMWF's Fujitsu VPP700. The resolution is T63L31 (triangular wave truncation up to wave number 63, and 31 vertical levels). The run is performed on six PE's. The trajectory runs and forecasts are performed at T213L31, but we will not discuss these steps here. We will focus on areas where the parallel implementation has performance problems.

Figure 6 shows the execution time for a 3D-Var run on 6 Fujitsu PE's.

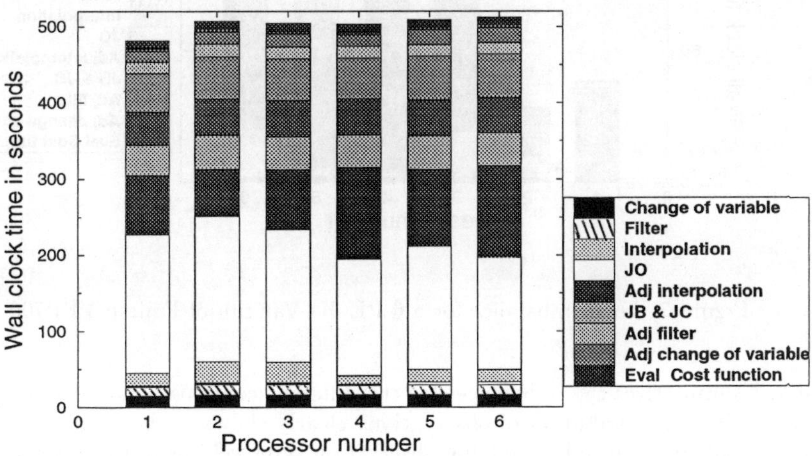

Figure 6: Run time for a 6 PE 3D-Var execution on Fujitsu VPP700

These numbers include computation time, communication time, and idle time waiting for other PE's. Load imbalances are only implicitly visible in this figure due to the implicit synchronization steps during an iteration. Most of the time is spent in J_O calculations, mainly on calculations in observation space, but a large fraction of time is also used on calculation of observation equivalents and their adjoints.

Figure 7 shows the load imbalance for the different computational phases on each of the 6 PE's. It is not possible to associate completely load imbalances with computational phases, but it is clear that there are large imbalances in the cost of processing observations (compare PE2 and PE6 in figure 7). The main reason for this is the method used in the present run for distribution of observations to PE's. Each PE is given an observation file of similar size, and we do not consider observation type or processing cost. We have used this default distribution method to show that it is important to cater for and to choose a more intelligent distribution strategy.

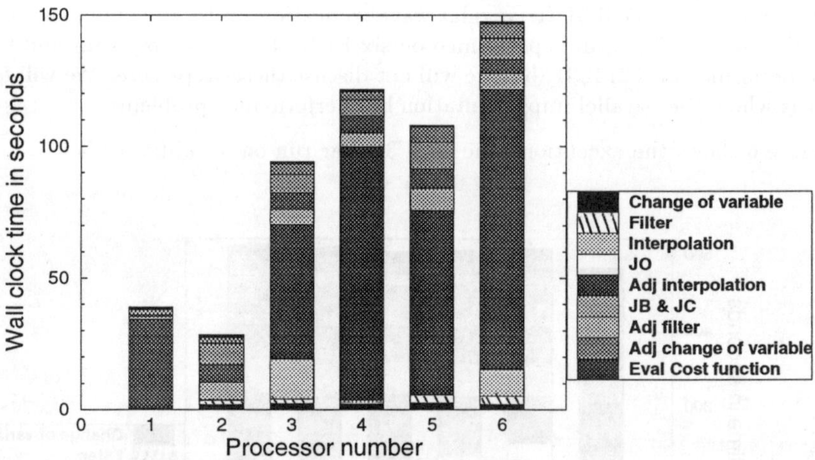

Figure 7: Load imbalance for a 6 PE 3D-Var run of Fujitsu VPP700

Table 1 summarizes the imbalances for the different computational phases. The imbalances in the distribution of observations clearly shows up here. It is possible to improve this imbalance by applying observation type weights in the observation distribution phase. For the other computational phases improvements can be achieved by allowing more messages to be sent simultaneously from each PE. Most of our message passing routines are able to do this.

Table 1 : Percentage load imbalance for each computational phase of a 3D-Var run of Fujitsu VPP700

Computational phase of 3D-Var	Imbalance in %
Change of variable	6.7
J_O calculation	22.7
J_B and J_C calculations	12.2
Adjoint change of variable	10.3
Evaluation of cost functions	3.1

Figure 8 shows the run time each PE uses for each of the 70 iterations for the computations associated with J_O.

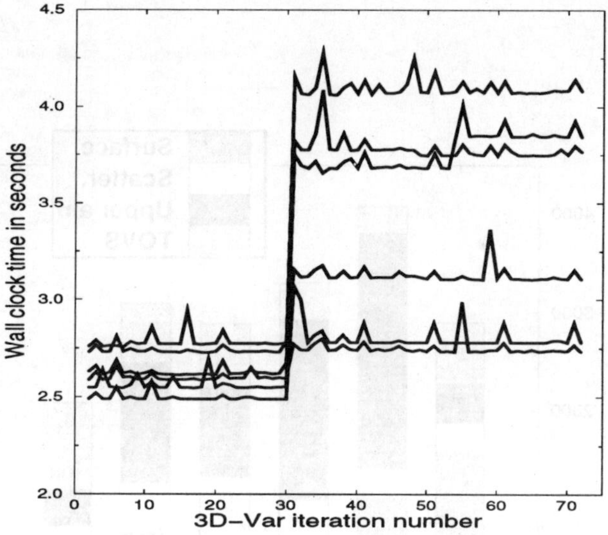

Figure 8: Cost of Observation Processing on 6 PE's. After 30 iterations expensive quality control starts

During the first 30 iterations the times vary between 2.5 and 2.8 seconds, that is a variation of 12 %. From iteration 30 onwards a quality control of observations is activated. In the present implementation it is very expensive to perform quality control on suspect radiosonde observations, and this is reflected in the times that now vary between 2.7 and 4.2 seconds per iteration (+55 % on the slowest PE). This imbalance can be reduced by optimising the quality control and by being aware of the cost when (radiosondes) are distributed to processors. However it is not possible to do an optimal static distribution of observation beforehand, because at that point it is not known whether an observation is suspect or not. A nice feature shown by figure 8 is the almost constant cost for the observation processing on each PE for iterations 1-30 and for iterations 31-70. This means an almost perfect load balancing can be achieved by reshuffling the data after iteration 1 and iteration 30. Such a method is being considered and is technically possible because the observation space calculations are separated from the grid point space calculations.

Figure 9 shows the number of observations of different types on each PE. On PE2 there are many radiosondes and on PE6 there are few, thus helping to explain the imbalances seen on figure 7. It does not explain everything because PE1 and PE6 have a similar number of radiosondes but the imbalance pattern on figure 7 differs quite a lot.

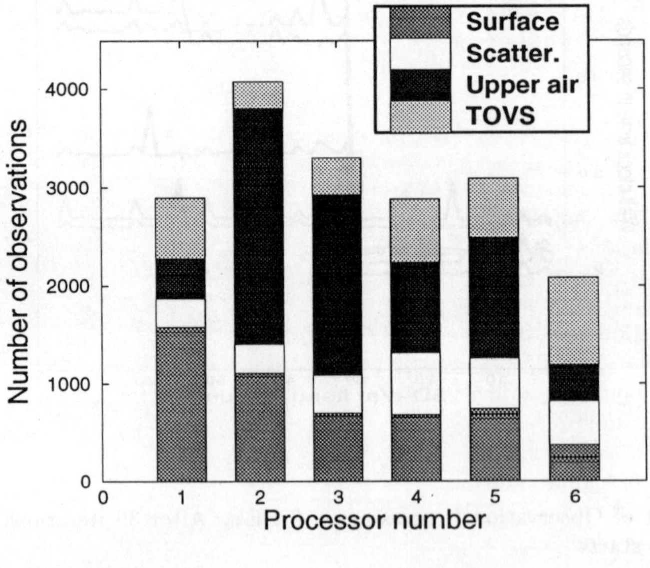

Figure 9: Number of observations on each PE

Figure 10 shows the run time for the observation space forward and adjoint calculations on the 6 PE's. The figure reflects the distribution shown in figure 9. We see a more even balance of the cost than could have been expected from figure 9. There is a good correspondence between figure 7 and figure 10, i.e. some of the load imbalance can be explained.

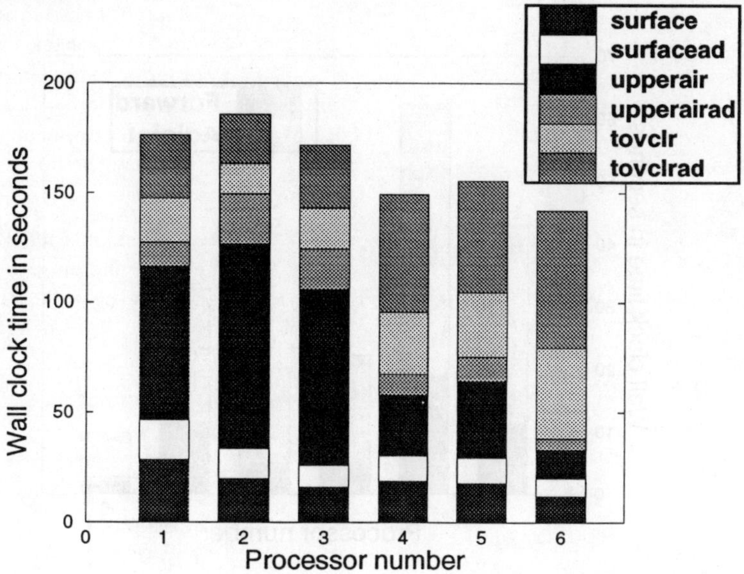

Figure 10: Time spent on observation space calculations for forward and adjoint calculations. Data shown for surface observation, upper air observations, and TOVS satellite observations

However figure 10 does not explain why PE2 and PE3 imbalances differ so much. This can be understood from figure 11 which shows the time spent on forward and adjoint interpolation to observation points. As discussed in section 2 the interpolation to observation points is performed by the PE that has the point inside its grid point area. This gives an unpleasant imbalance in the 6 PE run where we apply a grid point splitting only in the north-south direction. This means PE2 covers U.S.A., Europe and large parts of China where a large fraction of the total (conventional) observations comes from. PE6 covers the Antarctic region where few observations are available. In our present implementation interpolations have to be done in grid point space. It would be possible to outsource the forward interpolation work by restructuring the code, but it is difficult to see how reproducibility can be achieved in an outsourcing scheme for the adjoint calculations. We are investigating the use of linear instead of cubic interpolation for the interpolation to observation point as this would reduce the cost and also the imbalance substantially. When more satellite observations are used operationally this will also reduce the imbalances because these measurements will be more evenly distributed over the globe.

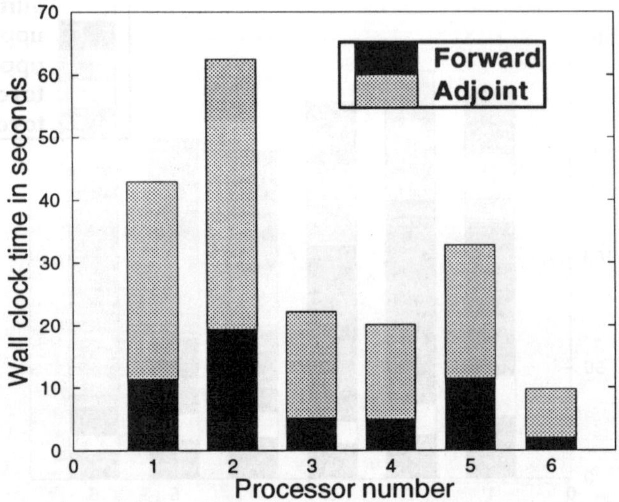

Figure 11: Time spent on interpolation to observation points on 6 PE's. Both forward and adjoint calculations shown.

Acknowledgements

We acknowledge the support from and fruitful discussions with David Dent, Otto Pesonen, and ECMWF's data-assimilation staff.

References

[1] S.R.M.Barros, D.Dent, L.Isaksen, G.Robinson, G.Mozdzynski, & F.Wollenweber : *The IFS model: A parallel production weather code.* Parallel Computing 21 (1995) 1621-1638.

[2] S.Barros, D.Dent, L.Isaksen & G.Robinson : *The IFS Model Overview and Parallel Strategies*, in Coming of Age, Proceedings of the 6th ECMWF Workshop on the Use of Parallel Processors In Meteorology, November 21-25, 1994.

[3] D.Dent & G.Mozdzynski : *ECMWF Operational Forecasting on a Distributed Memory Platform : Forecast Model*, in Making its Mark, Proceeding of the 7th ECMWF Workshop on the Use of Parallel Processors In Meteorology, December 2-6, 1996.

[4] Gilbert, J. C. and Lemaréchal, C. : *Some numerical experiments with variable storage quasi-Newton algorithms*, Mathematical Programming **B25** (1989), 407-435.

[5] L.Isaksen & S.Barros : *IFS 4D-Variational Analysis: Overview and Parallel Strategy*, in Coming of Age, Proceedings of the 6th ECMWF Workshop on the Use of Parallel Processors In Meteorology, November 21-25, 1994.

[6] S.Saarinen, D.Vasiljevic & H.Järvinen : *Parallelization of Observation Processing* , in Making its Mark, Proceeding of the 7th ECMWF Workshop on the Use of Parallel Processors In Meteorology, December 2-6, 1996.

ECMWF Operational Forecasting on a Distributed Memory Platform: Forecast Model

D.Dent and G.Mozdzynski
European Centre for Medium-range Weather Forecasts
Shinfield Park, Reading, RG2 9AX, U.K.
e-mail : david.dent@ecmwf.int, george.mozdzynski@ecmwf

Abstract

The development of a portable/parallel version of the IFS has been carried out over several years. The current version supports both macrotasking (for shared memory platforms) and message-passing. We describe the parallel I/O implementation for model input and post processing, along with work in progress to improve the message-passing aspects of the semi-Lagrangian scheme. Performance data from the Fujitsu VPP700 is presented.

1 Introduction

The IFS forecast model has been running operationally for many years at ECMWF on multi-processor shared memory computer systems from Cray Research. Parallelisation of the model on these systems was achieved by use of the Cray proprietary macro-tasking subroutine call interface. More recently the model has been ported to systems having a distributed memory architecture using message passing for communication between processor memories. The parallelisation scheme used on these systems is the transposition strategy and is described extensively in [1][2][3]. This port was initially confined to the forecast model, and focussed on achieving high scalability and performance for both vector and scalar/cache architectures. The work concentrated on exploiting the inherent parallelism in the IFS spectral model by using 2-dimensional (NPROCA,NPROCB) decomposition of model structures in grid-point space, fourier space and spectral space. However, to run operationally on distributed memory systems, other areas of the IFS forecast model required some development. In this paper, we present two of these areas, I/O processing and Radiation Interpolation and describe the adopted solutions. In both these cases, the new code had to support both distributed memory and shared memory as the IFS is required to run with both memory models.

The use of the semi-Lagrangian (SL) method in the IFS is another area where message-passing has been used. The present implementation uses a static scheme whereby processors communicate with neighbours at each time step to build a halo of data potentially required by the SL method. In section 4 some work in progress is described where we examine an enhancement which should reduce the SL communication overhead and further improve IFS scalability on large numbers of processors.

2 I/O Processing

An IFS forecast model execution needs to perform input once as part of its initialisation, and then output on a regular basis during a 10 day forecast. The format used for these is GRIB [5], or GRIdded Binary, which provides a very compact format for the representation of meteorological data. When considering possible approaches for I/O on a distributed memory system we obviously wanted scalability and performance, as we required a strategy that could work effectively on systems with large numbers of processors. Another important criteria was flexibility, as we preferred a solution that used exactly the same input files and produced the same output, irrespective of the number of processors used. These criteria were satisfied by implementing an I/O approach where the responsibility for the input and output of fields is assigned to specific processors at run-time. In this context a field represents a meteorological variable for the whole domain (grid-point or spectral) at a single level in the atmosphere or at the surface.

2.1 *Post Processing*

Post-processing is controlled by namelist input and performs output of model variables on a regular basis to a fields data base (FDB), which is simply a disk filesystem used to store fields for subsequent processing and archiving. For example, the T213L31 operational 10-day forecast model (time-step 15 mins) performs post-processing every 6 hours during the first 5 days and then every 12 hours for the last 5 days, a total of 1.7Gbytes being written. The number of fields written out today on a single post-processing timestep is shown in table 1.

An initial approach used for post-processing was for each processor to write its par-

Table 1: Fields written in a single post-processing time-step

	Spectral	Grid-Point	total
Model Level	124	125	249
Pressure Level	91	16	107
Surface	-	48	48
			404

tition of grid-point and spectral space to a separate file for the required fields. This seemed attractive as each processor could write out its data in parallel and then proceed with computation without delay. However, this approach required some other process to subsequently read these files, 'glue' the partial fields into complete fields, and encode into GRIB. This 'glueing' process was inefficient as it was serial, read a large amount of data, and processed a large number of files as NPROCA/B increased.

In contrast, the approach used today has none of the above problems as we always write out complete encoded fields which are gathered and encoded in parallel as shown in figure 1.

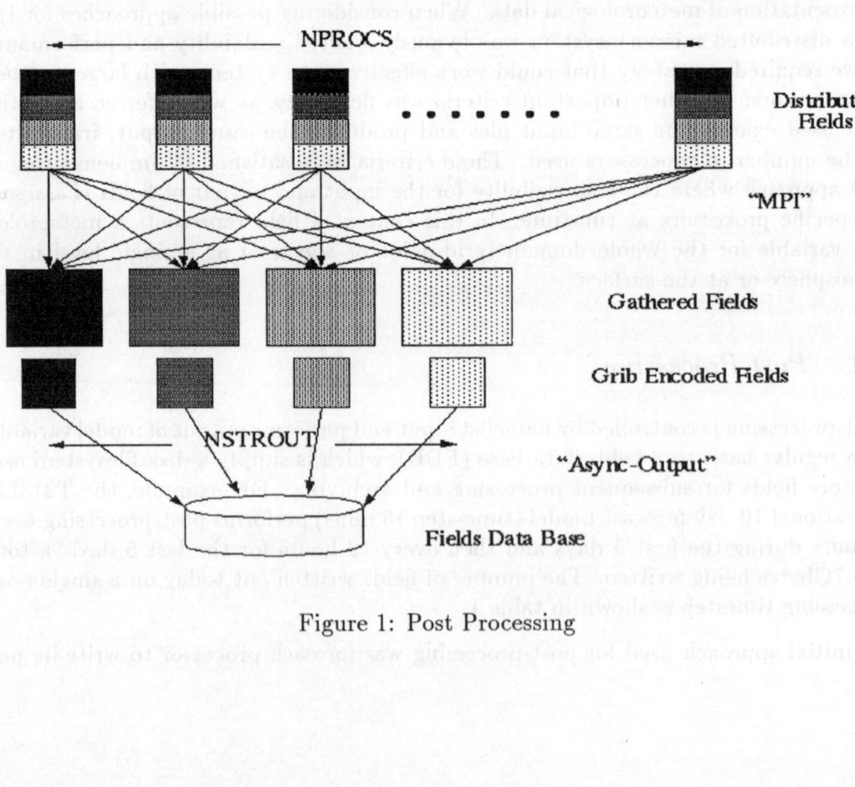

Figure 1: Post Processing

In more detail, each processor builds a work list of those fields that need to be post-processed, and then assigns processors to these fields using a simple round-robin scheme. The work list is then processed in two passes. In the first pass, those processors that are senders, initiate message-passing calls to send their partial fields to the receivers. In the second pass, those processors that are receivers simply receive their partial fields to build complete fields. The two passes are necessary to avoid any serialisation in the communication, as processors are typically both senders and receivers, but for different fields. When each field has been gathered, it is encoded into GRIB, and a 'write' operation issued to the FDB (asynchronously) or to a separate physical file. The actual implementation of the above includes namelist controls for the number of processors that will be used for I/O (NSTROUT) and the number of fields that are processed at a time (NFLDOUT) before a barrier operation. NFLDOUT is used to control how much message-passing buffer space is required. In addition the namelist logical LSPLITOUT is used in the case of non-FDB output to specify whether fields should be written as separate files or combined into a single file. The above strategy works well on distributed memory systems as the gathering, encoding and output operations can all be performed in parallel. On a system such as the Fujitsu VPP700 system at ECMWF, the gathering and encoding operations take advantage of the high performance interconnect and vector processors on that system. However, the output operation, although performed asynchronously, is relatively slow. Today, we have measured the I/O component of post-processing as taking about 1.5 minutes on an operational forecast lasting 35 minutes when using 38 processors. While this is acceptable today, more work will need to be done for execution on larger numbers of processors.

2.2 *Input Processing*

Although only performed once in a forecast run, the behaviour of input processing is still important for the large number of research experiments, many of which execute only a few time-steps. The distributed memory implementation for input processing is similar in many respects to that of post-processing, in that it uses a work list and has complementary namelist controls NSTRIN, NFLDIN and LSPLITIN. In particular, LSPLITIN specifies whether input fields are to be found together in a single file (LSPLITIN=F), or as separate files in a directory (LSPLITIN=T). For the latter case, a utility *splitgrib* was written to convert an input file (containing many fields) into a directory where each field is a separate file. Input processing can therefore run in parallel when LSPLITIN=T and the number of input streams NSTRIN is set greater than one. In this mode, each of the 'input' processors, reads a separate field, decodes it from GRIB format, and distributes it to the required processors. This approach produced a marked reduction in input processing time when used on a Cray T3D system, but this was not the case on the Fujitsu VPP700 system, due to the high cost of opening/reading the grib file. On the VPP700 system, large files are required for good performance, and our fields are only 0.25 Mbyte each at T213 resolution. So today on the VPP700 we run the IFS with LSPLITIN=F, which means that only one processor is used to read model input. This may change in the future as I/O performance is improved on this system.

2.3 *Checkpoint/Restart*

IFS checkpointing is mainly used to halt long-running experiments that will be restarted later. As the checkpoint and restart process should have no effect on the model results it is necessary that data be written out as full precision binary. Checkpointing can also be used as an aid to debugging by forcing a checkpoint at a timestep just prior to some failure point, and restarting the IFS with additional user diagnostics or with a debugger such as *totalview*. As the restart begins at the timestep just before the failure, much time can be saved in debugging as typically several debug runs are required. This is particularly important when a failure does not occur in the initial few timesteps of the model. The initial approach to checkpointing was for each processor to write its part of the IFS computational data structures to a separate file. This had the disadvantage that restart had to occur on the same processor configuration as the checkpoint, which was very restrictive. It had the further problem that the number of files produced became very large and the size of files small as the processor numbers increased. This was resolved by gathering the grid-point and spectral data structures into complete fields and writing these complete fields out to disk. Restart could then occur with any processor configuration with sufficient memory to run the problem. While the approach is serial, this is not such a problem as checkpoints are not frequent events.

3 Radiation Interpolation

The IFS radiation computations are very compute intensive, and for this reason are performed only every 3 hours (i.e. every 12 timesteps) during a 10 day T213 forecast. As a further means of reducing computing cost, the radiation computations are performed on a grid that is coarser than the reduced grid used for computations in grid-point space. To discriminate between the two grids we refer to them as the coarse grid and the fine grid. The coarse grid is typically one in every four points of the fine grid at equatorial latitudes, increasing gradually to every point at polar latitudes as shown in figure 2. Radiation interpolation is then the process of transforming variables between the two grids, using FFTs for this purpose in the shared memory implementation of the IFS.

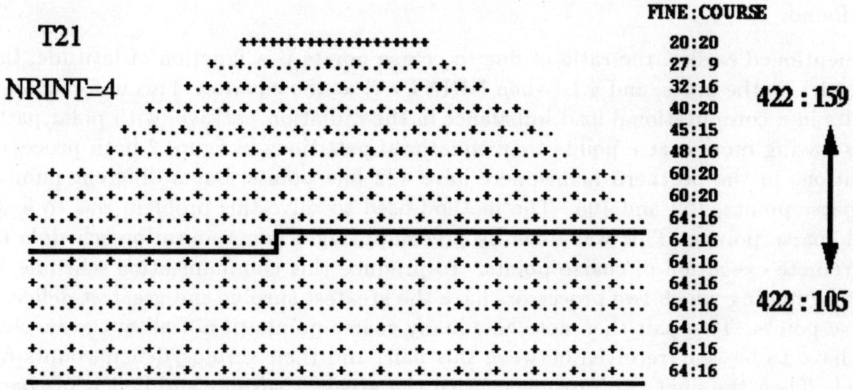

Figure 2: Fine and coarse grids used in radiation interpolation at T21 resolution. '+' denotes both a fine and coarse grid point, '-' denotes only a fine grid-point. The solid lines indicate partitioning over 2 processors.

FFTs are very suitable here, as the parallelisation strategy for shared memory has access to complete latitudes (scanned north to south). However, in our distributed memory implementation, complete latitudes are available only in very restrictive configurations (LAPPLE=T, LSPLIT=F), these being described in [1] and [2]. In general (NPROCA,NPROCB) configurations, only parts of latitudes are available in each processor. Therefore if we wanted to use FFTs we would require some additional transpositions which would involve significant communication and have a negative impact on scalability. This was compounded by the fact that the radiation interpolation routines were called within an outer NPROMA blocking loop, i.e. potentially many times within a single time-step depending on the value of NPROMA. For shared memory, NPROMA is the number of grid-points (vector length) for a set of complete latitudes, but on distributed memory represents a constant blocking factor in grid-point space, unrelated to latitude boundaries.

These problems were resolved by, firstly, relocating the radiation package to outside of the NPROMA blocking loop, and secondly, by supporting alternative ways of interpolating between fine and coarse grids. For transforming from the fine to the coarse grid, simple sampling was used, and in the other direction, Lagrangian cubic interpolation, which requires a maximum of 4 coarse grid-points from neighbouring processors for each latitude. As these clearly affected the results of forecasts, a meteorological evaluation was performed comparing the different radiation interpolation methods for a number of forecasts, and no significant degradation in forecast quality was found.

As mentioned earlier, the ratio of fine to coarse points is a function of latitude, this being 1:1 at the poles, and 4:1 (when NRINT=4) at the equator. This unfortunately results in a computational load imbalance in the radiation package, with polar partitions having more coarse points than equatorial partitions In figure 2 both processor partitions in the northern hemisphere have 422 fine points, but a different number of coarse points, 159 and 105. The method used to solve this problem was to load-level coarse points. This was done by precomputing a communication schedule for the remote execution of coarse points. To produce this communication schedule we start by finding which two processors have the greatest surplus and greatest deficit of coarse points. This pair then note how many coarse points (if not all of the surplus) will have to be sent/received between this pair, and their surplus/deficit counts updated. Then the next two processors with the greatest surplus and deficit of coarse points is found (taking into account the previous update), etc. This is repeated until load balance has been attained. What we have computed (at initialisation time) is a communication schedule for load balancing coarse points with the minimum number of messages. Radiation coarse grid load-balancing is essential for the IFS to scale on a distributed memory system. The method we have chosen to implement this works very well, but at the expense of adding yet more message passing.

4 Semi-Lagrangian Communications

The semi-Lagrangian (SL) calculation in the IFS consists of two parts, the computation of a trajectory from a grid point backwards in time, and the interpolation of various quantities at the departure point and at the mid-point of the trajectory [4]. The issue here for distributed memory systems is that both of these parts require access to grid point data held on neighbouring processors. This is resolved in the IFS today by constructing a halo of all grid columns from these processors based on the global maximum wind likely to be encountered. While this approach is relatively straightforward to implement it has a number of disadvantages. The first of these is that the use of a global maximum wind for determining the extent of the halo is a non-scaling factor as processor numbers increase. This is demonstrated in figure 3, where two examples of processor partitions are shown, a square partition and a long thin rectangular partition. It can be easily seen that the halo area does not reduce by the same factor as the decrease in core area.

Another disadvantage of the present halo scheme is that a large amount of halo data is communicated but never used. To quantify this, an experiment was performed to record SL halo area grid-point usage, for a number of processor configurations at T106 resolution. Figure 4 shows the results of one such run for an arbitrary choice of processor (1,1), where the number of grid points in the full (as derived by max wind)

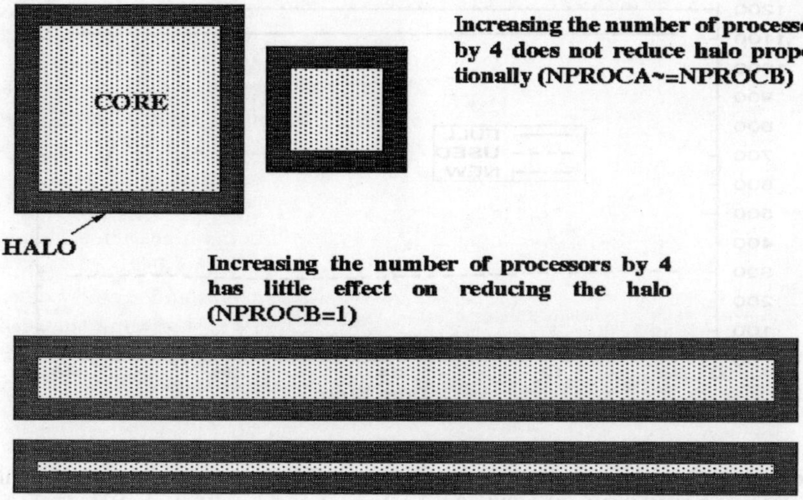

Figure 3: Semi-Lagrangian halo examples. This figure demonstrates the non-scaling effect of increasing numbers of processors (e.g. 4 times) on the SL halo area. In these two cases, the core region has reduced by a factor or 4, but the halo has only halved for square like partitions, while only a small reduction is observed for long thin partitions.

halo is compared against the number of grid points actually used at each time step. These experiments showed that between 18 and 30 percent of the communicated halo was used for all processors and time steps. This amount of usage was about what was expected. For a partition with high local maximum wind, we would expect less than 50 per cent of halo usage, as typically halo points from one direction would be required. In the case of low local maximum wind, then only a small amount of the halo would be used. As the number of halo points used changed very little between time steps, it was decided to measure how many new points were required between consecutive time steps. These new points are also shown in figure 4 for processor (1,1), while the distribution of such points over all processor/time-steps is shown in figure 5. This distribution is interesting as it suggests that a dynamic SL halo strategy could be used in the IFS. Clearly, if all halo points are communicated dynamically the overhead of such communication would be very high, as this would be dominated by a large number of relatively short transfers. However, if a high proportion of points could be communicated together based on halo usage from the previous time-step, then this would amortise the cost of a less efficient dynamic approach for the small number of extra points.

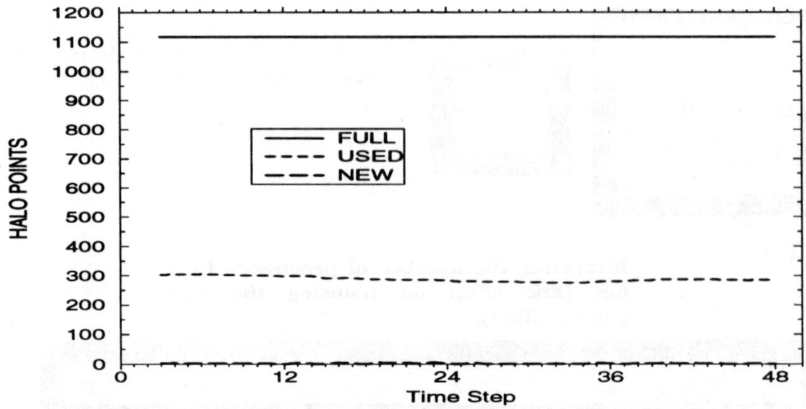

Figure 4: Semi-Lagrangian Halo Usage, T106 resolution on 64 nodes (NPROCA=16,NPROCB=4). This figure shows how many halo points are actually used by an arbitrary processor on a run of 48 time-steps. The halo communicated today (FULL) for this processor, and the number of halo points that are new over those on the previous timestep are also shown.

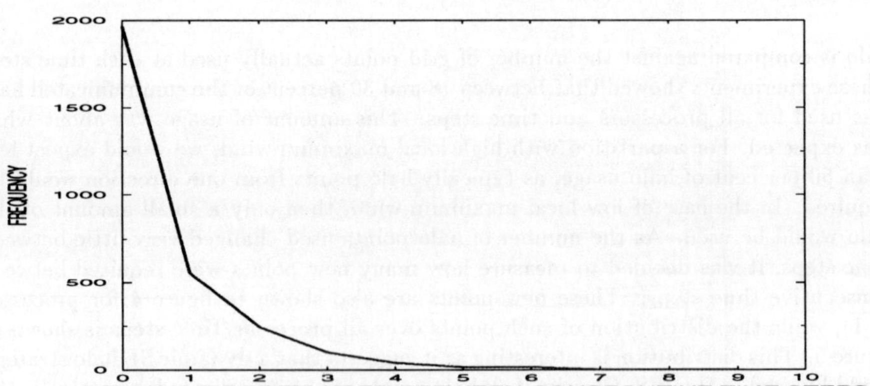

Figure 5: Distribution of numbers of new halo points, for all 64 processors and 48 timesteps, at T106 resolution.

5 Performance

The only performance metric which really matters is wall clock time. However, in order to investigate and optimise code performance, it is useful to utilise the hardware performance monitor facility in order to measure floating point execution rates. This can reveal, for example, poor vector behaviour due to short vector lengths, or due to memory bank conflicts. Figure 6 shows a display from such a performance monitor facility for a T106 execution, and illustrates the most expensive routines. It can be seen that there are 11 routines which each cost more than 2% of run time. These are components of convection, radiation, Legendre transformation and semi-Lagrangian advection. When we examine the execution rates within these routines (table 2), we see a great variation ranging from about 200 Mflops up to in excess of 1100 Mflops. The reasons for these variations are partly understood but in particular the conditional vector code within the physics routines needs to be studied to see if there are more efficient code structures and/or compiler directives that can be applied. The overall sustained rate for IFS is about 770 Mflops per processor, which is 35% of the theoretical floating point peak execution rate.

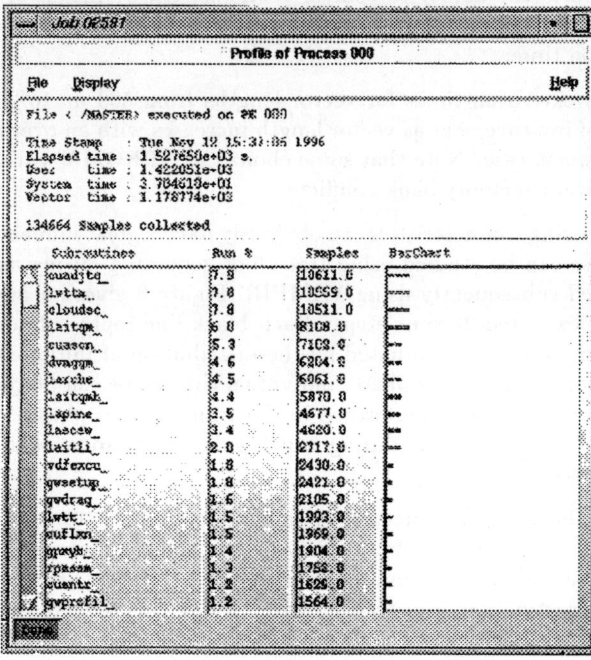

Figure 6: Profile of most expensive routines in IFS executing at T106 resolution.

Table 2: Floating point execution rates for the overall IFS and for a selection of the most expensive routines

	Mflops
CUADJTQ	756
LWTTM	1129
CLOUDSC	684
LAITQM	336
CUASCN	313
LARCHE	924
LAITQMH	345
LAPINE	385
LASCAW	209
LAITLE	273
Total IFS	770

Since IFS supports the ability to control vector length by means of a run time parameter (NPROMA), it is interesting to investigate the influence of this parameter on the execution time.

Figure 7 shows execution times for vector lengths from 500 up to 2000. There is a general trend of improvement as vector length increases with an overall gain of about 25% from the worst case. Note that some choices of NPROMA will almost certainly generate significant memory bank conflicts.

The message passing characteristics can obviously have a great influence on the overall run time. These can be studied using trace files generated during an IFS execution and interrogated subsequently using VAMPIR. Figure 8 gives an overall impression of a run which executed 24 time steps. Each black line represents one message and the initial 1.5 minutes are dominated by the distribution of initial data fields to the processor memories. After this phase, individual time steps can be readily identified, separated by communication performed in the transposition phases. The two larger time steps are 'radiation' steps during which substantial additional computation is performed (see section 4).

This trace can be examined in more detail by zooming in to any desired level of detail. Figure 9 illustrates this by identifying one specific message belonging to the SL halo communication phase and displays associated parameters, namely: message length, source, destination, and transfer rate. Using this tool, it should be possible to identify messages which are being communicated at sub-optimal transfer rates.

In order to quantify overheads, IFS has been instrumented so that, for example,

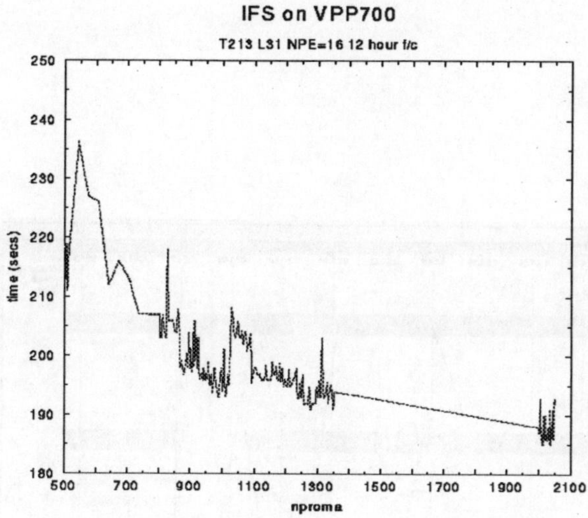

Figure 7: Forecast execution time as a function of vector length.

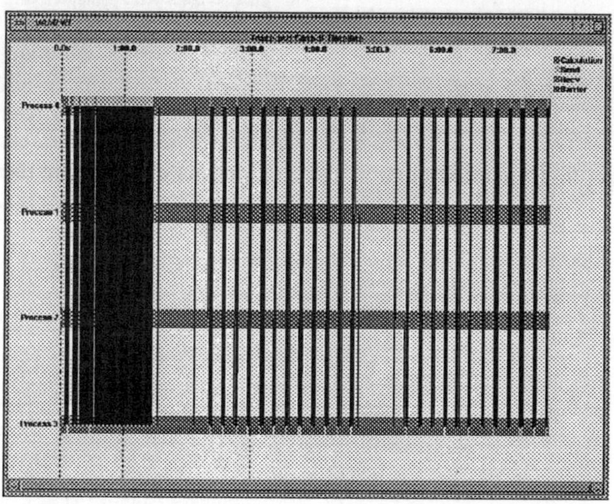

Figure 8: Message passing trace for initial phase of IFS (24 steps).

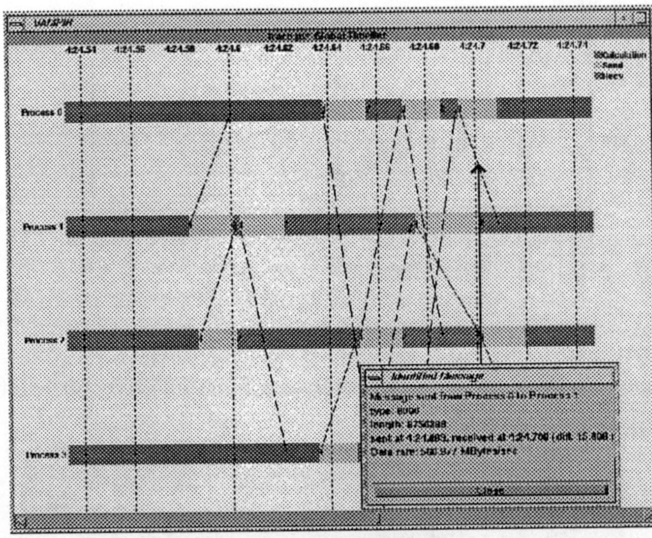

Figure 9: Expanded portion showing details of a specific message.

transposition cost, SL communication, and load imbalances can be separately identified. Figure 10 illustrates these for 3 differing model resolutions, showing that even at relatively low resolution, parallel execution utilising 46 PEs of a VPP700 exhibits acceptably low overheads.

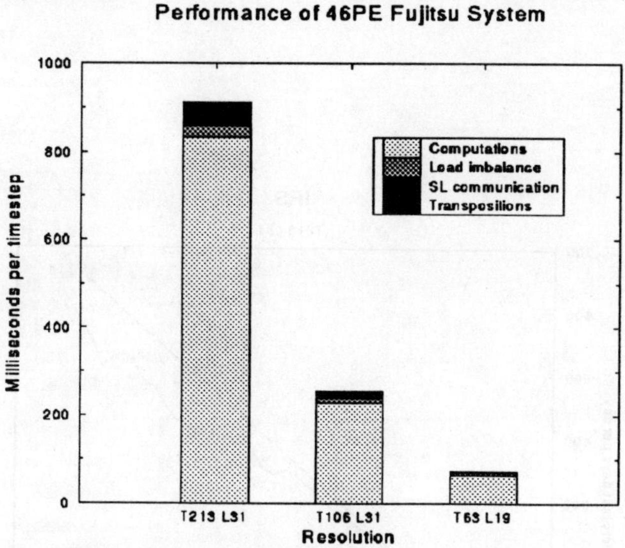

Figure 10: Cost of computation, work imbalances and message passing overheads for several model resolutions using 46 VPP700 processors.

To investigate the parallel performance when utilising a variety of machine partitions, the run time of IFS has been plotted against number of processors from a minimum of 4 (due to memory requirements) up to almost the full machine. Figure 11 shows these timings with Cray-C90 measurements for comparison. From 4 to 24 PEs the parallel efficiency is close to 100%. Even with the full machine, the parallel efficiency does not fall below 90%.

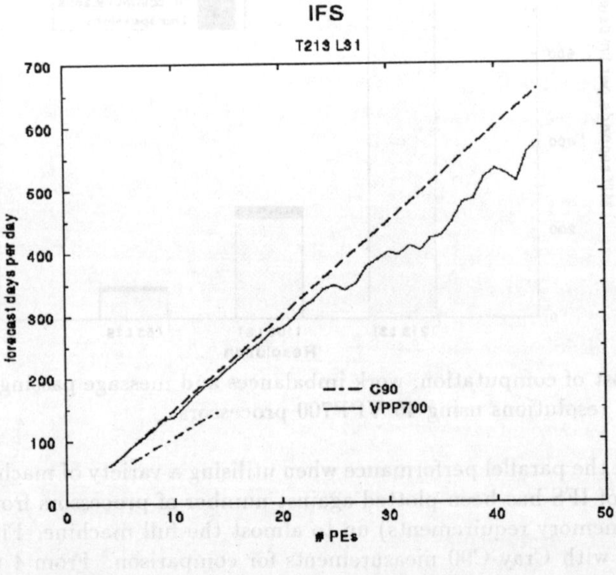

Figure 11: Forecast execution rates (forecast days per compute day) at T213L31 for various partition sizes of VPP700. Cray C90 rates are shown for comparison. The perfect speedup line assumes that the 4 processor efficiency is maintained without loss in larger configurations.

6 Summary

The current version of IFS runs in production on a VPP700 and can utilise any number of available processors. The parallelisation has been achieved using message passing, without sacrificing the shared memory parallel implementation. Performance measurements show that parallel scalability remains good when using more that 40 VPP700 processors. Load balancing efficiency has been maintained within radiation calculations at the expense of more complex code and message passing overheads. Semi-Lagrangian halo communication is much larger than is necessary and can be reduced in the future at the expense of more complex code.

References

[1] Barros, S. R. M., Dent, D., Isaksen, L., Robinson, G., Mozdzynski, G. and Wollenweber, F. : *The IFS Model: A parallel production weather code*, Parallel Computing **21** (1995), 1621-1638

[2] Barros, S. R. M., Dent, D., Isaksen, L. and Robinson, G. : *The IFS Model: Overview and Parallel Strategies*, Proceedings of the Sixth ECMWF Workshop on the use of Parallel Processors in Meteorology, G-R Hoffmann and N. Kreitz eds, World Scientific, 303-318.

[3] Dent, D., Isaksen, L., Mozdzynski, G., O'Keefe, M., Robinson, G. and Wollenweber, F. : *IFS Model – Performance Measurements*, Proceedings of the Sixth ECMWF Workshop on the use of Parallel Processors in Meteorology, G-R Hoffmann and N. Kreitz eds, World Scientific, 352-369.

[4] Ritchie, H., Temperton, C., Simmons, A., Hortal, M., Davies, T., Dent, D. and Hamrud, M. : *Implementation of the semi-Lagrangian method in a high resolution version of the ECMWF forecast model*, in Monthly Weather Review 123 (1995), 489-514

[5] WMO. WMO publication No. 306, Manual on Codes, Vol. 1, Part B, plus Supplements No. 1, 2 & 3. Technical report, World Meteorological Organisation, 1988.

Parallelization of Observation Processing

Sami Saarinen
Drasko Vasiljevic
Heikki Järvinen

European Centre for Medium-Range Weather Forecasts
Shinfield Park, Reading, Berkshire RG2 9AX, U.K.
e-mails: `firstname.lastname@ecmwf.int`

Abstract

The observation processing is an integral part of the meteorological data assimilation. Highly compressed BUFR-data sets are made available for the ECMWF variational data assimilation scheme through the flexible and computer independent CMA-format. However, the volume of data is growing rapidly mainly due to availability of new satellite data. Therefore, suitable data partitioning and parallelization is needed in order to maintain feasibility and scalability of the chosen assimilation scheme for the current and future parallel computer systems at ECMWF.

1 Introduction

The observational data is archived at ECMWF in a highly compressed BUFR-format [1]. Despite its compactness, this WMO-standard is less suitable for optimized numerical operations. Instead, its usage leads to a vast overhead due to unpacking and packing of data, which slows down the aggregate application performance.

At ECMWF, a new and flexible observational data format has been developed which is especially suitable for high performance computing. We describe this new Central Memory Array or CMA-format in more detail in section 3. The aim is to eliminate the BUFR-data while performing vector and parallel operations over observational data. Yet the corresponding BUFR-data are made available (for experiments, re-analysis, etc.) through standalone encoding routines. This is a necessary option because usually large CMA-files are not intended to be stored in the archive.

The new CMA-format could have introduced a problem in terms of enormous data volume, without the careful implementation of parallelization and data partitioning strategies[1]. This article describes the measures that have been taken into consideration in order to make handling of large observational data sets possible on virtually any modern parallel platform capable of message passing.

2 General dataflow

The following discussion applies to a single processor implementation only. In section 4, where the module OBSORT is introduced, the parallel aspects are covered in more detail.

A data assimilation cycle (see figure 1) starts with the retrieval of BUFR-data. Currently there are three different BUFR-files involved: conventional observations or GTS, and TOVS and SCAT satellite observations. Files are prepared for a six hour data assimilation period both in the 3D- and 4D-VAR [2] contexts. In the future, BUFR-files for six, twelve or twenty four hour periods will be prepared for the purposes of 4D-VAR. From the storage point of view, BUFR-files occupy up to tens of megabytes, depending on the chosen BUFR-compression scheme.

In the first step, called OBSPROC/MAKECMA (or shortly: MAKECMA), BUFR-files are read and decoded. All but formally erroneous observational data are transformed to the CMA-format and written to a so-called Extended CMA-file (ECMA). Furthermore, the unpacked BUFR-data are also written into another BUFR-format conforming file, where the high BUFR-compression rate is relaxed. The output BUFR-file maintains the same order of the observations as its counterpart CMA-file, which in turn contains only about 80% of the information that is present in the BUFR-file. The reduction in information content is the main reason to create a new BUFR-file. However, it should be emphasized, that the BUFR-data as such is not used in the true computations.

In the next step, the ECMA-file is passed to the SCREENING of observations that rejects undesirable observations using information of the meteorological fields within the IFS-code (Integrated Forecasting System [3]). At the startup time the current state of the atmosphere is known in terms of the observations and the best "a priori" estimate of the atmosphere, together with the estimate of its uncertainty (or error). The quality of observations is checked against a six hour forecast valid at the analysis

[1] Considerably different strategies from those presented recently in [9].

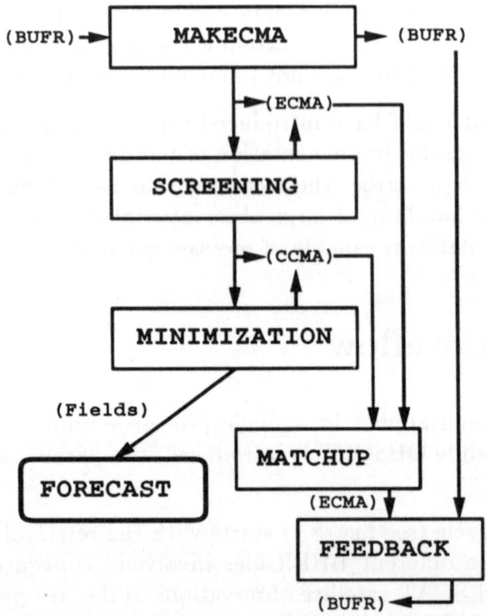

Figure 1: Dataflow in observation processing at the ECMWF on a single processor implementation.

time. The SCREENING picks up the best set of observations by rejecting erroneous, duplicated, blacklisted etc. observations from further processing.

All the qualified observations are written to a so-called Compressed CMA-file (CCMA). It still follows the CMA-format, but occupies much less space – about 1/10 of the original ECMA. The ECMA-file is also updated, since there is a need for data monitoring even of the rejected observations.

The observations for the actual data assimilation (MINIMIZATION [5]) are passed through a CCMA-file. The file is read into the memory in one large chunk to make subsequent operations on the data as fast as possible. Interpolation and extrapolation from the forecast model grid point space to the observation space can take place in a flexible manner without the need to carry out complicated data conversions present in older systems.

The MINIMIZATION provides the initial condition fields for the subsequent global forecast. It has also updated the CCMA-file with the observation departures from the analysis fields. At this point it is possible to start feeding information back to the archive. In practice this step is postponed until the time critical forecast model run has been finished.

The subsequent observation processing step to MINIMIZATION is called MATCHUP. It is part of the OBSORT-library, that will be described in section 4. The purpose of MATCHUP is to read the latest CCMA- and ECMA-data and return the extra information found in the CCMA-file back to the ECMA-file. This enables bringing the data that existed after MINIMIZATION only in a reduced set back to an extended CMA-set.

When the ECMA-file is up to date, the final part of the cycle begins. It is called OBSPROC/FEEDBACK (or shortly: FEEDBACK). Its purpose is to encode the ECMA-file back to a highly compressed BUFR-format. It also separates data back into the groups of GTS, TOVS and SCAT and retains the original input time periods (typically 6 hours). For this purpose FEEDBACK requires not only the updated ECMA, but also the (semi-)original BUFR-file from the MAKECMA output. This is a necessary burden since this BUFR-file contains some information not present in the ECMA. After the FEEDBACK-run the resulting BUFR-data are ready for archiving.

3 The CMA-format in a nutshell

This section explains the CMA-format in a concise way. Basically, a CMA-file consists of a stream of binary floating point numbers without logical file record delimiters. Each floating point number occupies eight (8) bytes (64-bits) and conforms to the IEEE 754-1995 -standard. Therefore, the format is fully portable to practically any modern computer system.

Low level I/O to the CMA-file is done via ECMWF's PBIO-library (Pure Binary I/O [7]). However, since the file is built up of a series of bytes, the standard C I/O-library calls could also be used. A unique feature of the CMA-format is that every logical section begins with the length of itself, which makes memory allocations and cross-checking relatively easy.

Logically a CMA-file is divided into two major parts: DDRs and BRD. The DDRs (or Data Description Records) give information about the contents of the actual data. Currently there are two DDRs of total size of 6144 floating point words. The BRD (or Bulk Report Data) consists of a number of variable length records – or observational reports. Each report in turn is divided into the header and body parts. The header contains common information about the report (like station identifier, latitude, longitude, observation type, etc.). The body parts contain actual values (observed, departure, error etc.) and data component type of an individual observation. There are usually multiple bodies per report.

The genuine integer numbers (count information, bitfields, etc.) and text strings are expressed in floating point numbers, since this is the only low level format available for CMA. In order to maintain accuracy, integers must contain only 32-bits of significant data, of which 31 bits fit easily into a 64-bit floating point number's large mantissa part. The remaining one bit is left for a sign bit. Integers are converted to the floating point numbers through regular substitution without explicit bit manipulation.

In the case of text strings, each "letter" is encoded into an extended ASCII number from 0 to 255 and treated as a regular integer. The scheme is portable compared to the more conventional style of programming where one attempts to pack eight bytes per floating point number[2]. Our scheme could have meant that a lot of space would have been wasted, if we had large amounts of text strings in the CMA-file. Fortunately, this is not the case since (almost) the only text string present is the eight character station identifier in the report header.

4 Parallelization and data partitioning

Due to the high data volume and the time critical scheduling in operations, it is necessary to parallelize all observation processing components. Most of the modules described in section 2 rely on availability of large memory, since the CMA-data are brought in-core for efficient data management reasons. This eliminates the need for slow random access I/O operations when the contents of a particular observation are required.

The fact that there is always a limit for the maximum available memory per processor enforces us to look at parallel processing almost immediately. Today a typical set of CMA-data for a six hour data assimilation period consume a quarter of a gigabyte of disk space, and more than twice as much in terms of in-core memory. When brought into memory the total CMA-array size is often doubled due to the shuffling algorithm (see section 4.3), where both incoming and outgoing CMA-data is kept in memory. Also some non-neglible in-core space is needed to hold certain global table information.

The disk space requirement may soon in fact exceed the gigabyte barrier, reaching multiple gigabytes in the not so distant future. This poses a serious problem on many 32-bit Unix-systems, which often support single file sizes "only" up to 2GB. Also excessive I/O-times would make runs unfeasible for a sequential computer.

From the discussion above it is clear that to accommodate such a vast amount of observational data would require a lot from a single processor computer system. Such systems are rare and expensive. Thus a parallel approach has to be adopted. As it will be shown later in this article, with careful partitioning of data one can solve the space problems rather easily using standard high performance solutions.

Although the SCREENING-module effectively reduces the size of the CMA-data by

[2]This "HOLLERITH"-packing is not portable across computer platforms.

an order of magnitude, it is that very module which requires the maximum amount of in-core memory as well as disk space time during the course of a data assimilation cycle. Fortunately, the SCREENING is a part of IFS, and thus the parallelization itself has been addressed in [4] and existed from the start. Only minor additions were needed to accommodate the handling of observational data.

Furthermore, decoding and encoding the BUFR-data in MAKECMA and FEEDBACK, respectively, although done only once in a cycle, is by nature very scalar. Fortunately, they are more or less "embarrassingly" parallel. This is to say, every processor can perform its unpacking or packing operations independently of each other, and therefore the performance ought to scale nicely[3].

4.1 *Data partitioning scheme*

At first sight, in the parallel implementation it might look feasible to split up the observational data set evenly (and geographically randomly) among the available processors. However, there is a variable amount of floating point operations per particular observation type when calculating contributions to the cost function in the MINIMIZATION. Furthermore, it is convenient to reach approximately the same geographical distribution of observations as in the grid point space in order to reduce communication between processors during interpolation and extrapolation phases. This led us to assign an observation type dependent weight per CMA-report [10], and have the same processor grid (`NPROCA` × `NPROCB`) as the MINIMIZATION.

We have found the following geographical partitioning of observational data to work well and is also practical for observation handling (see figure 2):

1. Set the origin for observational data space to the North Pole–Greenwich since this obeys the convention chosen in the GRIB-field origin definition.

2. Choose `NPROCA` to be a number of latitude bands from north to south, and `NPROCB` to be a number of longitude boxes within each latitude band.

3. Set the number of processors to `NPROC` = `NPROCA` × `NPROCB`.

4. Read local CMA-files and build up a table on the observational data, containing for each observation the geographical and processor location, unique sequence number, timestamp, 4D-VAR timeslot, relative weight, pointers to local CMA-data, etc.

5. Communicate the local table and sort locally the resulting global table with respect to timeslot (4D-VAR) and timestamp, and unique sequence number.

[3]Only if the I/O could be ignored.

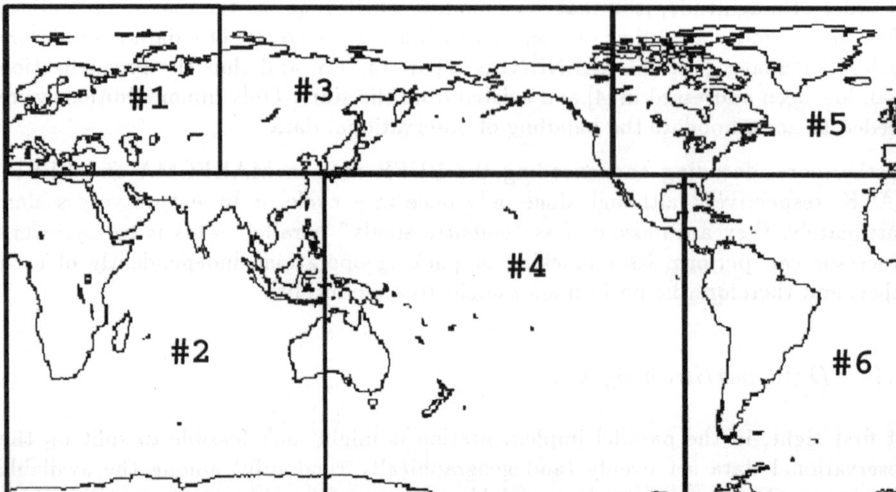

Figure 2: A typical split of the globe containing observational data, produced by OBSORT for a 2x3 processor grid. Each box contains approximately the same amount of observations (in a weighted sense). A number denotes the processor id to which observations in a particular box belong to. Partition boxes are different from those to be used in the grid point calculations.

6. For each timeslot, sort the locally available global information table with respect to latitude (from north to south). Subdivide the table into suitable parts, so that every latitude band contains about the same amount of observations in a weighted sense.

7. Continue for each latitude band to resolve the final longitudinal boxes.

8. Assign one box for each processor and update the destination processor information (the processor where a particular observation ends up) into the global table.

9. Shuffle the actual CMA-data based on the global information table. This step involves essentially all-to-all communication, where every processor (very likely) sends few CMA-reports to every other processor (including itself, but via local copies, not message passing). An efficient way to communicate data is to use parallel data transfer channels with a tournament table -like approach described in [6].

4.2 *Module OBSORT*

We have encapsulated the observational data partitioning scheme into a separate module called OBSORT. The source (mainly Fortran-90) is organised as one (but logically two) versions: a library version and a standalone program. Typically OBSORT is accessed via a library interface, because it brings parallelism into the observation processing in a modular and controlled way. The standalone mode is used only on special occasions, for example when particular CMA-data needs to be redistributed.

The parallelization scheme described earlier is suitable for BUFR-data partitioning also. Therefore, OBSORT has been extended to cope with BUFR-data alone, as well as with the CMA-driven kind where the counterpart BUFR-data is present. Furthermore, in MATCHUP, there is a need to bring updated and usually remote observational CCMA-reports back to the parent processor that owns the corresponding complete ECMA-report. Since there is a clear CCMA-data shuffle present, the functionality of MATCHUP is made available via OBSORT as well.

Summarizing OBSORT's functions, one can find four (4) different modes of operation under the shelter of the same program/library:

Table 1: Modes of operation for OBSORT functions.

Mode	Submodule	Description
0	BUFRsort	Partition and split the BUFR-data among the available processors (no geographical order).
1	CMA+BUFRsort	Geographically reorder the CMA-data in conjunction with the counterpart BUFR-data. So called CMA-data driven BUFR-sort.
2	CMAsort	Geographically reorder the CMA-data.
3	MATCHUP	Match up and update the ECMA-data present in one geographical distribution with the CCMA-data present in another distribution.

4.3 *Parallel dataflow*

For the parallel implementation we have to revise the dataflow diagram described in section 2 which was meant only for the single threaded execution. Program modules MAKECMA, SCREENING and FEEDBACK must be tied to OBSORT in order to run them in parallel with observations. It turns out that a call to the OBSORT-library entry point is the only notable change to introduce parallelism in observation processing.

Before going on, it is advisable to explain some internal details in reordering (or shuffling) of the observational data. The following helps us also to interpret the peculiar results obtained from the test runs (see tables 1 and 2). They show very little or even negative scalability in modules for which OBSORT is responsible. We concentrate on the CMAsort, where no BUFR-data is present. The extra cost of having BUFR-data in the CMA+BUFRsort would not alter the generality expressed here.

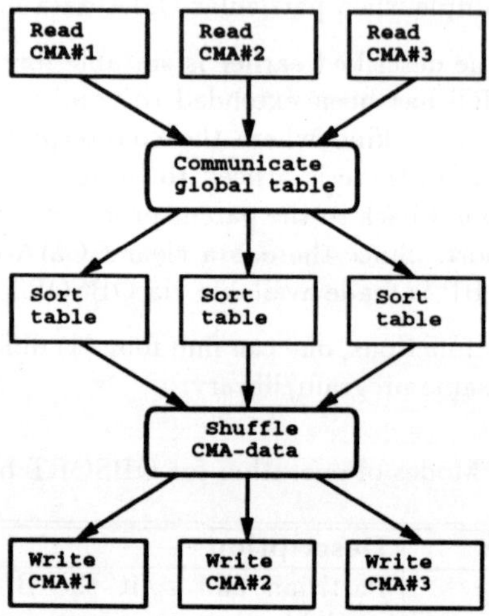

Figure 3: The five main stages when redistributing CMA-data among processors.

Figure 3 shows the five main stages in the geographical data reordering. Firstly, every processor reads independently the local CMA-files and establishes local tables which contain necessary information for the subsequent data shuffle. The second stage makes these tables global through an all-to-all communication step, and thus every processor obtains a local copy of a potentially large table. As a result, every processor knows all about the initial distribution of observations among processors. The third phase reorders the global tables (that are now physically local) independently on each processor. In the fourth step the actual shuffle of CMA-data takes place. This is also an all-to-all operation. In the worst case scenario every CMA-report would end up in a new owner processor. Finally, the new distribution is written out. Again there is one CMA-file per processor, and independent of all others.

The I/O in OBSORT to the CMA-file is done in two chunks. First, for the DDR-section, and then for the large BRD-part. Therefore, there is very little that can be improved in the high level programming.

Prior to the actual MAKECMA, we wanted to add functionality that redistributes the few input BUFR-files among the available processors. Despite an extra step, it boosts overall execution performance of the parallel MAKECMA, since all processors – not only the first few – will enter into the BUFR-data decoding phase. Currently each processor tries to read in one or more BUFR-files, in a cyclic fashion. If there were fewer files than processors, then only the few first processors would get a single file, others remaining idle. For each BUFR-message, an artificial latitude and longitude is assigned. Also the very basic contents of a BUFR-message is checked to get a proper weight assigned. After this the BUFR-data can be partitioned in a manner of the "geographical" reordering explained earlier. Despite the tricks involved, this method gives good results and offers nearly maximal reuse of the existing source code in OBSORT.

After receiving the redistributed BUFR-data, MAKECMA continues to perform its decoding functions, but now in a fully parallel mode. Every MAKECMA-process unpacks and writes a new CMA-report and the counterpart BUFR-message to a temporary file, one-by-one[4]. However, the geographical ordering is not optimal for SCREENING purposes. Therefore OBSORT needs to be consulted and a CMA-data shuffle followed by a CMA-driven BUFR-sort (i.e. CMA+BUFRsort) will be performed as a final substep in MAKECMA.

SCREENING proceeds in a more or less fully parallel way by reading local ECMA-files, one per processor, and performing its data validation functions. Once the resulting CCMA-data set is ready, it may happen that, in some of the processors, there will be very few CCMA-data left. Therefore, it is necessary to create a new geographical distribution of observational data, now applied to the CCMA-data. The new reordering is done as a final step in SCREENING.

There is a special function present in OBSORT for post-adjustment of clustered data. It was found that with a different number of processors more data was rejected in one case than in another one. Certain observation types, like AIREPs, tend to cluster around partition boundaries. In the first case a processor owned all the corresponding observations since there were no partition boundaries in the neighborhood. Whereas in the second case an AIREP-cluster was effectively split among two adjacent processors. For reproducibility reasons, it was necessary to check the existence of such clusters and move the clustered observations completely into the lowest numbered processor around the boundary of the processor in question.

For a better parallel performance in MINIMIZATION it is crucial to obtain a good load balance. Therefore, OBSORT's unique feature to assign weights for each observation report plays an important role. There are few built-in weight functions in OBSORT. Depending on the contents of the observation, a different weight is assigned. The weight functions are controlled via an external input file, where, for each

[4]We do not use I/O-to-memory buffering here, since we have very little idea in advance how much the heavily compressed BUFR-data will expand, when being transformed to the CMA-format.

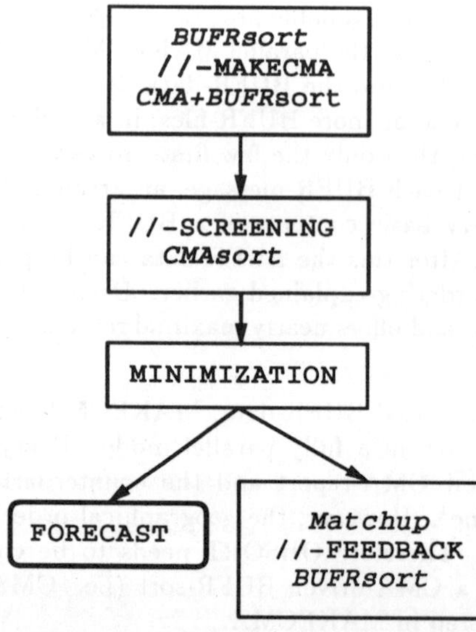

Figure 4: Dataflow in observation processing in the parallel scheme.

different observation type, separate weight coefficients can be assigned.

Finally, it was soon clear that MATCHUP could be integrated into FEEDBACK. As a result MATCHUP is performed quickly as the first parallel task in FEEDBACK. The updated local ECMA-data is then passed back in the form of a CMA-array to the actual FEEDBACK which encodes information back to a highly compressed BUFR-format. Furthermore, FEEDBACK benefits also from OBSORT's built-in feature to collect BUFR (or CMA) data into a single processor. This way we were able to go from final (but still distributed) BUFR-data back to the non-distributed and ready-to-archive BUFR-data in a rather elegant manner. Fortunately, it was not necessary to preserve the original BUFR-message order, thus simplifying programming.

5 Results

The parallel observation processing system has been tested on a 46-processor Fujitsu VPP700, Silicon Graphics server and Cray PVP systems at ECMWF. The current production machine (Fujitsu VPP700) is a distributed memory vector machine with a peak performance of 2.2Gflop/s and 2GB of local memory per processor. Processors are connected to each other via a fast crossbar interconnection network capable of more than 500MB/s point-to-point.

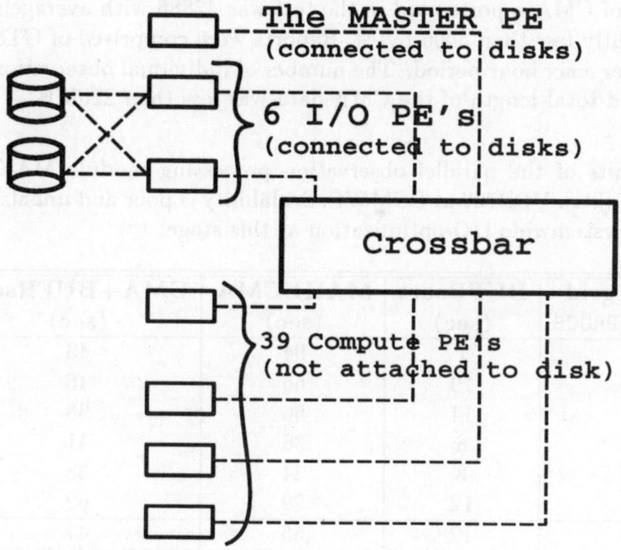

Figure 5: I/O arrangement on the Fujitsu VPP700/46 at the ECMWF.

Every CPU is capable of performing I/O-requests, but only 7 out of 46 are connected to disks. Actually only 6 out of 7 processors respond to I/O-transfer requests from other processors, since the remaining is busy in fulfilling its role as the master processor (see figure 5). Although the six actual I/O-processors are capable of performing computations similarly to other processors, large amounts of I/O-requests keep them continously busy. Therefore it was decided to free them from computations in all but special cases by using batch queue arrangements.

All processors run a full copy of the UNIX-kernel, thus leaving a bit more than 1.5GB for user memory and message buffers. Part of the user memory can be assigned to a memory resident file system to which only the local processor has access to. Also, its contents are deleted upon completion of a batch job, thus making its usage less convenient than the SSD-hardware on Cray PVP systems.

In the tests (see tables 1 and 2) we concentrated on measuring performance of OBSORT and speed of MAKECMA. The FEEDBACK-module was not tested since many internal tests have shown that it takes roughly twice the time of a corresponding MAKECMA-execution. Furthermore, SCREENING is an IFS-module, and since it performs mostly operations other than those connected to OBSORT, it was not included in the test.

Total number of CMA-reports used in the test was 47886 with average report length occupying slightly less than 5000 bytes. Reports were comprised of GTS, TOVS and SCAT-data over a six hour period. The number of individual observations ("bodies") was 808683 and total length of the CMA-data was less than 220MB.

Table 2: Results of the parallel observation processing module MAKECMA with OBSORT on Fujitsu VPP700 at ECMWF. Scalability is poor and unbalanced because of the lack of systemwide I/O-optimization at this stage.

Processor grid NPROCA × NPROCB	**BUFRsort** (sec)	**MAKECMA** (sec)	**CMA+BUFRsort** (sec)	**Total** (sec)
3 × 1	7	94	43	159
6 × 1	19	66	46	156
8 × 1	14	56	38	129
10 × 1	8	36	41	102
12 × 1	8	34	38	97
16 × 1	12	29	62	124
2 × 3	12	55	47	140

5.1 *Interpretation of results*

The tests were carried out on a non-dedicated Fujitsu-system with non-optimized I/O. Therefore, I/O-speeds tend to be relatively low, up to 4MB/s for write, and slightly more for read. More importantly, the I/O was in serious imbalance, as can be seen from the table 2. There, the time spent in a CMA-data read or write was measured for three different processor configurations. Although the size of the local data scale down by the increase of processors, the corresponding local I/O-times do not decrease at all. Timings behave unpredictably and seem to have more or less constant cost independent of the local data size.

Bad scalability can perhaps be understood from the I/O-setup of the system in figure 5. Also the fact that the I/O-subsystem was not dedicated for these runs certainly played some role. Non-scalable I/O hurts dramatically the aggregate performance. However, it should be noted that the kernel operation – decoding the BUFR-data or MAKECMA itself – shows scaling, despite all the I/O it performs at the same time. But in MAKECMA the I/O is perhaps less synchronized than in OBSORT, and benefits from the current system setup.

It is expected that when the number of observations go up, the remaining major operations (BUFR- and CMA+BUFR-sorts) will not increase their time proportionally,

Table 3: I/O-performance for local CMA-read and write operations in the parallel implementation on Fujitsu VPP700. Despite decreasing size of the local data, the timings remain more or less unchanged.

Processor grid	Proc #	CMA-read			CMA-write		
		Size (MB)	Time (sec)	Rate (MB/s)	Size (MB)	Time (sec)	Rate (MB/s)
3 × 1	1	62.4	5.7	10.9	60.8	16.2	3.8
	2	77.5	7.1	10.9	71.4	18.1	3.9
	3	75.5	8.2	9.2	83.1	19.6	4.2
6 × 1	1	30.7	4.2	7.3	38.5	19.2	2.0
	2	38.5	5.4	7.1	22.3	7.5	3.0
	3	38.0	4.2	9.0	33.0	11.1	2.9
	4	31.7	4.2	7.5	38.5	14.6	2.6
	5	37.6	5.0	7.5	37.6	14.1	2.7
	6	37.5	5.9	6.4	45.5	19.1	2.4
10 × 1	1	20.7	5.7	3.6	28.0	11.7	2.4
	2	21.8	7.0	3.1	24.4	5.3	4.6
	3	21.8	8.1	2.7	13.5	4.9	2.8
	4	21.2	8.5	2.5	16.1	6.1	2.6
	5	21.4	7.9	2.7	21.8	13.3	1.6
	6	21.6	6.7	3.2	22.0	14.0	1.6
	7	21.4	7.1	3.0	23.2	18.8	1.2
	8	21.9	5.4	4.1	23.7	19.0	1.2
	9	22.2	5.8	3.8	20.7	17.1	1.2
	10	21.4	5.5	3.7	32.0	15.0	2.1

since all operations (except I/O) in OBSORT are fast and mainly vectorized. This can be achieved in practice if certain I/O-optimizations (memory resident I/O, disk preallocation, larger I/O buffers) for Fujitsu-runs are implemented.

Some good news is that the all-to-all operation when performing geographical re-ordering of CMA-data takes only a fraction of a second. Total aggregate bandwidth varies between 100 and 300 MB/s with data to be exchanged being – in the worst case – about the total size of CMA-data. Relatively low performance of the exchange operation is explained because it includes the time for gathering (scattering) the data from (to) the large random access CMA-arrays into (from) a message buffer that is currently set to a "safe" 2MB. This leads to multiple transfer transactions between processor pairs, and thus extra latencies. Optimally the size of the message buffer should be dynamically adjustable and equal to the total size of CMA-data divided by

NPROC^2. However, the scheme may lead to memory shortages even on the VPP700, as the CMA-data size keeps growing.

All message passing in OBSORT is based on the ECMWF message passing library MPE. The library is built on top of Fujitsu's native MPlib to minimize delays. A portable MPI-version is also available.

It should be noted that there is room for dramatic performance improvement by use of memory resident files. However, at the time of writing its usage is less convenient for parallel jobs, since all data is lost upon completion of a parallel batch job, unless copied to disk thus moving the I/O-performance problem elsewhere. Some unofficial tests have shown, though, that the performance boost is about a factor of 5 when using memory resident file system, and OBSORT seems to benefit most from this setup.

6 Conclusions

An efficient partitioning scheme to parallelize observation processing has been presented. It operates on unpacked BUFR-data called CMA (Central Memory Array) and enables fast vector and distributed memory -parallel operations over a vast amount of observational data.

Although the parallel implementation works well, it suffers from serious I/O bottlenecks existing in the present VPP700 -system setup at ECMWF. Therefore, a thorough scalability study is less relevant at the moment. Use of the memory resident file system will improve the performance, but requires more support from the system software side.

This article covered only one tip of the iceberg of what is going on currently in the ECMWF data assimilation program development. It gave a picture of the complexity while trying to balance between the flexible data management and ever growing size of the data sets on a parallel platform. The software systems presented have been in successfull operational use at ECMWF on the Fujitsu VPP700 -system since the end of September 1996.

Acknowledgements

Authors would like to thank all the staff at ECMWF who participated directly or indirectly in the project during 1995-96. Special thanks for all ideas suggested by Mats Hamrud, Lars Isaksen, Otto Pesonen and Per Undèn.

References

[1] M.Dragosavac : *BUFR User Guide and Reference Manual*, ECMWF Meteorological Bulletin M1.4/4 October 1994.

[2] L.Isaksen & S.Barros : *IFS 4D-Variational Analysis: Overview and Parallel Strategy*, in Coming of Age, Proceedings of the 6th ECMWF Workshop on the Use of Parallel Processors In Meteorology, November 21-25, 1994.

[3] S.Barros, D.Dent, L.Isaksen & G.Robinson : *The IFS Model Overview and Parallel Strategies*, in Coming of Age, Proceedings of the 6th ECMWF Workshop on the Use of Parallel Processors In Meteorology, November 21-25, 1994.

[4] D.Dent & G.Mozdzynski : *ECMWF Operational Forecasting on a Distributed Memory Platform : Forecast Model*, in Making its Mark, Proceeding of the 7th ECMWF Workshop on the Use of Parallel Processors In Meteorology, December 2-6, 1996.

[5] L.Isaksen & M.Hamrud : *ECMWF Operational Forecasting on a Distributed Memory Platform : Analysis*, in Making its Mark, Proceeding of the 7th ECMWF Workshop on the Use of Parallel Processors In Meteorology, December 2-6, 1996.

[6] T.Kauranne, J.Oinonen, S.Saarinen, O.Serimaa & J.Hietaniemi : *The Operational HIRLAM 2 Model on Parallel Computers*, in Coming of Age, Proceedings of the 6th ECMWF Workshop on the Use of Parallel Processors In Meteorology, November 21-25, 1994.

[7] J.D.Chambers : *Encoding and decoding GRIB data*, ECMWF Meteological Bulletin M1.9/3, October 1995.

[8] *Fujitsu VPP700 series : Hardware User's Guide.*

[9] Hong Q.Ding & Robert D.Ferraro : *Climate Data Assimilation on a Massively Parallel Supercomputer*, Jet Propulsion Laboratory, Caltech, Pasadena, CA. In Supercomputing'96 proceedings, Pittsburgh, November 17-22, 1996.

[10] Erik Andersson (ECMWF) : *Personal communication.*

Parallelising the Unified Model for the Cray T3E

P. Burton , A.Dickinson
Meteorological Office, Bracknell, Berks, UK.

December 20, 1996

Abstract

The UK Met Office is in the process of replacing its shared memory Cray C90/16 with a distributed memory Cray T3E. The T3E is being delivered in stages, with the final configuration of 696 PEs being available by mid 1997.

The main modelling code used by the Met Office, known as the Unified Model (UM), is being modified to allow it to work on massively parallel distributed memory machines. The UM is a grid point model supporting both prediction and data assimilation for the atmosphere and ocean. It is used for operational global and limited area forecasting as well as climate modelling.

The parallelism is being achieved by use of a regular domain decomposition. Interprocessor communication is accessed via the portable GCOM library which provides a high level common interface to a range of message passing languages.

1 Introduction

The UK Met Office's (UKMO) suite of modelling codes known as the Unified Model (UM) [1] has been in use for both operational Numerical Weather Prediction (NWP) and climate prediction since 1991. The UM consists of over half a million lines of Fortran 77 code, which include atmospheric and oceanic prediction models (which may be coupled), and data assimilation for both of these models. A wave model is currently being ported into the UM. The models can be run in many different configurations and resolutions. Table 1 shows the most commonly used configurations at UKMO, and table 2 shows new configurations for the atmospheric forecast which are planned for introduction during 1997.

In contrast to many other forecasting centres, which use semi-implicit spectral models, the UM uses an explicit grid point formulation [2] in both its global and regional configurations. The global configurations use a regular latitude-longitude grid which causes a convergence of points towards the poles, requiring a zonal Fourier filter to be applied at high latitudes to enable a longer timestep to be used. Limited area configurations use a rotated latitude-longitude grid, such that the centre of the integration domain effectively contains the equator, so minimising the distortions in the grid.

Application	Mode	Grid (Km)	Dimensions
Global Climate	Atmosphere	300	96x73x19L
Global ocean	Ocean	300	96x73x20L
Global NWP	Atmosphere	90	288x217x19L
Regional NWP	Atmosphere	50	229x132x19L
Mesoscale NWP	Atmosphere	17	92x92x31L

Table 1: Current configurations for the Unified Model

Application	Mode	Grid(Km)	Dimensions
Global NWP	Atmosphere	60	432x325x30L
Mesoscale NWP	Atmosphere	10	150x150x38L

Table 2: Future configurations for the Unified Model

The code was originally developed to make use of the vector, shared memory architecture of supercomputers, such as the the Cray C90. Over the past few years, however, scalable parallel distributed memory supercomputer architectures (commonly referred to as Massively Parallel Processors, or MPPs) have emerged as a viable alternative to the more traditional, modestly parallel, shared memory designs. This shift in hardware design, has required a corresponding change in software design to exploit the increasing parallelism; namely the introduction of message passing techniques to facilitate communication between the machine's processors.

Over the last four years, the UKMO has been investigating the potential of MPP machines, by developing message passing versions of key components of the code. This work eventually led to a message passing version of the main component of the UM — the atmospheric prediction code and this formed a part of the benchmark suite used during the recent competitive procurement process to find a replacement for the UKMO's Cray C90.

2 The UKMO goes MPP

The result of the procurement was that the UKMO decided to purchase a 696 processor Cray T3E, installation of which began in October 1996, with the full system due to be in place by mid 1997. The majority of the machine will consist of the 450MHz version of the DEC Alpha EV5 RISC processor, each processor being capable of a theoretical peak performance of 900Mflops, and addressing a local memory of 128Mb. Eight of the machine's processors are configured with 512Mb of memory - these are intended for running single processor jobs with large memory requirements. A comparison between the Cray T3E and UKMO's existing C90 is shown on table 3.

	Cray C90	Cray T3E
Number of PEs	16	696
Peak Mflops/PE	1000	600/900
Expected Max Sustained Mflops/PE	400	80/100
Memory GBytes	2	92
Disk GBytes	140	720

Table 3: Comparison between the Cray C90 and T3E

3 Overview of the UM Parallelisation

3.1 Domain Decomposition

The fundamental concept in the parallelisation of the atmospheric and oceanic models, and their data assimilation schemes is "Regular Domain Decomposition". A regular grid is overlaid on the model grid, such that each processor is responsible for computing a rectangular geographical subsection, containing all the vertical levels of the model grid. In the atmospheric model and data assimilation, a two dimensional decomposition is used, illustrated in figure 1, but the ocean model uses a one dimensional decomposition, such that each processor is responsible for a number of complete latitude rows. This means the maximum number of processors that can be used for running an ocean model is limited by the number of latitude rows in the configuration being run.

The decomposition is performed such that the subdomains contain, as far as possible, equal numbers of points. Unfortunately, this condition is not always satisfied perfectly, because the standard UM resolutions often contain awkward sizes (such as 73 rows in the climate model) which cannot be equally subdivided, so there can be a small variation in the sizes of the subdomains. This can cause a slight load imbalance in the dynamics, where the work is generally spread evenly over all points. The effect is not significant in the physics, where the natural load imbalance of physical processes, such as convective activity which is highly concentrated around equatorial regions but decreases markedly towards the poles, has a much larger effect.

The rectangular subdomain on each processor is extended by one grid point on each side to provide a "halo" area. This contains data copied from the edges of adjacent processors' subdomains, allowing the processor to update all the points in its subdomain when a second order operator is applied. The halo areas are updated as required by calling the routine `SWAPBOUNDS`.

3.2 Message Passing

There are many message passing libraries available, and while some such as PVM and MPI are available on most MPP platforms, others such as Cray's SHMEM are proprietary systems available only on a certain platform. The proprietary systems are often preferable as they

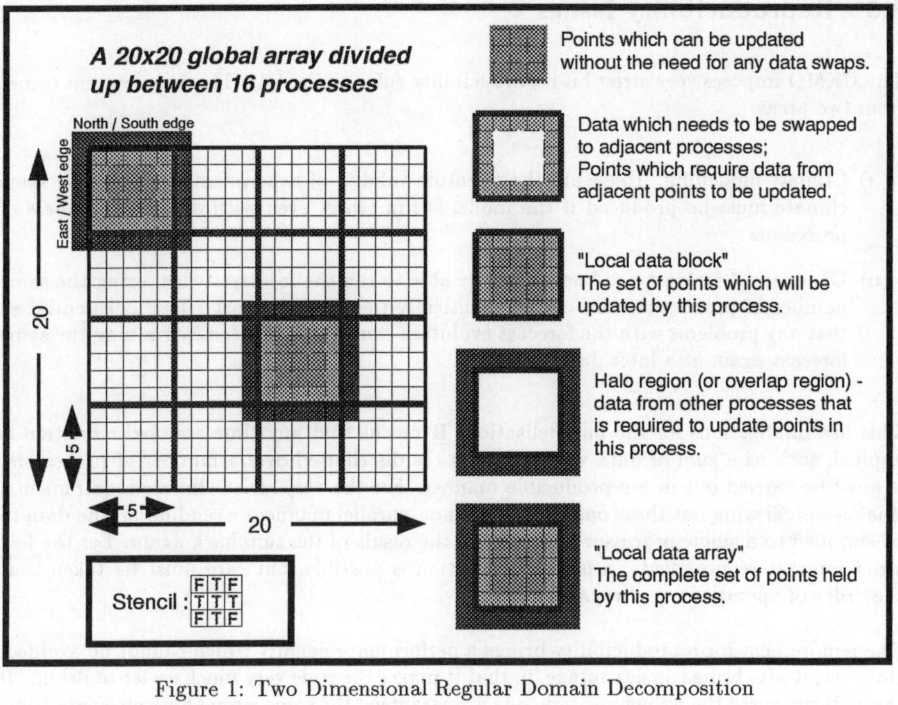

Figure 1: Two Dimensional Regular Domain Decomposition

offer the highest performance, but undesirable as they make the code non-portable.

To overcome the problem with portability, but retain the high performance of the proprietary libraries, the UKMO has elected to use a general interface library, GCOM [3] (Generalised COMmunications), which offers a consistent user interface to a number of message passing libraries, both portable and proprietary. GCOM was developed by SINTEF Applied Mathematics, with the primary aim of providing a fixed interface to many different message passing languages, with minimum performance overhead.

GCOM provides all the common basic message passing operations, such as send/receive, synchronisation, max/min and sum. SINTEF have also created an additional library, called GCG [4] (Generalised Communications for Groups), which allows global operations such as sums and max/min to be performed over subsets (groups) of processors. GCG also supports some more complex operations, such as vector sums, with and without reproducibility and an all-to-all transpose which enables a user definable movement of data between any set of processors.

3.3 Reproducibility Issues

The UKMO imposes very strict bit reproducibility rules on the UM. This requirement comes from two areas:

i) Climate modelling. To ensure the scientific validity of experiments, exactly the same climate must be produced if the model is run twice, even with different numbers of processors.

ii) Operational forecasting. Users must be able to run the forecast twice, using the same number of processors, and obtain identically the same forecast. This is essential so that any problems with the forecast evolution can be investigated by running the same forecast again at a later date.

This has implications for the parallelisation. It means that any time a global reduction is applied, such as a sum of data where the data is distributed over a number of processors, it must be carried out in a reproducible manner. For the very strict climate requirements, this means carrying out these operations in a non-parallel manner — sending all the data to be summed to a single processor which sends the result of the sum back again. For the less strict forecast requirements, a parallel operation is possible, but care must be taken that the order of operations is always identical.

The requirement for reproducibility brings a performance penalty which cannot be avoided. However, it also brings an advantage, in that it makes the code very much easier to debug. If the code produces the wrong answers, one is guaranteed the same wrong answers every time, no matter how many processors are used, so the code can simply be re-run and debugged using a smaller number of processors.

3.4 Code Maintenance Issues

The UM is a portable code, capable of being run on any platform, from a desk top PC (with enough memory!) to a supercomputer, which supports Fortran-77 plus a small set of standard extensions. The MPP version was coded so that the portability of the code was not compromised.

The UM uses the Cray *nupdate* source code manager to control the development and compilation of the code. *Nupdate* allows conditional compilation of code via an `*IF,*THEN,*ELSE` type construct, and this is used to select the MPP or non-MPP code. The scientific code has been converted to MPP so that the differences are as small as possible.

3.5 Scalability of the Parallel Code

The scalability of the model is limited by a number of factors:

i) The large number of communications that are required through each timestep (see section 5.1).

ii) The relatively small number of gridpoints per processor for small configurations, such as the climate model shown in table 1, when large numbers of processors are used.

iii) The load imbalance, especially in the physical parameterisations.

The lack of scalability to very large numbers of processors is not a real problem though, as we do not wish to use the whole machine to run a single job. Instead, a mix of jobs will be run simultaneously on the machine, as currently happens on the C90. Climate configurations will probably use less than a hundred processors each, as will ocean models, where the maximum number of processors is limited to the number of latitude rows. NWP models will use a few hundred processors each, the exact numbers being determined when the full configuration machine is available and the code optimised for the T3E.

4 Examples of UM Parallelisation

This section will examine a number of areas of the UM's functionality, to illustrate how the parallelisation has been achieved.

4.1 Finite Difference Code

The atmospheric model uses an explicit grid point formulation in the dynamics, so the code contains many loops evaluating finite difference type equations. Figure 2 gives an example of such a loop, where the array `new` is calculated by applying a finite difference function on the `old` array. A feature of the code is the use of 1D arrays in the horizontal, which increase the vector performance on machines such as the C90, at the expense of carrying out redundant calculations at the edges, which must be recalculated in a separate loop, where a global wrap-around is applied.

Figure 3 shows the same code after it has been parallelised. The `new` and `old` arrays are now just a subdomain of the entire field, and carry an extra halo around their edges, as illustrated in figure 1. The changes to the code are:

i) Use of *nupdate* `*IF DEF,*ELSE,*ENDIF` to allow conditional compilation - the functionality is very similar to the `#ifdef` type construct used by the *cpp* preprocessor.

ii) Two new `*CALL` files (these are the *nupdate* equivalent of the `#include` files used with the *cpp* preprocessor). They contain variables, most of which are passed into subroutines via argument lists, and a small number of which are carried in `COMMON` blocks. The first file (`MPPINFO`) contains general information about the decomposition that is in use, such as local sizes and position in the decomposition, and the second file (`LOOPLIM`) contains commonly used loop limits. These loop limits are defined in the control code, and are set appropriately for the processor in question, taking into account its position in the decomposition.

```
REAL new(p_field)
REAL old(p_field)

DO i = row_length+1 , p_field-row_length

  new(i) = 0.25*(
    old(i-row_length) + !  north
    old(i+1) + !  east
    old(i+row_length) + !  south
    old(i-1) ) !  west

ENDDO

DO i = row_length + 1, p_field - row_length ,
  row_length

     !  update left and right boundaries enforcing
     !  wrap around for global models

ENDDO
```

Figure 2: Original Finite Difference Code

iii) Call to SWAPBOUNDS - this subroutine updates the halo area of array old, ready for the finite difference calculations which follow.

iv) The code to apply the global wrap-around is not required for the MPP code, as the halos contain wrap-around data for processors at the edge of the LPG. The finite difference loop itself will update these points correctly, but still perform redundant calculations in the halo regions.

4.2 Filtering at High Latitudes

The convergence of the grid towards the poles when the model is run in the global configuration results in the explicit integration scheme generating computationally unstable high frequency oscillations at high latitudes. These are removed by applying a zonal FFT filter at high latitudes. Typically around a third of all rows need to be filtered. The exact number is calculated at each timestep by finding the first row at which the wind speed exceeds the CFL stability criterion — every row from this row poleward is set to be filtered.

There are two problems to be overcome in the parallelisation of the FFT filtering:

i) Generally there will be a decomposition in the East-West direction, so processors will not contain whole rows of data, while the FFT routines require complete rows of data

```
*IF DEF,MPP
*CALL MPPINFO
*ENDIF
*CALL LOOPLIM

  REAL new(p_field)
  REAL old(p_field)

*IF DEF,MPP
  CALL SWAPBOUNDS(old,row_length,p_rows,halo_size)
*ENDIF

  DO i = first_p_point , last_p_point
    new(i) = 0.25*(
      old(i-row_length) + !  north
      old(i+1) + !  east
      old(i+row_length) + !  south
      old(i-1) ) !  west
  ENDDO

*IF -DEF,MPP
  DO i = first_p_point , last_p_point
    row_length

    !  update left and right boundaries enforcing
    !  wrap around for global models

  ENDDO
*ENDIF
```

Figure 3: Parallelised Finite Difference Code

to operate on.

ii) Only around one third of the rows require filtering, so there is a load balancing problem. Processors around the equatorial regions have no work to do.

The resolution of these two problems is to do a redistribution of data, such that processors contain complete rows of data, and that all processors have approximately the same number of rows to filter. Once this transposition has been done, the FFT filtering code can be called exactly as it is in the non-MPP code. The filtered data is then re-transposed back to where it originally came from.

The process is illustrated in figure 4. Once the position at which filtering starts has been calculated, the transposition can be defined. This is done by calculating a **send_map** array which defines where all of the sub-rows requiring filtering need to be sent to, and a **receive_map** array which defines where all the rows this processor will be filtering are coming from. The GCOM/GCG routine **RALL_TO_ALL** is then called, which uses these map arrays to perform the transposition of data, moving data from the **source_data** array to the **fft_data** array, ready to be filtered. Once the data has been filtered, the **RALL_TO_ALL** routine is called again, but with the **send_map** and **receive_map** arrays interchanged, so that the data is returned to its orginal location in the **source_data** array.

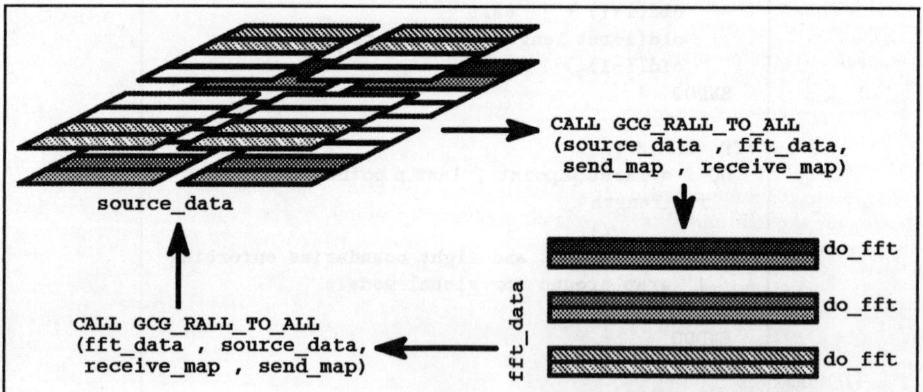

Figure 4: Data transpose for the FFT filtering

Timing results from this code on the T3E show that the complete redistribution of data over all processors is beneficial, resulting in a significantly better scalability than could be obtained by just doing an East-West transposition of data and ignoring the problem of load imbalance.

4.3 Atmospheric Data Assimilation

The analysis correction data assimilation scheme [5] is called as a subroutine from within the atmospheric model. The parallelisation of this scheme has therefore used the same 2D domain decomposition as is used in the model.

The initial implementation of the parallel scheme works by each processor being responsible for all the observations within its local subdomain. Each processor adds the increments from its local observations to an array which covers the full grid (not just the local subdomain). This full grid sized array is needed because the increments generated from an observation can sometimes spread over a very large area, well outside of the local subdomain. When all processors have finished processing their observations, the full grid increment arrays from each processor are combined to create a final global increment array which contains the contributions from all observations. This global increment array is then decomposed back over the 2D atmospheric model decomposition before the rest of the scheme is executed.

This implementation was found to have an unacceptable level of load imbalance. Observations are not spread evenly over the globe, and different types of observations have different spatial localities. For example surface pressure observations are mostly over densely populated land masses, which cover a large proportion of the Northern hemisphere but very little of the Southern hemisphere.

To overcome the problem of load imbalance, the code has been modified so that observations are spread more evenly between the processors; those not having many observations to process are given some extra observations from processors which have a large number of observations to process. This modification is not too difficult to implement since each processor already has a globally sized increment array, so can easily process observations at any point on the model grid. It should be possible to tune the distribution of observations such that the load imbalance is removed.

4.4 Some Comments on I/O

The I/O has been modified to work on MPP systems, but has mostly not been parallelised. The parallelisation of the I/O will be examined during future optimisation work.

The I/O in the UM can be broadly broken into three areas:

i) Fortran I/O for reading namelists and formatted data during the model initialisation. This has generally been left untouched, so that every processor reads in the namelists or data.

ii) Reading/writing model dumps and diagnostic data. This code required some work in order for it to function correctly in MPP mode. The data files concerned can be thought of as consisting of some header information, followed by a set of 2D meteorological fields.

The general approach is for the I/O to go through a single processor. For the headers a single processor reads the headers, and broadcasts them to all the other processors

— and when writing, the single processor just writes its copy of the header. The meteorological fields are more complicated to handle, as the decomposition means that although on disk each 2D field is a contiguous set of data, within the machine it is spread over all the processors. Special scatter and gather routines (which use the GCOM/GCG routine `RALL_TO_ALL`) allow the data to be spread out after it has been read in by a single processor, and gathered back to a single processor when it must be written.

iii) Writing/reading temporary files containing model fields — such as atmospheric data fields that are written out to disk whilst oceanic timesteps are being executed, and then read back in for the next set of atmospheric time steps.

As these files are not accessed outside of the model, there is no requirement for a single file of contiguous data, as there is for dumps and diagnostic data. It is quite satisfactory for each processor to write out its own independent file of temporary data, and then read it back in again when it is required — no communication or synchronization between processors is required. The only change required to the code is to make sure that each processor generates unique filenames for its temporary files.

Future parallelisation work will centre on the optimization of reading/writing model dumps and diagnostic data. One possibility would be to build on the code already written, by allowing a user definable number of processors to perform the I/O. This would allow us to take full advantage of the T3E's architecture which gives equal access to I/O devices from all processors, but also retain portability to other machines where this is not the case.

A T3E specific optimization would be to use asynchronous writes, so removing the significant cost of writing the large amount of diagnostic and dump data written during the model integration.

4.5 Small Executables

Small executables are short Fortran codes that typically run in a few seconds on the C90, and do some preprocessing of data before it is used in a following model run or data assimilation. The approach taken with most of these codes is to run them as single processor jobs, but run them simultaneously rather than sequentially, as at present. This means that the parallelisation is within the scripts that call the executables, rather than within the executables themselves.

5 Porting and Optimising the code for the Cray T3E

There were no major problems porting the code to the T3E, due to the portable nature of the UM, and also the fact that the T3E's user environment is very similar to that on the UKMO's current Cray C90. One of the main differences between the two systems is the move to IEEE floating point format on the T3E, requiring a conversion of any binary data moved from the C90. Another difference was the transition from the F77 compiler to the T3E's F90 compiler, but this presented no real problems.

The main effort was in tuning the code for performance on the T3E, which can be broken into two areas: improving the parallelisation aspects of the code, and improving the single processor performance. Much of the work detailed here is still in progress at UKMO, but the techniques discussed were proven in the benchmarking carried out during the procurement.

5.1 Improving the parallelisation aspects

The easiest performance improvement was gained by switching to the fast Cray SHMEM library to do the message passing, but generally still using GCOM as the interface. In key areas of the code using SHMEM directly allows even greater performance to be achieved.

The atmospheric model's integration scheme, can be described by equation 1, which shows the value of some variable x at time $n+1$, based on its value at time n:

$$x^{n+1} = \mathbf{P}(\mathbf{D}^3(\mathbf{Ad}^2(\mathbf{Aj}^3(x^n)))) \tag{1}$$

where **Aj** represents an adjustment timestep, **Ad** an advection half timestep and **D** a single pass of the ∇^2 diffusion operator. **P** represents the effects of the physical parametrizations which include treatments of convection, large-scale precipitation, radiation, boundary-layer exchanges and gravity wave drag.

Each adjustment timestep, advection half timestep and pass of the diffusion operator, require the application of the `SWAPBOUNDS` routine to each variable being updated, so its halos can be refreshed before the operator is applied. The `SWAPBOUNDS` is also called every time an interpolation is carried out between the staggered temperature and wind grids. Such interpolations occur many times during each timestep.

The frequent use of `SWAPBOUNDS` makes it an obvious target for performance improvement. The routine has been rewritten to use the one-sided SHMEM get/put operations directly, without using GCOM. This, together with the resulting reduction in the number of barriers (synchronisations), produced a useful increase in performance.

5.2 Improving the single processor performance

The UM was written to be highly efficient on vector based machines, such as the Cray C90, so it is not perhaps suprising that some effort is required to achieve good performance on a cache based machine such as the T3E.

The work required, is however, not perhaps as arduous as might at first be imagined. A complete re-write of the code is not necessary! The good news is that a lot of the optimisation can be performed very satisfactorily by the compiler which makes good use of the cache and streaming mechanisms on the T3E to allow vector code to run efficiently. Three of the most common optimisations are discussed briefly below.

i) **Loop unrolling** This ensures that the maximum amount of data in cache is used in each iteration of the loop. Unrolling is done automatically by the compiler. A few loops required directives to the compiler in order to achieve maximum performance.

ii) **Loop merging** By merging two or more loops into one single loop, memory traffic can be significantly reduced, and by unrolling the resulting loop even more performance improvement can be obtained. The compiler is capable of carrying out some loop merging, but manual loop merging in key sections of code gives many advantages:

a) increased performance

b) intermediate temporary arrays can be removed

c) the code becomes more readable - and often more closely resembles the equations it is evaluating

d) performance is also improved on vector shared memory machines.

iii) **Vector maths functions** The T3E provides a library of vector maths functions. When a standard mathematic function (such as `log`, `sqrt`, `exp`) must be applied to all members of an array, the use of the vector library gives a substantial performance increase. This can be improved further by using the reduced precision version of the library, at the expense of slightly less accurate answers. A future release of the compiler will be capable of doing this optimisation automatically.

6 Future Work

The UKMO currently has the atmosphere and ocean models running on its T3E, but there is still much work to do. Firstly in improving the performance of these codes, and secondly in the parallelisation of new algorithms that will be introduced over the coming years.

i) **Performance Improvement**

The single processor performance optimisation described in the previous section is an ongoing piece of work, but there are other performance improvement measures that are still to be addressed.

One of the most major of these is to solve the problem of load imbalance in the physics code. This is particularly a problem in the short-wave radiation, convection and gravity wave drag routines, where the physical processes are spread very unevenly over the model domain. The solution we plan to adopt is to perform a dynamic redistribution of data within the physics routines, based on the time taken during the previous timestep, to try and ensure that each processor has a similar amount of computation to perform.

ii) **Parallelisation of new algorithms**

Two major new algorithms will be introduced into the UM in the next few years, a new semi-implicit dynamics scheme using semi-Lagrangian advection, and 3D variational assimilation. Both of these projects are putting considerable effort into parallelisation issues.

Semi-Lagrangian advection poses an interesting problem in parallelisation, once again, due to the convergence of the grid towards the poles. The closeness of grid points around polar regions, combined with the long time steps possible with semi-Lagrangian schemes, means that the departure point calculations will often require data that is not within the subdomain of the processor containing the arrival point. In fact the departure point may not even be contained in any of the neighbouring processor's subdomains.

A solution to this would be to use a very large halo region, and/or reduce the timestep. Both of these solutions are not optimal - the first results in an increased communications and memory cost, while the second results in an increased computational cost. The solution currently being investigated, called "*computation on demand*" allows a moderate, tunable, halo size, and no decrease in the time step. A halo of moderate size, which will probably mean a halo four points wide, will contain much of the data required for the departure point calculation. When the departure point is not within a processor's subdomain, or its halo region, the processor sends the relevant information (departure point location and interpolation weights) to the processor containing the departure point, which carries out the computation, and sends the result back. Initial tests show this method to have very good scalability on the T3E.

7 Conclusions

The UKMO has completed the initial port of the atmospheriv model, atmospheric data assimilation and oceanic model components of the UM to the Cray T3E, by using regular domain decomposition and message passing via a portable interface library, GCOM.

We have shown that it is possible to run grid point models which use split explicit integration schemes and staggered grids on the Cray T3E. Parallelisation of such integration schemes generates an interprocessor communication pattern consisting mostly of large numbers of fairly short messages, to update the halo regions. The low latency high bandwidth SHMEM communications on the T3E are ideally suited for this type of application.

We have found that the T3E's Fortran 90 compiler performs much useful optimisation. Manual optimisation, such as loop merging, actually benefits the readability and conciseness of the code, as well as increasing performance on both the T3E and vector shared memory machines.

There are still problems to be solved, which currently limit the scalability of the UM to a few hundred processors. Achieving better load balance in the physics, and reducing the cost of I/O will contribute significantly to improving this.

8 Acknowledgements

The authors wish to thank SGI/Cray Research (UK), particularly Deborah Salmond and Bob Carruthers for helping to optimise the UM for the T3D/E. We would also like to thank Roar Skålin at SINTEF for his assistance with the FFT filtering and the parallelisation of the semi-Lagrangian advection scheme.

References

[1] Cullen M.J.P. 1991 A unified forecast / climate model. *ECMWF Seminar Proceedings - Numerical Methods in Atmospheric Models*, **2**, 233-260

[2] Cullen M.J.P., Davies T. 1991 A conservative split-explicit integration scheme with fourth order horizontal advection. *Q.J.R. Meteorol. Soc.*, **117**, 993-1002.

[3] Amundsen J. , Skålin R. 1996 GC User's Guide Release 1.1 *SINTEF Applied Mathematics Report STF42 A96504*

[4] Amundsen J. , Skålin R. 1996 GCG User's Guide Release 1.1 *SINTEF Applied Mathematics Report STF42 A96505*

[5] Lorenc A.C. , Bell R.S. , Macpherson B. 1991 The Meteorological Office Analysis Correction Data Assimilation Scheme. *Q.J.R. Meterol. Soc.*, **117**, 59-89

MODEL DEVELOPMENT FOR PARALLEL COMPUTERS AT DWD

ULRICH SCHÄTTLER
Deutscher Wetterdienst, Geschäftsbereich für Forschung und Entwicklung
Postfach 100465, 63004 Offenbach, Germany

and

ELISABETH KRENZIEN
Deutscher Wetterdienst, Geschäftsbereich für Technische Infrastruktur
Postfach 100465, 63004 Offenbach, Germany

DWD develops the 4th generation of its numerical weather prediction system. It consists of a global grid point model (GME) based on a triangular grid and a non-hydrostatic local model (LM). The new global model replaces today's global model and the regional "Europa-Modell" and provides the lateral boundary fields for the local model. The LM will be the central weather forecast model for the area of Germany. For both models there is a very high computational demand which can only be met by parallel computers.

The parallelization strategy used in both cases is the domain decomposition. While this is straightforward for the regional LM with a rectangular domain, a more sophisticated partition is proposed for the global model. We will report on the structure of the models, the numerical algorithms and the parallelization approaches.

1. Numerical Weather Prediction at DWD

1.1. Today's Model Generation

28 years after introducing the first numerical model for weather forecasting, DWD now is in the process of developing the 4th generation of its numerical weather prediction (NWP) system. The current operational model chain consists of three different models. The spectral "Global Modell" is used for large-scale predictions. Its spectral horizontal representation is T106 with 320×160 grid points in the Gaussian grid. The results of the GM are provided to the "Europa-Modell" (EM) as lateral boundary conditions. The EM is a regional grid point model for the synoptic and meso-α scale covering the Northern Atlantic and Europe. The grid size is 55 km, 181×129 grid points and 20 vertical levels.

In 1993 the high resolution meso-β scale "Deutschland-Modell" (DM), developed in close cooperation with the Swiss Meteorological Institute, was added as the main short range weather forecasting tool for Germany. The DM is running with the same code as the EM, but on an operational domain size of 109×109 grid points and 20 vertical levels with a mesh size of 14 km. A more detailed up-to-date description of today's operational models can be found in the Quarterly Report of the Operational NWP-Models[11].

The three models are run with highly vectorized and optimized code on a Cray Y-MP4 and a Y-MP2. For the EM/DM there exists a standard Fortran77 version that is used by 6 meteorological services and about 20 research institutes which operate it on workstations. The code of this version has also been parallelized for distributed memory computers using the message-passing library PARMACS 6.0. This program was used as a benchmark for the procurement of the new computer system at DWD. The basic parallelization has been described by Schättler and Krenzien[10].

1.2. Objectives of the New Models

In the last years not only the meteorological but also the computer sciences have shown developments that provide new opportunities to numerical weather prediction. In addition, the political and economical requirements regarding weather forecasting have changed. This ranges from modelling the transport and deposition of chemical and radioactive pollutants, for which global and regional models provide the meteorological data base, to detailed weather forecasts for all parts of the world, e.g. for aviation or touristic business. Besides, the timely prediction of severe weather events always was and still is one of the most important goals in numerical weather prediction.

The main objectives for DWD's new system can be characterized as follows:

- Produce operational weather forecasts for up to 7 days globally and for the area of Germany on the meso-γ scale locally (including clouds, fog, precipitation, local wind systems, severe weather events).
- Provide a global and regional meteorological data base for various post processing modules (computing trajectories, prediction of water levels, ocean waves).
- Simulate regional and local climate scenarios (urban climate with a resolution of up to 100 meters).
- Serve as a research tool for various scientific applications (parametrizations, data assimilation, ensemble prediction, long range forecasts, climate).

The new model chain will consist of only two models. The global model GME (its complex physical equipment is based on the EM/DM) will replace GM and EM and should produce forecasts at least in the quality of the EM. The local model LM will be used for weather forecasts over Germany. Its lateral boundary conditions are provided by GME.

Table 1.1 summarizes the numerical characteristics of the models and gives a hint of the computational requirements.

Table 1.1: Characteristics of the New Models

Model		Resolution			Flops / 24 h forecast
		horizontal		vertical	
		km	grid points		
GME	1998:	50-60	160.000	40 layers	
	2000:	25-30	650.000	50 layers	ca. $300 \cdot 10^{12}$
LM	1998:	8-10	40.000	40 layers	
	2000:	2-3	640.000	50 layers	ca. $300 \cdot 10^{12}$

It is clear that these requirements can only be met by parallel computers. Therefore it is necessary to construct not only efficient but also scalable algorithms that can execute on hundreds or even thousands of processors.

1.3. Phases of the Development

The following milestones have been set up for the development:

Table 1.2: Milestones for the Development

	GME	LM
1996	sequential semi-implicit version with full physics of EM/DM	parallel Eulerian version with full physics of EM/DM
1997	parallel version with improved physical packages	selection, implementation and testing of high resolution physical packages
1998	quasi-operational run of both models parallel to the old system	
1999	switch over to new computer system and operational run of both models	
$\geq$ 2000	further improvements and work on the models	

For both models projects have been launched and working groups have been established at DWD. Collaborations have been started with some national and international research institutes and universities. The LM is available within the Short Range Numerical Weather Prediction (SRNWP) initiative of the EUMETNET.

2. The Nonhydrostatic Local Model LM

The basic features of the present first version of the limited area model LM are the nonhydrostatic dynamics in a similar form as in the MM5 (Grell et al.[6], 1994 and Dudhia[5], 1993), the organizational framework and the model physics of the EM/DM and the parallelization methods of the DM. Modifications have been made with respect to the grid structure, physical parameterizations and especially the programming language in order to guarantee the main design goals of portability, reproducibility and easy source code management for the sequential and parallel version of the model.

2.1. Basic Structure of the LM

The model is based on the set of equations for a nonhydrostatic full compressible atmosphere. Introducing a spherical coordinate system (λ, ϕ) with a rotated pole and a terrain-following vertical coordinate σ the prognostic equations for momentum (u, v, w), pressure perturbation (p'), temperature (T) and moisture (q) take the form

$$\frac{\partial u}{\partial t} + \mathbf{v}\cdot\nabla u - \frac{uv}{a}\tan\varphi = fv - \frac{1}{\rho\, a\, \cos\varphi}\left(\frac{\partial p'}{\partial \lambda} - \frac{\sigma}{p^*}\frac{\partial p^*}{\partial \lambda}\frac{\partial p'}{\partial \sigma}\right) - \left(\frac{\nabla\cdot\underline{\mathbf{F}}}{\rho}\right)\cdot\mathbf{e}_\lambda,$$

$$\frac{\partial v}{\partial t} + \mathbf{v}\cdot\nabla v + \frac{u^2}{a}\tan\varphi = -fu - \frac{1}{\rho\, a}\left(\frac{\partial p'}{\partial \varphi} - \frac{\sigma}{p^*}\frac{\partial p^*}{\partial \varphi}\frac{\partial p'}{\partial \sigma}\right) - \left(\frac{\nabla\cdot\underline{\mathbf{F}}}{\rho}\right)\cdot\mathbf{e}_\varphi,$$

$$\frac{\partial w}{\partial t} + \mathbf{v}\cdot\nabla w = \frac{g\,\rho_0}{\rho\, p^*}\frac{\partial p'}{\partial \sigma} + g\frac{\rho_0}{\rho}B - \left(\frac{\nabla\cdot\underline{\mathbf{F}}}{\rho}\right)\cdot\mathbf{e}_z, \tag{2.1}$$

$$\frac{\partial p'}{\partial t} + \mathbf{v}\cdot\nabla p' - g\,\rho_0\, w = -\gamma\, p\, D + \frac{\gamma\, p}{T}\frac{Q}{c_p}, \tag{2.2}$$

$$\frac{\partial T}{\partial t} + \mathbf{v}\cdot\nabla T = \frac{1}{\rho\, c_p}\left(\frac{\partial p'}{\partial t} + \mathbf{v}\cdot\nabla p' - g\,\rho_0\, w\right) + \frac{Q}{c_p}, \tag{2.3}$$

$$\frac{\partial q^k}{\partial t} + \mathbf{v}\cdot\nabla q^k = -\frac{1}{\rho}\left(\nabla\cdot\mathbf{J}^k + \nabla\cdot\mathbf{F}^k\right) - \frac{1}{\rho}I^k, \tag{2.4}$$

with the buoyancy term

$$B = \frac{T - T_0}{T} - \frac{p'\, T_0}{p_0\, T} + \left(\frac{R_v}{R_d} - 1\right) q^v - q^l - q^f, \tag{2.5}$$

the advection operator

$$\mathbf{v}\cdot\nabla = \frac{1}{a\,\cos\varphi}\left(u\frac{\partial}{\partial\lambda} + v\cos\varphi\frac{\partial}{\partial\varphi}\right) + \dot{\sigma}\frac{\partial}{\partial\sigma}, \tag{2.6}$$

the contravariant vertical velocity

$$\dot{\sigma} = -\frac{\sigma}{p^* a\cos\varphi}\frac{\partial p^*}{\partial\lambda}u - \frac{\sigma}{p^* a}\frac{\partial p^*}{\partial\varphi}v - \frac{g\,\rho_0}{p^*}w, \tag{2.7}$$

and the threedimensional wind divergence

$$\begin{aligned} D = \nabla\cdot\mathbf{v} = &\frac{1}{a\,\cos\varphi}\left(\frac{\partial u}{\partial\lambda} - \frac{\sigma}{p^*}\frac{\partial p^*}{\partial\lambda}\frac{\partial u}{\partial\sigma}\right) \\ &+ \frac{1}{a\,\cos\varphi}\left(\frac{\partial}{\partial\varphi}(v\,\cos\varphi) - \cos\varphi\frac{\sigma}{p^*}\frac{\partial p^*}{\partial\varphi}\frac{\partial v}{\partial\sigma}\right) - \frac{g\,\rho_0}{p^*}\frac{\partial w}{\partial\sigma}. \end{aligned} \tag{2.8}$$

The vertical coordinate σ is defined by the relation

$$\sigma = \frac{p_0 - p_{top}}{p_{sealevel} - p_{top}} = \frac{p_0 - p_{top}}{p^*}. \tag{2.9}$$

Here subscript 0 denotes the reference state and primes the deviations from the reference state. The constants a, c_p, f, g, R_v, R_d and γ represent radius of the earth, heat capacity of dry air at constant pressure, Coriolis parameter, acceleration due to gravity, gas constants for vapour and dry air and the ratio of heat capacities for air at constant pressure to constant volume. The remaining terms Q, I, $\nabla\cdot\underline{F}$ and $\nabla\cdot\underline{J}$ describe the diabatic heating rates, the cloud microphysical sources for the moisture components q^k and the divergence of the turbulent and the diffusive fluxes (Doms[4] 1995).

2.2. Numerical Aspects

Finite difference methods are employed to solve the LM model equations (2.1) - (2.4). The equations are discretized on an Arakawa-C grid where pressure, temperature and the moisture variables are located at the centre of a gridbox whereas the momentum components are defined at the gridbox interfaces. The spatial discretization uses second-order centered differences for both horizontal and vertical differentials. To control small scale noise a linear fourth order diffusion scheme is applied.

While in hydrostatic models fast moving external gravity waves limit the time step of the integration method, the full compressible nonhydrostatic equations permit high frequency acoustic waves which require also a short time step for numerical stability. To increase the efficiency of the program, the splitting method described by Klemp and Wilhelmson[7] (1978) and Dudhia[5] (1993) is used for the time integration.

The idea of the time splitting is to treat the fast terms describing sound and gravity wave propagation with small time steps while doing a large step for the slow terms (advection, physics). This is possible because the terms responsible for these waves can be separated from the slowly varying terms:

$$\begin{aligned}
\frac{\partial u}{\partial t} &= -\frac{1}{a\rho\cos\varphi}\left(\frac{\partial p'}{\partial\lambda} - \frac{\sigma}{p^*}\frac{\partial p^*}{\partial\lambda}\frac{\partial p'}{\partial\sigma}\right) + S_u, \\
\frac{\partial v}{\partial t} &= -\frac{1}{a\rho}\left(\frac{\partial p'}{\partial\varphi} - \frac{\sigma}{p^*}\frac{\partial p^*}{\partial\varphi}\frac{\partial p'}{\partial\sigma}\right) + S_v, \\
\frac{\partial w}{\partial t} &= g\left(\frac{\rho}{\rho_0}\right)\frac{1}{p^*}\left\{\frac{\partial p'}{\partial\sigma} + p^*\left(\frac{T'}{T} - \frac{T_0}{T}\frac{p'}{p_0}\right)\right\} + S_w, \qquad (2.10) \\
\frac{\partial p'}{\partial t} &= -\gamma p\left(\nabla_h\cdot D_h + \frac{g\rho_0}{p^*}\frac{\partial w}{\partial\sigma}\right) + g\rho_0 w + S_{p'}, \\
\frac{\partial T}{\partial t} &= -\frac{\gamma p}{\rho c_p}\left(\nabla_h\cdot D_h + \frac{g\rho_0}{p^*}\frac{\partial w}{\partial\sigma}\right) + S_T.
\end{aligned}$$

$\nabla_h\cdot D_h$ stands for the horizontal divergence and S_u, S_v, S_w, $S_{p'}$ and S_T represent the slowly varying terms. These are held constant for a large time step interval Δt, while the integration of (2.10) is performed with small time steps Δt_s.

Because only a subset of terms in the model equations is integrated with a small time step Δt_s whereas the computational expensive slow terms have to be updated less frequently, the algorithm becomes numerical efficient. The method is illustrated in Figure 2.1.

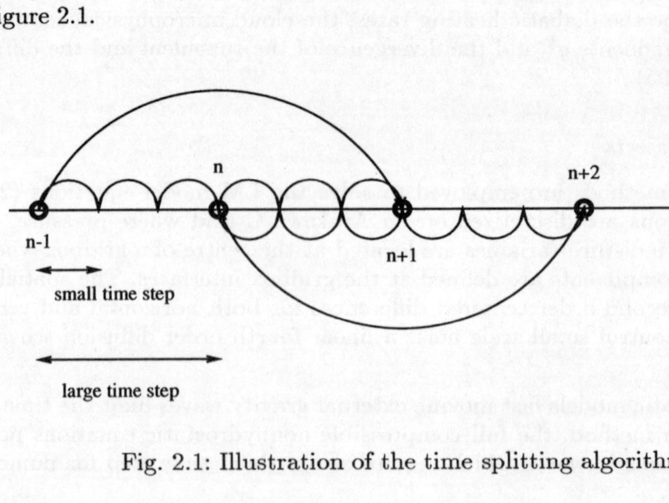

Fig. 2.1: Illustration of the time splitting algorithm

According to a proposal by Skamarock and Klemp[12] (1992), the time splitting scheme has been extended by including the terms for vertical gravity wave propagation, i.e. the complete bouyancy terms, in the subset (2.10). Thus the stability criterion for the leapfrog scheme used for the integration of the slow terms becomes independent of the gravity wave restriction $\Delta t < N^{-1}$, where N is the Brunt-Väisälä frequency. This restriction would severly limit the time step Δt in meso-β applications with a large horizontal grid spacing Δs (for $\Delta s \geq 10$ km, N^{-1} is about 30 - 40 seconds). Δt is now limited only by the Courant-Friedrichs-Levy criterion for horizontal motion:

$$\Delta t \leq \frac{\Delta s}{\sqrt{2}\,|\mathbf{v}|_{\max}}, \tag{2.11}$$

where $|\mathbf{v}|_{\max}$ is the maximal horizontal wind speed.

In the small step the horizontal wave propagation is computed explicitly using a foreward-backward scheme and the vertical acoustic and gravity waves are treated implicitly using the Crank-Nicolson method. The implicit vertical differencing guarantees that the short time step Δt_s is independent of the vertical resolution and thus limited only by the acoustic wave speed c_s and the horizontal grid spacing according to

$$\Delta t_s \leq \frac{\Delta s}{\sqrt{2}\,c_s}. \tag{2.12}$$

An important and critical part of limited area models is the adaption of the model variables to lateral boundary conditions. For the moment the Davies relaxation scheme for one-way nesting is implemented. Periodic boundary conditons can be used, too. At the upper boundary a Rayleigh damping layer is introduced.

For the model physics, the formulations developed for the EM/DM are implemented into the LM as a first step.

2.3. Parallelization and Implementation

Because the organizational framework of the LM was taken from the EM/DM, the structure of the parallel LM is similar to the parallel DM. The parallelization strategy is the 2D domain decomposition where each processor gets the appropriate part of the data to compute the forecast. The difference to the DM is in the time splitting integration method described above. All other features (explicit Eulerian time step, horizontal diffusion, boundary relaxation) are the same as in the parallel DM and are explained by Schättler and Krenzien[10].

In the DM, the computation of the semi-implicit corrections requires the solution of an elliptic partial differential equation and hence global communication in the parallel program. In the splitting method used for the LM the system of equations (2.10) is treated explicitly in the horizontal and therefore involves only a nearest neighbour exchange, i.e. local communication. The structure of the adiabatic time stepping can be characterized as follows:

Loop over large time steps: $t = 0, 1, 2, \ldots$

1: Compute the tendencies for the horizontal advection and diffusion.

2: Compute the tendencies for the slow waves with a large time step Δt.

3: Loop over small time steps: $\tau = 0, 1, 2, \ldots$

- Compute the variables for the new small time step $\tau + 1$.
- Exchange boundaries for the necessary variables.

4: Compute the boundary relaxation and the Asselin time filtering, hence the prognostic variables for time step $t + 1$.

5: Exchange boundaries for the prognostic variables.

The LM has first been implemented in a serial Fortran 77 version to gather experience in nonhydrostatic modelling at an early stage of the development. Later a transformation to Fortran 90 took place to benefit from the new features of this language like dynamic memory allocation, modules and interface constructions. The Fortran 90 version now is parallelized using the Message Passing Interface (MPI).

The original Fortran 77 code is separated into data and source `MODULE`s which are chosen in accordance with the different tasks and functions they have. While data `MODULE`s replace the `COMMON` block feature, source `MODULE`s are ordered in three hierarchical levels:

First level:	Main program and task managing.
Second level:	Model set-up and different tasks (dynamics, physical packages).
Third level:	Auxiliary utilities and libraries.

This procedure allows for a modular program development and facilitates the possibility of managing the parallel and the sequential version of the model in only one program code. All subroutines that need MPI-calls are gathered in the third-level `MODULE environment`. This module exists twice, first as `parallel_environment`, and also as `sequential_environment` having the same routines as `parallel_environment` but mostly as dummy subroutines only.

On a sequential computer, where MPI is not available, the model has to be linked with `sequential_environment` (so the linker will not complain about unsatisfied external references), while on parallel machines with MPI it has to be linked with `parallel_environment`. The code itself is not affected by switching the platform.

3. The New Global Model GME

3.1 Basic Concepts for GME

The system of equations solved in the GME is explained in the simple framework of the shallow-water equations, given here in local η, χ-coordinates.

$$\begin{aligned} \frac{du}{dt} &= -\frac{g}{a}\frac{\partial h}{\partial \eta} + fv - K_4 \nabla^4 u = R_u, \\ \frac{dv}{dt} &= -\frac{g}{a}\frac{\partial h}{\partial \chi} - fu - K_4 \nabla^4 v = R_v, \\ \frac{dh^*}{dt} &= -\frac{h^*}{a}\left(\frac{\partial u}{\partial \eta} + \frac{\partial v}{\partial \chi}\right) - K_4 \nabla^4 h^* = R_{h^*}; \end{aligned} \tag{3.1}$$

u and v are the horizontal coordinates of the speed, $h = h^* + h_s$ is the fluid height and h_s the height of the orography. g, f and a are the gravity of the earth, the Coriolis-parameter and the radius of the earth. K_4 is the diffusion coefficient (linear fourth order). R_u, R_v and R_{h^*} are the right-hand sides of the prognostic equations.

These equations are integrated using a semi-implicit algorithm. The explicit part is solved with a three time-level semi-Lagrangian scheme that can be written as

$$\begin{aligned} u^{n+1} &= I_d\left(u^{n-1}\right) + 2 \cdot \Delta t \cdot I_m\left(R_u^n\right), \\ v^{n+1} &= I_d\left(v^{n-1}\right) + 2 \cdot \Delta t \cdot I_m\left(R_v^n\right), \\ h^{*n+1} &= I_d\left(h^{*n-1}\right) + 2 \cdot \Delta t \cdot I_m\left(R_{h^*}^n\right), \end{aligned} \tag{3.2}$$

where I_d denotes the horizontal interpolation of the prognostic variables at the previous time-level $n-1$ to the departure point of the trajectory and I_m the interpolation of the right hand side at the current time-level n to the midpoint of the parcel trajectory. Only the horizontal diffusion is taken at the previous time-level.

To enlarge the time step which is restricted by the explicit algorithm, semi-implicit corrections are computed. For this purpose a Helmholtz-equation has to be solved for the height field.

$$\left(\nabla^2 - d\right) s = f\left(h_{exp}^{n+1}, u_{exp}^{n+1}, v_{exp}^{n+1}, h^n, u^n, v^n, h^{n-1}, u^{n-1}, v^{n-1}\right), \tag{3.3}$$

where s is the solution and d a positive constant. The right hand side of (3.3) depends on the new values of the prognostic variables after the explicit step and on the values of time level n and $n-1$. The corrections for the values of h^{n+1}, u^{n+1} and v^{n+1} can be derived from s.

3.2 Grid Generation

For the discretization of the equations and the practical implementation a grid has to be introduced. The one that is used for the GME will be explained in more detail. It was first introduced in a meteorological model by Sadourny[9] et. al. (1968) and Williamson[13] (1969). The approach outlined here is based on the work of Baumgardner[1] (1983).

To construct the grid, the sphere is divided into 20 spherical triangles of equal size by placing a plane icosahedron into it. The 12 vertices of the icosahedron touch the sphere, one vertex coincides with the north pole and the opposite one with the south pole.

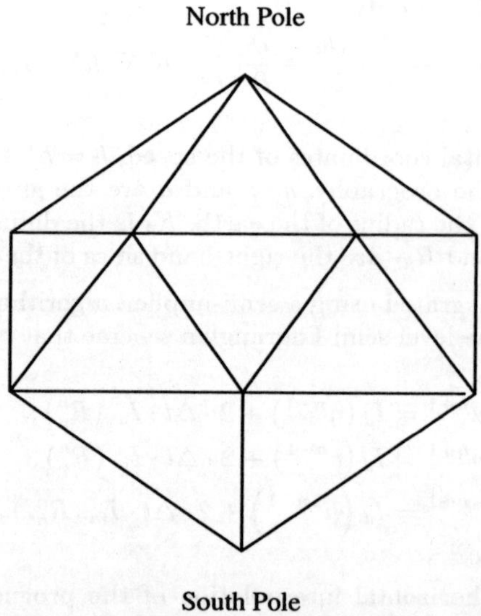

Fig. 3.1: Plane icosahedron consisting of 20 plane triangles

The spherical triangles are defined by the great circles connecting two vertices respectively. Each of the 12 vertices then is surrounded by 5 spherical triangles. Two adjacent triangles are combined to form a "diamond", i.e. a logically square block. Five of these diamonds originate from the north pole and five from the south pole (see Figure 3.2).

For further grid generation, the sides w of the 20 main triangles are subdivided iteratively into ni equal parts to form subtriangles. Each point in a main triangle is surrounded by six triangles and accordingly is in the centre of a hexagon. However, the points which form the vertices of the icosahedron are surrounded by only five triangles and therefore are the centres of pentagons. For the first subdivision, the triangle sides may be divided into three parts, later only bisections are allowed. This restriction is due to the use of a multigrid solver for the Helmholtz-equation in the semi-implicit time stepping. These solvers work efficiently with this kind of mesh refinement. Some resulting grids are illustrated in Figure 3.3.

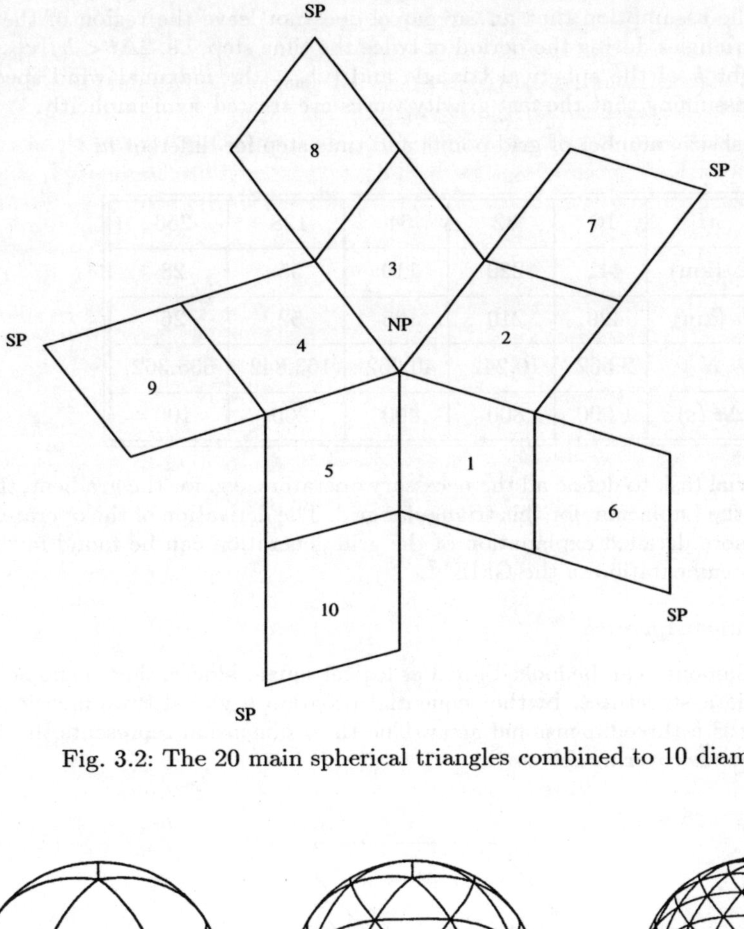

Fig. 3.2: The 20 main spherical triangles combined to 10 diamonds

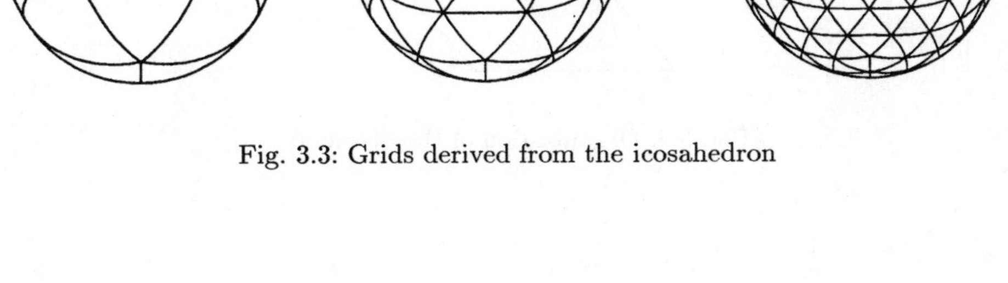

Fig. 3.3: Grids derived from the icosahedron

Table 3.1 gives the minimal meshsize Δ within a diamond (Δ varies within a diamond by 20% at most), the number N of grid points and the time step Δt for different values of ni, if only bisections are performed. The time step Δt is calculated under the assumption that an air parcel does not leave the region of the 6 surrounding triangles during the period of twice the time step, i.e. $2\Delta t < h/|\mathbf{v}|_{\max}$ with the height h of the spherical triangle and $|\mathbf{v}|_{\max}$ the maximal wind speed ($\approx 125\ m/s$) assuming that the fast gravity waves are treated semi-implicitly.

Table 3.1: Meshsize, number of grid points and time step for different ni

ni	16	32	64	128	256
Δ (km)	441	220	110	55	28
h (km)	420	210	105	52	26
N	2.562	10.242	40.962	163.842	655.362
Δt (s)	1.600	800	400	200	100

It is a non-trivial task to define all the necessary operators, e.g. for the gradient, the divergence or the Laplacian, for this triangular grid. The derivation of the operators as well as a more detailed explanation of the grid generation can be found in the preliminary documentation of the GME[8,2].

3.3 Parallelization Approach

Because the diamonds can be looked upon as logical square blocks, there is no need for complex data structures. In the sequential program a global twodimensional field is stored as a threedimensional array. The third dimension represents the 10 different diamonds covering the earth.

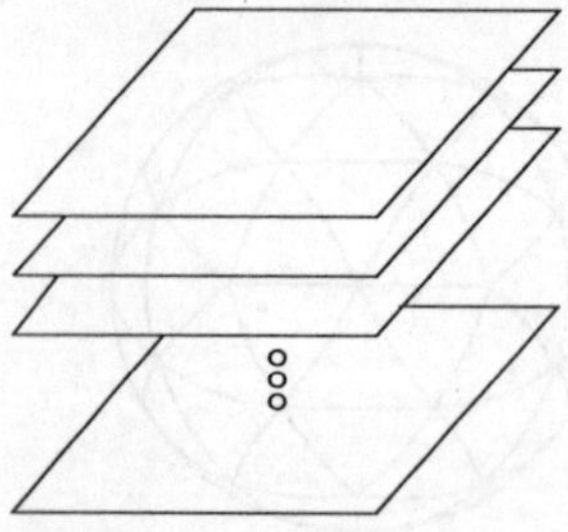

Fig. 3.4: Organization of the diamonds

To ease the implementation, each diamond is extended by additional rows and columns taken from the appropriate neighbouring diamonds. This allows to employ the stencils for the operators without using values from neighbouring diamonds directly.

Because the four vertices of the diamond core have only 5 neighbours, the structure of the extended diamond becomes rather complicated with mirrored and even undefined points. For $ni = 4$ this is shown in Figure 3.5.

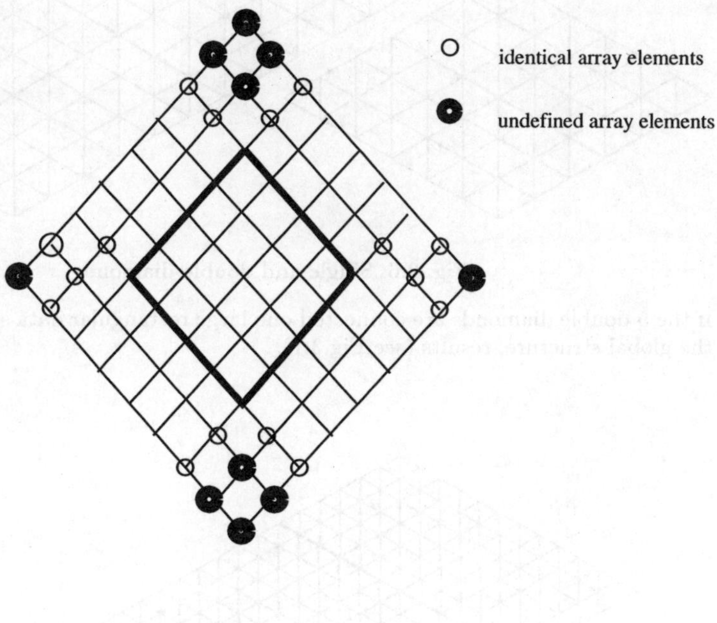

Fig. 3.5: Extended array with mirrored and undefined grid points

The parallelization approach again is the domain decomposition. But while this is straightforward for a regional model a more sophisticated strategy has to be used here, where every diamond can be regarded as a regional model.

Grouping 2 diamonds together gives 5 double-diamonds, reaching from the North to the South Pole. Both, the 10 diamonds and the 5 double-diamonds are called diamond-local structures (see Fig. 3.6).

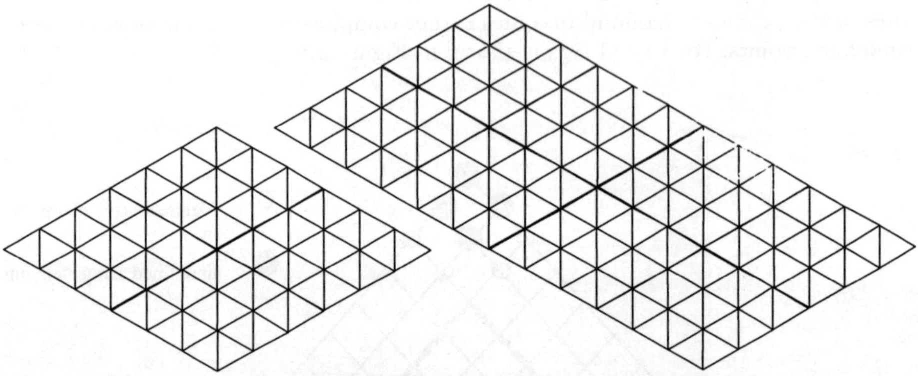

Fig. 3.6: Single and double-diamond

If the 5 double diamonds are connected one large rectangular data structure, called the global structure, results (see Fig.3.7).

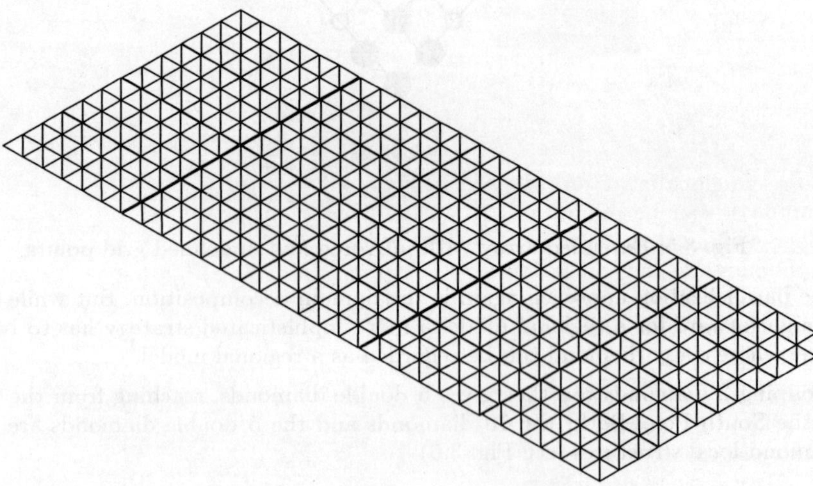

Fig. 3.7: Global view of the data structure

From this global structure decompositions not reflecting the geometrical background of the grid generation nor the distribution of the single diamonds to the sphere can be derived. Possible partitionings of the global structure are rowwise, columnwise, twodimensional (the same as in regional models) and the so-called orange-peeling (see Fig. 3.8).

The decompositions of both, the local and the global structures, have been theoretically investigated by GMD[3] regarding static load balancing, communication patterns and the amount of communication.

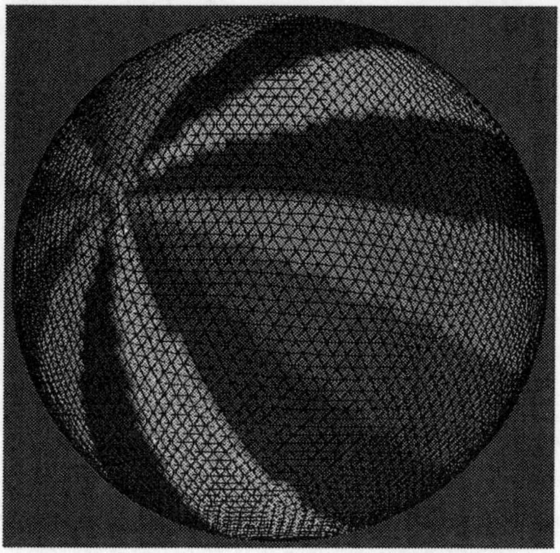

Fig. 3.8: Orange-peeling of the sphere for 16 processes

Using the global structure the amount of data that has to be exchanged between boundaries can be minimized. However, the organization and implementation is rather complicated, because different subdomains can have different numbers of neighbours and the communication patterns are not the same for all subdomains. For the global structure the communication patterns can be regular only if the number of processors is a multiple of 10 (for the single diamonds) or of 5 (for the double-diamonds).

A practical way for the implementation is based on the diamond-local structures and related to an idea of John Baumgardner[2]. Every diamond can be partitioned in the chessboard manner like the domain of a regional model. Since all diamonds are of equal size, their decomposition is identical. If the processors are arranged in a twodimensional grid corresponding to the decomposition of the diamonds, every processor gets a part of each diamond. This kind of decomposition is shown in Figure 3.9 for 4 processors and in Figure 3.10 (left).

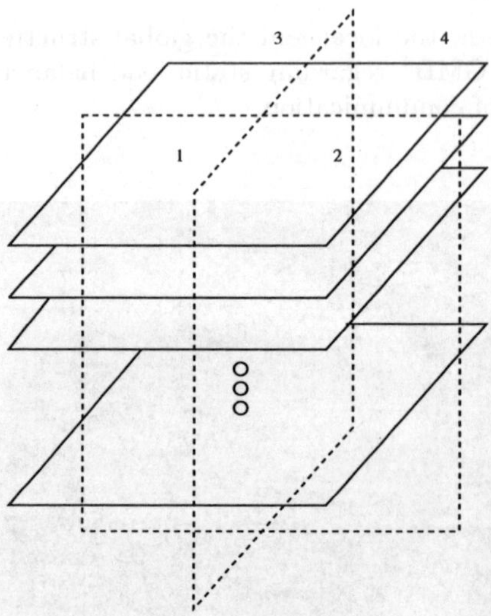

Fig. 3.9: Logical view of the decomposition of the diamond-local structures

In spite of the shortcoming that a larger amount of data has to be transferred, this decomposition has a certain potential in NWP-models. There are several advantages:

- The implementation (organization of neighbours, data exchange between subdomains) is as straightforward as in a regional model.
- A processor gets a rectangular block from all 10 diamonds. Therefore the parallel and the sequential versions of the program have the same structure.
- The 10 blocks are distributed regularly over the earth. From a statistical viewpoint there is a chance to get a rather balanced load distribution, regarding the computations in the physical packages (day-night radiation, land-water distribution; see also the left picture of Figure 3.10).

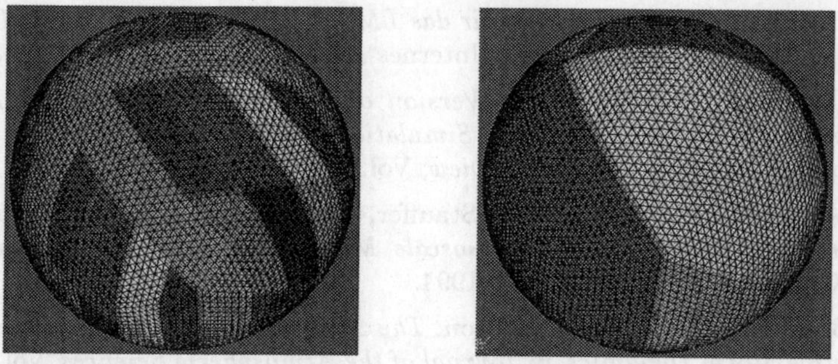

Fig. 3.10: Diamond-local and global 2D decomposition of the sphere

In a way this decomposition can be compared to the methods used in models for computational fluid dynamics, e.g. for the modelling of spacecraft or airplanes. The difference is that these codes use unstructured grids and therefore get blocks of different sizes. The decomposition of the blocks into subdomains and the distribution of the subdomains to the processors is not as straightforward as in the GME.

Acknowledgements

This article is based on the work of two development groups at DWD. We thank the GME group (Axel Diehl, Michael Gertz, Dörte Liermann, Detlev Majewski, Peter Prohl, Bodo Ritter, Ulrich Schättler) and the LM group (Axel Diehl, Günter Doms, Erdmann Heise, Annemarie Link, Maud Minotte, Ulrich Schättler, Jürgen Steppeler) for providing their material. The work on GME was enriched by two visits of John Baumgardner at DWD in 1995/96. Most important, we owe thanks to G. Doms and D. Majewski for their careful review of the manuscript.

References

1. J. Baumgardner, *A Three-Dimensional Finite Element-Model for Mantle Convection*, Ph. D. thesis, The University of California at Los Angeles, 1983.
2. J. Baumgardner, Los Alamos National Laboratory. Private communication.
3. O. Bröker, K. Cassirer, R. Hess, C. Jablonowski, W. Joppich and S. Pott, *Forschungs- und Entwicklungsarbeiten im Rahmen des neuen Global-Modells (GME) des DWD*, Gesellschaft für Mathematik und Datenverarbeitung, Internes Arbeitspapier, 28. November 1996.

4. G. Doms, *Basis-Gleichungen für das LM — 1. Arbeitspapier zur LM-Entwicklung*, Deutscher Wetterdienst, Internes Arbeitspapier, Februar 1995.
5. J. Dudhia, *A Nonhydrostatic Version of the Penn State / NCAR Mesoscale Model: Validation, Tests and Simulation of an Atlantic Cyclone and Cold Front*, in *Monthly Weather Review*, Vol. 121, 1993, pp. 1493-1513.
6. G.A. Grell, J. Dudhia, D.R. Stauffer, *A Description of the Fifth Generation Penn State / NCAR Mesoscale Model (MM5)*, NCAR Technical Note (NCAR/TN-398+STR), June 1994.
7. J.B. Klemp and R.B. Wilhelmson, *The Simulation of Three-dimensional Convective Storm Dynamics*, in *Journal of the Atmospheric Sciences*, Vol. 35, 1978, pp. 1070-1096.
8. D. Majewski, *Preliminary Documentation of the New Global Model GME*, 1996, Deutscher Wetterdienst.
9. R. Sadourny, A. Arakawa and Y. Mintz, *Integration of the Nondivergent Barotropic Vorticity Equation with an Icosahedral-Hexagonal Grid for the Sphere*, in *Monthly Weather Review*, Vol. 96, No. 6, 1968, pp. 351-356.
10. U. Schättler and E. Krenzien, *Experiences on Porting the "Deutschland-Modell" of DWD to Parallel Platforms and First Results*, in *Coming of Age – Proceedings of the Sixth ECMWF Workshop on the Use of Parallel Processors in Meteorology*, eds. G.-R. Hoffmann, N. Kreitz (World Scientific, 1995), pp. 223-239.
11. R. Schrodin (Ed.), *Quarterly Report of the Operational NWP-Models of the Deutscher Wetterdienst*, No. 4, October 1995, Geschäftsbereich für Forschung und Entwicklung, Deutscher Wetterdienst.
12. W. Skamarock and J.B. Klemp, *The Stability of Time-Splitting Numerical Methods for Hydrostatic and Nonhydrostatic Elastic Systems*, in *Monthly Weather Review*, Vol. 120, 1992, pp. 2109-2127.
13. D. Williamson, *Numerical Integration of Fluid Flow over Triangular Grids*, in *Monthly Weather Review*, Vol. 97, No. 12, 1969, pp. 885-895.

ADVANCES IN METEO-FRANCE MODELS PARALLELISATION

J.-F. ESTRADE
METEO-FRANCE SCEM/TTI/DEV/ST 42, av G. Coriolis 31057 TOULOUSE CEDEX
e-mail : jean-francois.estrade@meteo.fr

Liviu DRAGULANESCU
Institutul National de Meteorologie si Hidrologie
Sos. Bucuresti-Ploiesti 97
Sector 1 BUCAREST
ROUMANIE
dragulanescu@meteo.inmh.ro

and
Denis PARADIS
METEO-FRANCE SCEM/TTI/DEV 42, av G. Coriolis 31057 TOULOUSE CEDEX
e-mail : denis.paradis@meteo.fr

For two years, Météo-France has been porting the ARPEGE/IFS code to distributed memory architectures. Now, the code uses Fortran 90. New applcations have been introduced as the stretched schemes with two levels of parallelisation and the digital filter initialisation scheme. The ARPEGE part of the code has been tested on Fujitsu VPP700 and Cray C90. At the same time, we started some investigations on the ALADIN model. The ALADIN model is a limited area model which use part of subroutines from ARPEGE. For the moment being, the ALADIN code can run on a distributed platform and on the C90. Only one set of the distributed environment has been developed for the Eulerian scheme. The semi-Lagrangian scheme is under development.

1. INTRODUCTION

Two main models have been implemented on a distributed platform : the global spectral model ARPEGE/IFS and the limited area model ALADIN, coupled with ARPEGE. For the ALADIN model, a part of the subroutines is common with the ARPEGE/IFS model. First, ARPEGE/IFS model was parallelised and then, we adapted the parallelisation strategy of the ARPEGE/IFS model to the ALADIN model.

In the case of ARPEGE/IFS, the first configuration has been running in distributed mode for several years as this part has been well implemented by ECMWF. At Météo-France, we only needed to implement some special schemes (Digital filter initialization (DFI), stretched grid, input/output and physics) used in our operational suite.

In the first part, we will talk about the implementations of the stretched grid with two levels of distribution and the DFI in ARPEGE/IFS. In the second part, we discuss more about the ALADIN parallelisation and in the last part, the integration of the ARPEGE/ALADIN suite is described.

2. ARPEGE/IFS

2.1. General overview

As the ARPEGE/IFS code is the most important operational code at Météo-France, we consider the study of its parallelisation as a strategic development.

ARPEGE/IFS stands for "Action de Recherche Petite Echelle Grande Echelle" (in English : "large scale fine scale research action") and designates a project aiming at replacing most of the operational and research numerical simulation tools of Météo-France by a single fully integrated code. This huge code has been developed in full collaboration with ECMWF where it has the name IFS

(Integrated Forecasting System). The main common part between IFS and ARPEGE/IFS consists in the dynamics, but other parts (such as physics, analysis and post-processing) are quite different.

In 1995, we built a new version of ARPEGE written in Fortran 90 that we tested on a distributed memory platform but also on a shared memory platform. For the distributed version of the code, we can only execute the first configuration of ARPEGE on a reduced grid. So we need to implement two new schemes to be very near to the operational suite : the stretched grid and the DFI.

In this part, we describe these two new schemes which have been tested on a distributed memory platform. For the description of the ARPEGE/IFS parallelisation, you can refer to [2 to 10].

2.2. The digital filter initialization

The digital filter initialization is a very simple initialization method based on the concept of digital filtering(references [1],[11]). This method allows to eliminate spurious high-frequency oscillations from the forecast.

In fact, for the ARPEGE case, we achieved :

* first, a backward short-range adiabatic integration of the forecast model with a reduced time-step T/2 (to avoid explosions).

* second, a forward short-range diabatic integration of the forecast model with a time-step T close to the time-step of the forecast.

We don't encounter any problem of parallelisation of the DFI as this scheme uses the same tarnsposition strategy developed by the ECMWF. Unfortunately, we couldn't test this scheme with more than 20 processors because we haven't got access to any such machine. But we achieved different tests with the two levels of distribution and checked the validity of the results. Nevertheless, it might be interesting to study the impact (in time, memory and accuracy of the results) of the DFI scheme.

2.3. The stretched grid

At Météo-France we use a rotated, stretched grid. The pole of this grid is located in the middle of France. Then, we use a stretching coefficient of 3.5. It means that we get more grid points near one pole (in our case the center of France) and less values on the opposite pole in order to get more accurate forecast on France. The mesh of the grid point near France is about 40 km and 400 km near Australia. Also, the computation in the spectral space is not exactly the same between a stretched grid and a non-stretched grid due to new dependencies.

In the ARPEGE/IFS model (without stretching), we can use two levels of parallelisation. The first level of distribution consists in splitting the grid point space over the latitude and the spectral space over the total wave number. The second level of parallelisation is a little bit more complicated. In the grid point space, we split the data over latitude and longitude. Each processor has full verticals for a packet of latitudes and longitudes. In the spectral space, we split the data following the wave number. Computations are independent between each spectral point on an horizontal plan but we get a dependency on the vertical level. So, we need to get all the data on a vertical.

For a stretched grid, the dependency of the data is not exactly the same in the spectral space. In this space, for some computations we have always some dependencies on the vertical level but for other computations we have a dependency on the total wave number. So, we need to transpose the data between each computation with respect to the different dependencies.

In a first approach, it was easy to adapt the spectral computations for the stretched grid in the case of the first level of parallelisation because we respect all the dependency. But, as we get a vertical dependency in the semi-implicit computation, we need to develop a new strategy for implementing the second level of parallelisation. First, we separate the computation of the horizontal diffusion from the computation of the semi-implicit part. At the beginning of the semi-implicit calculations, we achieve a transposition from the distribution set (SET_N) (cf figure 1) with full total wave numbers (n) to the distribution set (SET_L) with full verticals only for the divergence, temperature and surface pressure fields.

Figure 1 : spectral grid

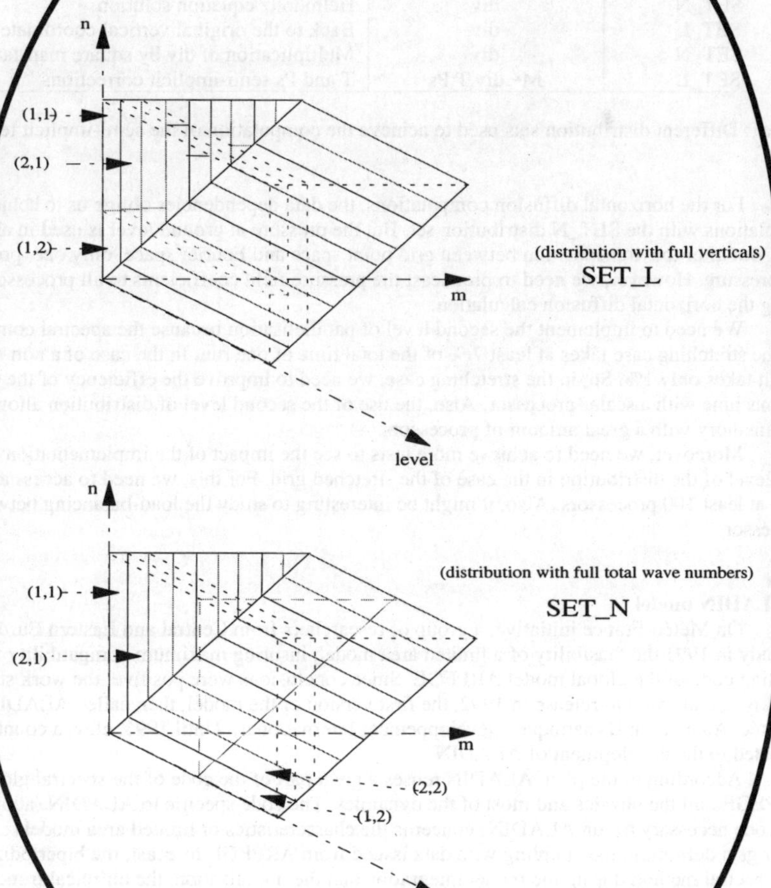

(i,j)=proc (i*j), i in the A-set and j in the B-set.

Then, the projection on the vertical eigenmodes is achieved on the SET_L distribution set. Afterwards, the Helmholtz equation is solved on the SET_N distribution in the stretching case. To return to the original vertical coordinate, we have to transpose the divergence field back to SET_L distribution. Then, the multiplication of the divergence by the square map-factor must be done on SET_N distribution whereas the last calculations to get temperature and surface pressure terms must be done on the SET_L distribution. In fact, we use four transpositions inside the semi-implicit computations (see Tab.1).

Distribution set	Fields	Computation
SET_L	div,T,Ps	Projection of div on vertical eigenmodes
SET_N	div	Helmholtz equation solution
SET_L	div	Back to the original vertical coordinate
SET_N	div	Multiplication of div by square map-factor (M^2)
SET_L	M^2.div,T,Ps	T and Ps semi-implicit corrections

Table1: Different distribution sets used to achieve the computation of the semi-implicit terms.

For the horizontal diffusion computations, the data dependencies oblige us to achieve the calculations with the SET_N distribution set. But the pressure at ground level is used in all processors. But after the transposition between grid point space and Fourier space, only one processor get the pressure. However, we need to broadcast the pressure field coefficients to all processors before doing the horizontal diffusion calculation.

We need to implement the second level of parallelisation because the spectral computation for the stretching case takes at least 7 % of the total time of one run. In the case of a non-stretched run, it takes only 1%. So, in the stretching case, we need to improve the efficiency of the code for the machine with a scalar processor. Also, the use of the second level of distribution allow us to use less memory with a great amount of processors.

Moreover, we need to achieve more tests to see the impact of the implementation of the second level of the distribution in the case of the stretched grid. For this, we need to access a machine with at least 100 processors. Also, it might be interesting to study the load-balancing between each processor.

3. ALADIN model

On Météo-France initiative, a group of researchers from Central and Eastern Europe began to study in 1991 the feasibility of a limited area model, insuring maximum compatibility with the existing code of the global model ARPEGE. Since conclusions were positive, the work started quickly and allowed to release in 1992, the first version of the model, then called ALADIN (Aire Limitée, Adaptation Dynamique, Développement International). Until 1995, eleven countries participated to the development of ALADIN.

According to the plan, ALADIN reuses a great part of the code of the spectral global model ARPEGE : all the physics and most of the dynamics. The code specific to ALADIN (about 30% of the code necessary to run ALADIN) concerns the characteristics of limited area models : the rectangular grid definition, the coupling with data issued from ARPEGE forecast, the biperiodization to use spectral method during the model integration and the initialization, the elliptical truncation.

ALADIN model has been defined as a dynamical adaptation model : its initial conditions and its boundary conditions are given by the global model ARPEGE. Also, the format of the INPUT files is the same than in ARPEGE except that the domain is not the same.

As the distributed environment has been implemented in ARPEGE model, we need to implement this environment in the ALADIN model. Nevertheless, we need to achieve some adaptation in the different space.

3.1. Distribution in the spectral space

On the figure 2, you can see the distribution in the spectral space. This distribution is a 2D distribution. For the moment, we have only implemented and tested a 1-D distribution for the eulerian case. The strategy used for the 1D distribution is quite the same than this used in ARPEGE. For example, if we take an elliptical truncation (NMSMAX=15,NSMAX=13) with a distribution on 4 processors, the first zonal wave number (m=0) is dedicated to the processor 1, m=1 to processor 2, m=2 to processor 3, m=3 to processor 4, m=4 to processor 4, m=5 to processor 3, m=6 to processor 2, m=7 to processor 1, m=8 to processor 1, m=9 to processor 2 ,..., until m=15. If we achieve the sum of all the elements computed by the processor we get quite the same amount of elements between each processors.

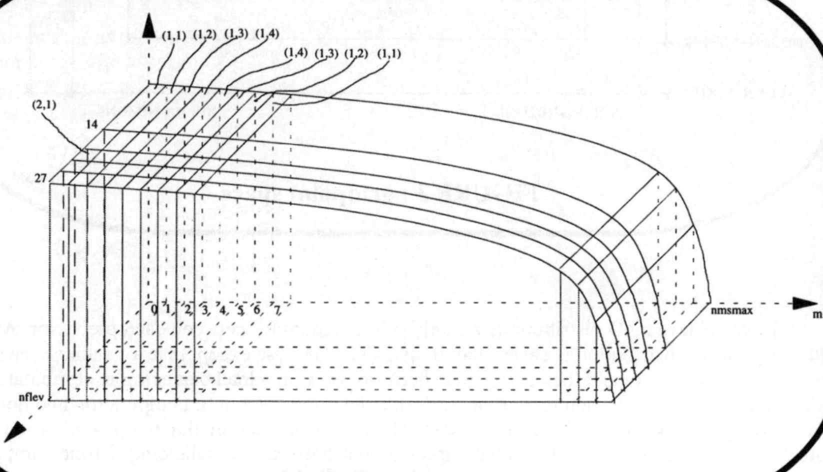

FIGURE 2 : spectral space

We can also achieve a 3D transposition (i. e. 2D distribution between vertical levels) In fact, even with an elliptical truncation, the total amount of elements for each processor is quite the same. We are thinking that the load imbalanced will only appear with a great number of processors. As the truncation is not too big (for the time being), a massively parallel machine is not useful for the ALADIN case and in fact we need less than (NMSMAX+1)/2 processors to correctly run ALADIN.

3.2. Distribution in the grid point space

In the gridpoint space we have three different areas : the extension zone (used for the FFT) the coupling zone (used for the coupling with the ARPEGE files) and the central zone (where we achieve all gridpoint computations). (see figure 3).

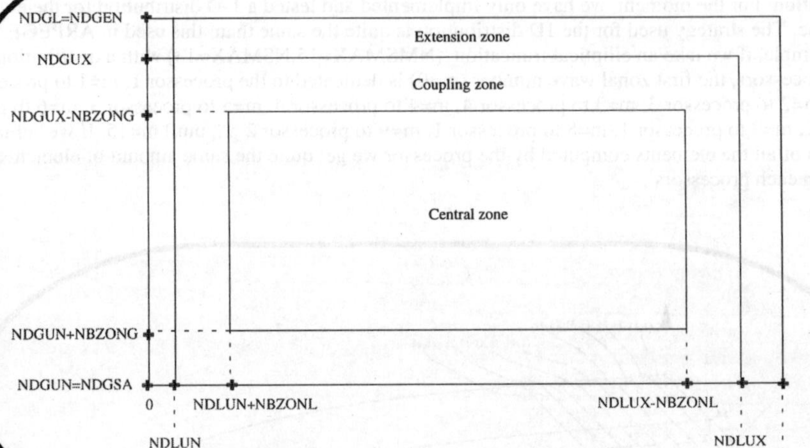

FIGURE 3 : gridpoint space

If we achieve a 1D distribution, we only split the latitudes between each processor. As the grid is regular the distribution is easier than in the ARPEGE case except that we achieve the splitting only in the coupling and the central zone because we want a load balancing in computation and not in amount of data. So each processor gets quite the same amount of computation and not the same amount of data. (see figure 4). In the case of figure 4, we can see that the processor 4 has more data than the processor 3. It is very important to achieve a load balancing in time computation between processors because the physics takes about 50 % of the total time of computation.

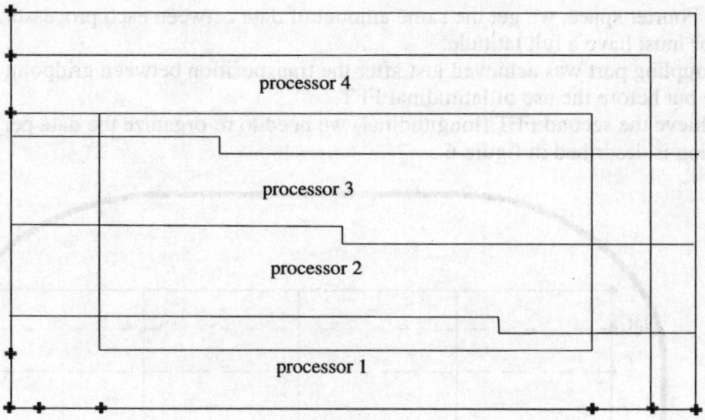

FIGURE 4 : gridpoint distribution

3.3. Transposition between gridpoint space and spectral space

We use quite the same transpositions than in ARPEGE except that we need to re-organize the data before doing the FFT because in the Fourier space the full domain is concerned on the contrary to gridpoint space where the calculations are achieving only on the coupling and central zone. The new distribution in the Fourier space is described in figure 5 :

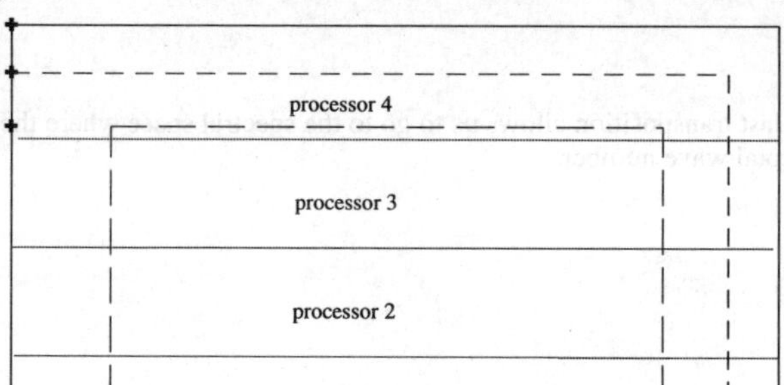

FIGURE 5 : Fourier distribution

In the Fourier space, we get the same amount of data between each processor. Moreover, each processor must have a full latitude.

The coupling part was achieved just after the transposition between gridpoint space and the Fourier space but before the use of latitudinal FFT.

To achieve the second FFT (longitudinal) we need to re-organize the data per longitude. The new distribution is described in figure 6 :

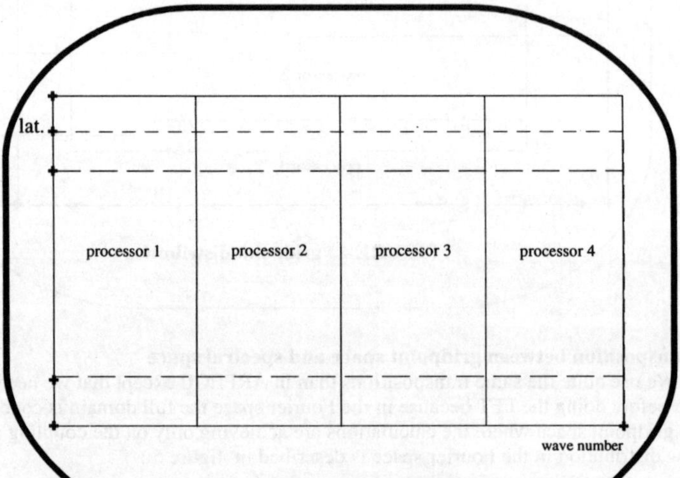

FIGURE 6 : Fourier distribution

Then the last transposition allows us to go to the spectral space where the distribution is achieved on the total wave number.

3.4. The semi-Lagrangian case

For the semi-Lagrangian computation, it is necessary to build a halo. Inside the coupling and central zone, the problem is the same than in ARPEGE. For the computation outside these zones, we project the points from the extension zone to the border of the coupling zones. So we don't need any halo on the extension zone. Also, if we use a 2D distribution, the halo is useless in the extension zone. In the figure 7, the dashed zone is the halo zone for processor 4 and for processor 2.

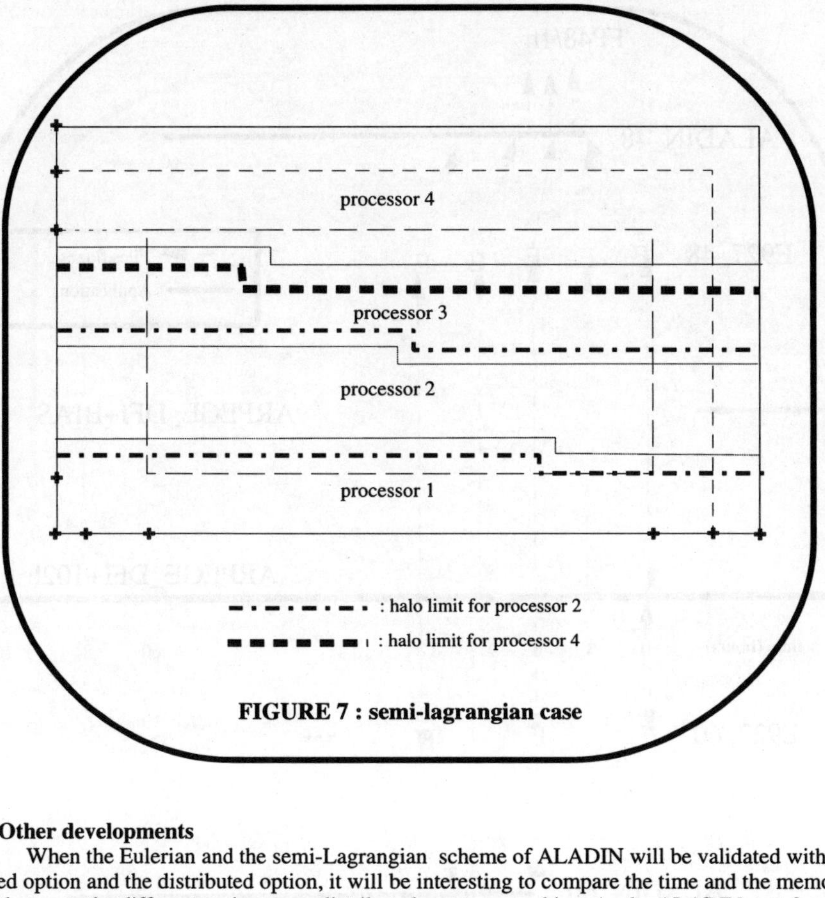

FIGURE 7 : semi-lagrangian case

3.5. Other developments

When the Eulerian and the semi-Lagrangian scheme of ALADIN will be validated with the shared option and the distributed option, it will be interesting to compare the time and the memory used between the different options on a distributed memory machine. As the ALADIN part for the Eulerian case was included in the benchmark of our ITT, we'll get some results next year on different platforms.

4. The use of ARPEGE/ALADIN applications in a simulation of an operational context

In fact in an operational suite, we need to get ARPEGE and ALADIN forecasts. So, in our benchmark, we achieved a test including both ALADIN and ARPEGE models. In a first approach, we remark that it is not so easy to run efficiently both ARPEGE and ALADIN models plus the building of coupling files and the post-processing on the same platform because each application use a lot of memory and the scheduling between each application is not easy to define.

The simulation of the forecasting suite is described in the figure 8.

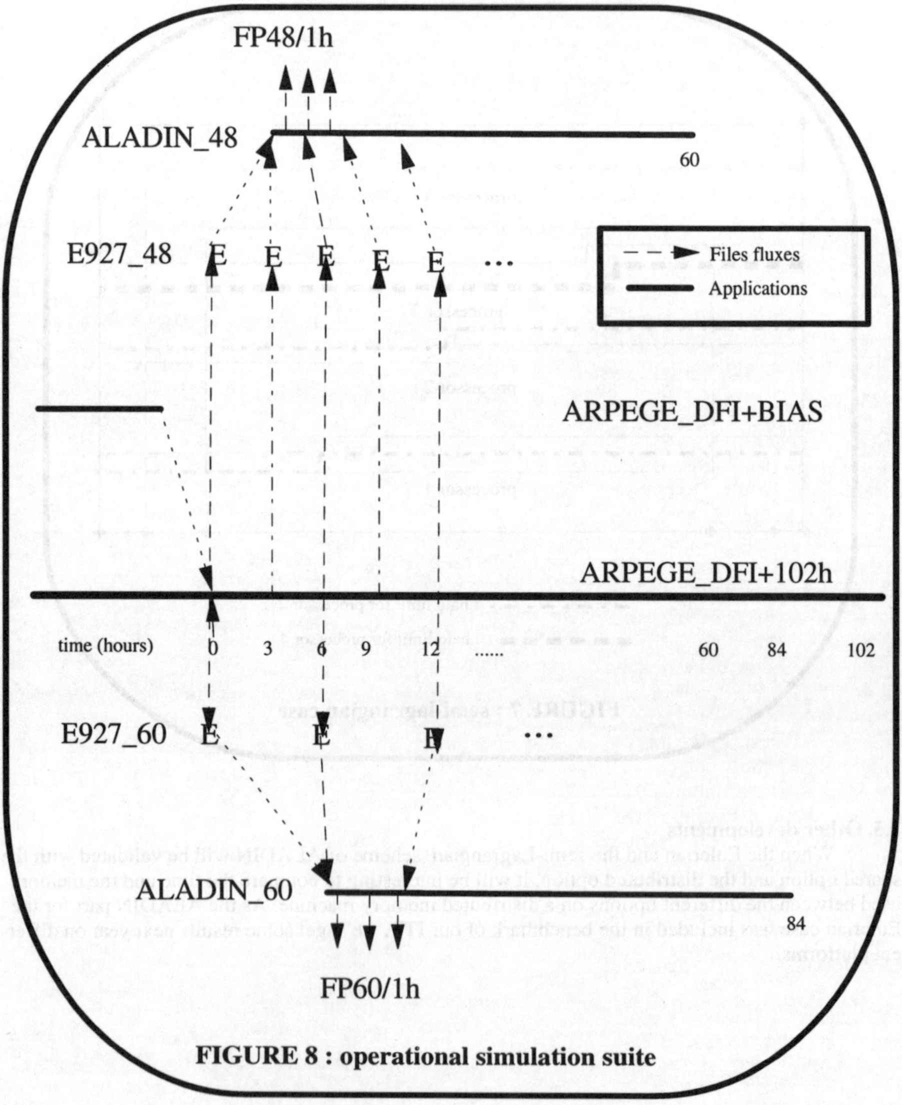

FIGURE 8 : operational simulation suite

First, we execute a DFI application on the guess file. This run builds a new file that will be used in the main run of ARPEGE.

The second step is the main run of ARPEGE which uses the analysis file and the output file of the latest configuration. Each 3 hours, this configuration builds some output files. These output files are useful for the building of the coupling files for ALADIN.

So, every three hours, we run an ARPEGE/ALADIN application (E927) which creates a coupling file. This run is not distributed because it doesn't take too much memory and time. So we need 36 runs of E927 to be able to run the two ALADIN configurations. If ARPEGE is very fast, then we can run several E927 at the same time.

To start ALADIN, we only need the two first coupling files. But after three hours (or six hours) of simulation of ALADIN, we must get a new coupling file in the ALADIN run. Also, ALADIN generates an output file, each hour, which is post-treated by a Full-pos of ALADIN.

The Full-pos of ALADIN runs on only one processor because it doesn't take too much time and memory. With the two ALADIN, we generate 144 output files which are post-treated by 144 Full-pos of ALADIN.

In this suite, we can remark that not all the applications are parallelised (only ARPEGE and ALADIN models). But we have a potential of parallelism between the applications. However, it is not easy to build a script because we have to wait for the availability of some files (for example, the coupling files for the ALADIN case). Also, the input/output files are splitted in the case of the distributed applications. So, before using the non-distributed applications (Full-pos) we have to unsplit the files. All the parameters of this suite have been chosen in such a way that it takes more than 70 minutes of total elapsed time of execution (on an YMP C90, 8 processors). At the end, it is not obvious to determinate which run finishes last.

5. CONCLUSION

During these last two years, we tried to build a simulation of the operational suite for the benchmark of the ITT of Météo-France which is able to run on different platforms. This simulation will allow us to find the best machine for an operational context. Of course, this simulation doesn't represent the full operational suite used at Météo-France because the integration of all the runs for a distributed machine takes a lot of time. But, we have learnt that it is not so easy to use a distributed machine in an operational context and a lot of work is still to do.

It is obvious also that we need to achieve more tests for ALADIN because some optimization is needed. Also, it will be very interesting to compare the efficiency between the multitasked version and the distributed version of ALADIN.

Acknowledgments :

David Dent, Laars Isaksen, Saulo Barros, Fritz Wollenberg, Guy Robinson are thanked for helpful collaboration on the parallelisation of ARPEGE/IFS. Besides, Jean-François Geleyn, Patrick Lemoigne, Radmila Bubmova, Gabor Radnoti, Mehdi El Abed, Ryad El Khatib, Dominique Giard, Eric Sevault, Jean Philippe Piedelievre, Bouchaib Radi, Vijay Saravane, Dominique Birman are thanked for helpful conversations, for the encouragement provided in our project and for some helps on the parallelisation of ARPEGE/IFS and ALADIN.

Jean Latour from CRAY RESEARCH FRANCE is thanked for his help on the parallelisation of ARPEGE/IFS on the C90 and the adaptation of the code to the T3D.

References :

1. P. Lynch, X. Huang : Initialization of the HIRLAM model using a digital filter , June 1992, Monthly Weather Review, p 1019

2 . S. R. M. Barros, T. Kauranne : On the parallelisation of Global Spectral Eulerian Shallow-Water Models, Parallel supercomputing in atmospheric science, Nov. 1992, ECMWF, World Scientific, p 36
3. D. Dent : Parallelisation of the IFS Model, Parallel supercomputing in atmospheric science, Nov. 1992, ECMWF, World Scientific, p 73
4. U. Gartel, W Joppich, A. Schuller : Parallelisation of the IFS : Status report and first results, Parallel supercomputing in atmospheric science, Nov. 1992, ECMWF, World Scientific, p 152
5. L. Isaksen : Parallelizing the ECMWF optimum interpolation analysis,Parallel supercomputing in atmospheric science, Nov. 1992, ECMWF, World Scientific, p 240
6. S. R. M. Barros, T. Kauranne : Scalablity estimates of parallel spectral atmospheric models, Parallel supercomputing in atmospheric science, Nov. 1992, ECMWF, World Scientific, p 312
7. D. Dent, L. Isaksen, S. Barros, F. Wollenberg, G. Robinson : The message passing version of ECMWF's weather forecast model, High Performance Computing and Networking, Apr. 1994, Lecture notes in computer science,Springer-Verlag, p 299
8. U. Gartel, W Joppich, A. Schuller : Parallel computing for weather prediction, High Performance Computing and Networking, Apr. 1994, Lecture notes in computer science,Springer-Verlag, p 305
9. J. F. Estrade, D. Birman : Adapting Parallel IFSARPEGE to Météo-France implementation Nov. 1994, ECMWF, World Scientific, p 206
10. S. R. M. Barros, D. Dent, L. Isaksen, G. Robinson : The IFS model - Overview and parallel strategies, Nov. 1994, ECMWF, World Scientific, p 303
11. P.Lynch, D. Giard, V. Ivanovici : Improving the efficiency of a digital filtering scheme for diabatic initialization, Nov. 1996 , Internal notes (Météo-France)

OPERATIONAL IMPLEMENTATION OF PARALLEL MODELS AT THE NORWEGIAN METEOROLOGICAL INSTITUTE

DAG BJØRGE
Norwegian Meteorological Institute
P.O. Box 43, Blindern, N-0313 Oslo, Norway

PER CHRISTIAN MOAN and ROAR SKÅLIN
Academic Computing Service
Norwegian University of Science and Technology
N-7034 Trondheim, Norway

ABSTRACT

The experience gained when parallelizing the forecasting part of Hirlam has been utilized when the other operational models at DNMI were modified for use on parallel platforms. Some details of the parallel semi-Lagrangian code is shown, and some ideas on more efficient parallel I/O is briefly discussed.

1. Introduction

The Norwegian Meteorological Institute (DNMI) will use a 64 node CRAY T3E recently installed at the Norwegian University of Science and Technology (NTNU) for operational weather forecasting and other compute intensive simulations. To take advantage of this system, it was decided to parallelise serveral of the operational models at DNMI. The project "Massive Parallel Processing in Numerical Weather Prediction" was initiated at DNMI in 1993 finalized by the end of 1995. As a result of this project we now have Parlam, a parallel version of the forecasting parts of the Hirlam model. Using the same techniques as in Parlam, the EULMOD chemical tracer model has successfully been parallelised. A parallel version of the operational ocean model is soon to be released.

The Hirlam model[1,2] can be described as an Eulerian, or optionally a semi-Lagrangian gridpoint model, with semi-implicit time integration. At the previous ECMWF workshop on the "Use of parallel processors in meteorology", the parallelisation efforts put into the Eulerian code was covered[3]. The parallel version of the Eulerian code is also described and analyzed in other publications[4,5,6]. In this paper a brief overview of the parallel semi-Lagrangian code will be given. We will also show some recent measures of the performance of the semi-Lagrangian code on the Cray T3E.

The operational models at DNMI applies the same I/O scheme and file structure. So far, this scheme has been ported to parallel machines in a straigthforward way, by using only one processor for reading and writing data. On powerful computers like the T3E, this is clearly unsatisfactory, as I/O may account for up to 30% of the total elapsed time. Thus, a project has been initiated to develop a common, parallel I/O scheme for all the models. Our initial ideas on how to design this scheme are presented briefly.

2. The semi-Lagrangian scheme in the Hirlam model

The semi-Lagrangian scheme in the Hirlam model is a two time level, three dimensional semi-implicit grid point scheme with sixth order simplified horizontal implicit diffusion to maintain stability for large time steps.

2.1. General parallelisation strategy

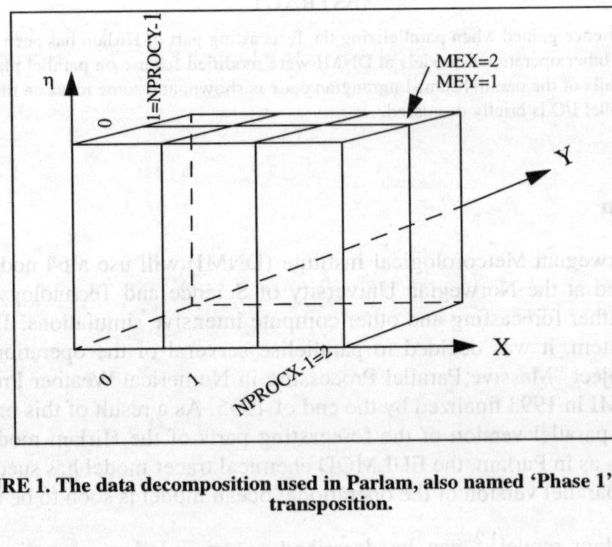

FIGURE 1. The data decomposition used in Parlam, also named 'Phase 1' in the data transposition.

In Parlam the following parallelisation strategy has been chosen:

- To use a two-dimensional latitude-longitude domain decomposition, with subdomains consisting of contiguous vertical columns extending from the surface to the top of the atmosphere, as shown in Figure 1.
- To keep the same data structure as in the serial code, with the obvious exception that each processor now holds only the data for its subdomain.

- To apply data transposition in the implementation of the original Helmholtz solver, which is used in the semi-implicit time scheme, and in the implementation of the implicit horizontal diffusion used in the semi-Lagrangian scheme.

Using the two dimensional longitude by latitude decomposition with NPROCX * NPROCY processors (Phase 1) as a starting point, we can use data transposition to redistribute the data into latitude by level (Phase 2) and longitude by level (Phase 3) decompositions. In this way both the Fourier transforms and the Gaussian elimination in the Helmholtz solver, and the Gaussian elimination for the low pass sine filters in the implicit horizontal diffusion scheme, can all be done locally on each processor.

In order to reduce the amount of communication, we have introduced the additional constraint of only redistributing the data in one direction when going from one phase to another. As an example, the latitude decomposition is kept constant when going from a longitude by latitude to a latitude by level decomposition (Phase 1 to Phase 2), and NPROCX processors is used in the vertical direction. To avoid unnecessary load imbalance, NPROCX should be of the same size as, or smaller than, the number of levels. This is however not considered to be a critical constraint, as NLEV typically is larger than 30 in operational setups. Furthermore, increasing the number of processors probably leads to an increase of the number of levels, in which case this will never become a real constraint. Figure 1 together with Figure 2 show an example of data distribution for the three phases, using NPROC = 6 processors with NPROCX = 3 and NPROCY = 2.

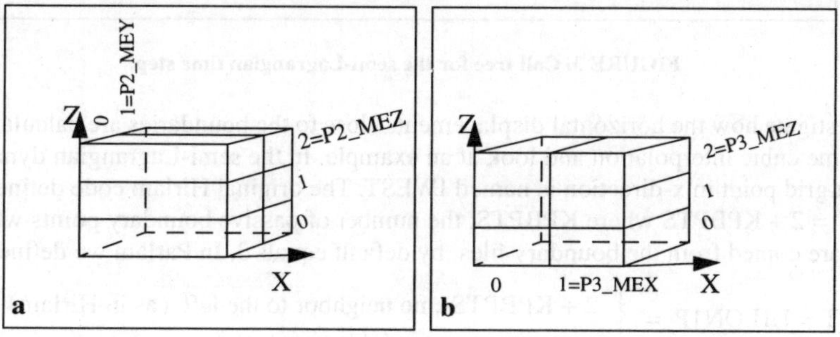

FIGURE 2. a) Phase 2, and b) Phase 3 in the data transposition.

2.2. Parallelisation of the semi-Lagrangian scheme

A schematic overview of the calling tree in a semi-Lagrangian time step is given in Figure 3 below.

```
SL2TIM
  COMPFX      first guess interpolation in time
  SLDYN       explicit dynamics
  SLDYNM
    preparations
    vertical interpolation
    NNCOM     communicate velocities for destaggering
    DESTAG    bicubical destaggering
    vertical integration
    NNCOM     communicate work arrays
    CALPQR    calculate displacements
    NNCOM     communicate all prognostic variables
    BIXINT    interpolation to departure points
    NNCOM     communicate displacements for staggering
    INTPQR    staggering
    BIXINT    interpolation of velocities
  SLEXPA      preparations for the Helmholtz solver
  HHSOLV      semi-implicit time stepping
  IMPADJ      the implicit adjustment
  PHCALL      physical parameterisation
  DIFFH       implicit horizontal diffusion
  BNDREL      boundary relaxation
```

FIGURE 3. Call tree for the semi-Lagrangian time step.

To investigate how the horizontal displacements close to the boundaries are calculated, let us assume cubic interpolation and look at an example. In the semi-Lagrangian dynamics the first grid point in x-direction is named IWEST. The original Hirlam code defines IWEST = 2 + KPBPTS where KPBPTS, the number of passive boundary points where values are copied from the boundary files, by default equals 3. In Parlam we define

$$\text{IWEST} = \text{L1LON1P} = \begin{cases} 2 + \text{KPBPTS} & ;\ \text{no neighbor to the left (as in Hirlam)} \\ 1 & ;\ \text{a neighbor to the left} \end{cases}$$

If the wind speed exceeds $v_{max} \approx \frac{2\Delta x}{\Delta t}$, where Δx and Δt are the grid size and time step length, the necessary values for interpolation will be symmetrically extrapolated around JX=2.

In the case of a global boundary, the procedure is shown graphically in Figure 4.

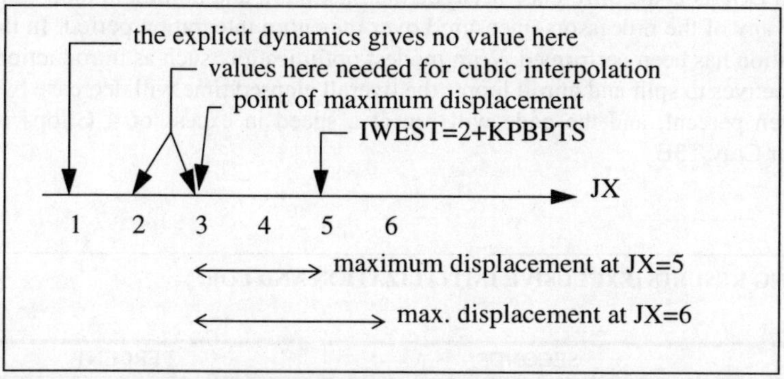

FIGURE 4. Displacements at a western global boundary.

For an internal boundary, the procedure can be described by Figure 5.

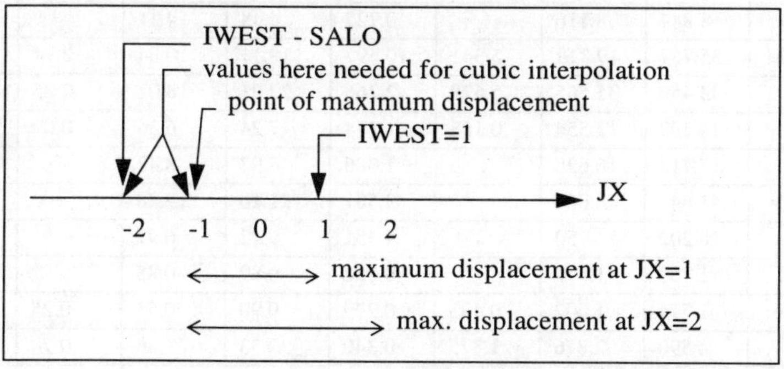

FIGURE 5. Displacements at a western internal boundary.

Where the semi-Lagrangian halo, called SALO for short, can be defined as SALO = KPB-PTS to obtain the same 'accuracy' on internal boundaries as on the global boundaries. If the wind speed exceeds v_{max} somewhere in the interior of the computational area, one must redefine SALO larger than KPBPTS to reproduce the behaviour of the serial code exactly.

2.3. Semi-Lagrangian timings

Timings of Parlam on the new Norwegian 64 node CRAY T3E for a 48 hour run with a semi-Lagrangian time step of 20 minutes and full physics at every time step is shown below. In this test run the model has been integrated on the current operational grid

(188 * 152 grid point and 31 levels), with boundary files from ECMWF every sixth hour. The term LOAD is the difference between the maximum and minimum total elapsed time taken by any of the processors, measured over the entire integration period. In this run no optimization has been performed. With modest optimization, such as introduction of compiler directives to split and unroll loops, the overall elapsed time will decrease by approximately ten percent, and the code will run at a speed in excess of 4 Gflops on the 64 processor Cray T3E.

TIMING RESULTS (EXCLUSIVE INITIALIZATION AND I/O):

	SECONDS				PERCENT			
	TOTAL	COMP	COMM	LOAD	TOTAL	COMP	COMM	LOAD
GEMINI:	4.283	2.882	1.021	0.381	2.17	1.46	0.52	0.19
SL2TIM:	3.689	3.036		0.653	1.87	1.54		0.33
SLDYN:	8.844	8.116		0.727	4.48	4.11		0.37
SLDYNM	35.757	19.819	5.341	10.597	18.11	10.04	2.71	5.37
DIFFH:	43.459	35.565	5.628	2.266	22.01	18.02	2.85	1.15
PHYS:	14.297	12.554	0.168	1.574	7.24	6.36	0.09	0.80
VDIFF:	17.714	16.628		1.086	8.97	8.42		0.55
SUNDQ:	41.647	38.066		3.581	21.10	19.28		1.81
RADIA:	18.202	13.750		4.451	9.22	6.97		2.25
BNDREL:	1.753	1.672		0.082	0.89	0.85		0.04
EXPADJ:	1.774	1.002	0.493	0.279	0.90	0.51	0.25	0.14
HHSOLV:	4.590	2.876	1.375	0.340	2.33	1.46	0.70	0.17
IMPADJ:	1.409	0.779	0.204	0.427	0.71	0.39	0.10	0.22
TOTAL:	197.419	156.726	14.232	26.461	100.00	79.39	7.21	13.40

DIFFH, the sixth order implicit horizontal diffusion, is parallelised by calling the data transposition routines described above. The diffusion is applied on velocity components, temperature and humidity, so that the communication seen in DIFFH will be four times larger than the HHSOLV communication. Together with the semi-Lagrangian dynamics and the Sundqvist condensation scheme, the diffusion routines are obvious candidates for further optimization.

3. Parallel I/O

The operational models in question all use the same sort of input and output files. Each sequential Unix file consists of a number of two-dimensional horizontal fields. The data in each field are packed, either as the currently used scaled 16 bits DNMI records, or as variable precision GRIB records. In Parlam the most straight forward method where one node reads everything and distributes, has been used so far. In the table below, elapsed times for the routines involved in the input of 9 boundary files are shown.

BASIC TIMING RESULTS OF INPUT (seconds):

	COMP	COMM
Distribution:	0.135	6.118
Reading / unpacking:	7.416	

As input and output together seems to account for more than 20 percent of the total time on the T3E, several other more sophisticated I/O schemes have been considered.

A simple refinement of the one-to-all strategy is to organize the distribution of data in a tree-like fashion. Another straight forward approach is to parallelise the I/O with respect to the number of 2D-fields. Many nodes accessing the disc hardware simultaneously will cause blocking effects because of excessive disc access, and it is therefore important to tune the algorithm. When the data are packed as in our case, file I/O can be overlapped with data manipulations and distribution in a pipelined fashion. If enough memory is available, asynchronous I/O should also be considered.

4. Conclusions

The operational models operated by the Norwegian Meteorological Institute have been parallelised, and are ready for operational use on the new Cray T3E computer. Experiments with the forecast model indicates that the 64 node machine delivers in excess of 4 Gflops. This implies that the model will be run on a 10 km grid with 332 by 350 grid points and 31 levels for the new operational setup. Because of limited storage resources, postprocessed files will not immediately be of the same resolution. As boundary files are received from ECMWF on a 0.5 by 0.5 degree grid, and the interpolation to the runtime grid can be parallelised, the I/O will probably not be a bottleneck. If, or when, the amount of I/O grows larger, some of the improved I/O schemes discussed in this paper will be implemented.

5. Acknowledgments

The MPP project and the Parallel I/O project are supported by The Research Council of Norway, grants 100955/432 and 116829/431.

6. References

1. *The HIRLAM Documentation Manual, System 2.5*, ed. E. Källén (1996). Available from SMHI, S-60176 Norrköping, Sweden.

2. A. McDonald, *The HIRLAM two time level, three dimensional semi-Lagrangian, semi-implicit, limited area, grid point model of the primitive equations.* (Hirlam Technical Report No. 17, 1994.Available from SMHI, S-60176 Norrköping, Sweden.)

3. D. Bjørge, *Implementation of the Semi-Implicit Scheme in a Message Passing Version of HIRLAM.* (Proceedings of the Sixth ECMWF Workshop on the Use of Parallel Processors in Meteorology: Coming of Age. Eds.: G-R Hoffmann and N. Kreitz)

4. D. Bjørge and R. Skålin, *PARLAM the parallel HIRLAM version at DNMI.* (DNMI Research Report No. 27, 1995)

5. R. Skålin, *Scalability of Parallel Grid Point Limited Area Atmospheric Models. Part I: Explicit Time Integration Schemes.* Journal of Atmospheric and Oceanic Technology, to appear.

6. R. Skålin, *Scalability of Parallel Grid Point Limited Area Atmospheric Models. Part II: Semi-implicit Time Integration Schemes.* Journal of Atmospheric and Oceanic Technology, to appear.

OPTIMIZING THE HIRLAM SYSTEM TO NEC-SX4

J. U. Jørgensen
Danish Meteorological Institute,
Lyngbyvej 100, DK-2100 Copenhagen, Denmark

Abstract

The transition to message passing and fully distributed memory has been postponed to the next generation of computers at DMI. The NEC SX4 fulfills DMI's requirements for both memory bandwidth and scalability, even though it is a vector parallel system with shared memory. A single NEC processor is more than 15–20 times faster than the fastest scalar processor on the HIRLAM reference forecast model [1], but the price of a 20 processor MPP system is normally higher than a single processor NEC SX4. We consider the NEC SX4 to be a very efficient system in the range 1–30 Gflops for the purpose of running the HIRLAM system operationally.

1. Introduction

The HIRLAM system has been run operationally at DMI since 1990 on CONVEX vector parallel computers. The first 3 years on C2 system and the latter 3 years on a C3880, which is an 8 processor system with a peak performance of 2 Gflops. A CRAY C90 processor is 5 times faster than a CONVEX 3880 processor on the HIRLAM forecast model, thus the C90 is more efficient, probably due to the limited memory bandwidth on the CONVEX system.

The *straightforward* solution to the memory bandwidth limit is to use fully distributed memory. However, a message passing version of the full HIRLAM system does not exist yet.

The HIRLAM analysis has been ported to four cached based systems with acceptable performance. Two of them were shared memory systems with enough performance for running the program operationally. Work spent on parallelizing the analysis for distributed memory would be wasted since three or four dimensional data assimilation will replace the current OI analysis in a few years time.

Several groups have parallelized the forecast model with good performance, but the performance on scalar cache based systems is far from peak, leaving room for improvements in the performance. Such improvements require major code changes and they are difficult to maintain in a portable code like the HIRLAM forecast model.

1.1. Selection of a new Supercomputer

A portable set-up for running the HIRLAM forecast model has been

Table 1:

Company	Processor	Compile flags	Mflops/s
NEC	SX4	-float1 -b -stack	330.0
NEC	SX3	-float1 -b -stack	297.0
CRAY	C90-240MHz	-Zv -Wd-ei -Wf -o.	179.6
CONVEX	C4	-O2 -re -f90 -uo	154.7
FUJITSU	VPP500	?	114.2
CRAY	C90-240Mhz	-Zv	106.1
CONVEX	3800-66Mhz	-O2 -re -f90 -uo	74.2
SGI POWER	R8000-90	-O3 -TARG=fp_preci.	67.5
IBM590	pwr2-66Mhz	-O3 -qfloat=HSFLT	61.9
SGI POWER	R8000-90	-O3	46.4
CONVEX-SP	HP-PA-100	-O2 -re -f90 -uo	45.0
CONVEX	3200-33	-O2 -re -f90 -uo	28.0
SGI CHAL.	R4400-150	-O2 -mips2 -Olimit.	24.3
SGI INDY SC	R4000-100	-O2 -mips2 -Olimit.	17.3
DELL	586-120	-O4	17.3
SGI INDY	R4000-100	-O2 -mips2 -Olimit.	16.0
TOSHIBA	586-100	-O	16.0
SGI INDY PC	R4600-100	-O2 -mips2 -Olimit.	14.4
DIGITAL-NT	586-60	-O4	10.5
OLIVETTI	586-60	-O4	9.8
TOSHIBA	486-75	-O	6.2
INTEL	486-66DX2	-O4	4.3
INTEL	i486-33DX	-O4	2.5
SUN(OS 5.3)	SPARC-25	-O2	1.1
POWERNOTE	486(+7)SLC-2	-O4	0.7
IBM P70	i386-20	-O4	0.4

developed and executed on a number of single processor machines with different grid-dimensions and resolutions. The smallest size could execute using one Mword of memory. The vector length for most loops was 624, long enough to gain efficiency from vectorization, but small enough in theory to fit into cache on cache based systems. The purpose of this test was to gain some insight into the performance of scalar systems for a typical meteorological application instead of just comparing the idealized Lin-pack test, which most compilers happen to know too well.

All results are scaled to Mflops based on the C90 hardware performance monitor (*hpm*). Wall clock rather than user time was used for scaling to Mflops, because different operating systems divide system and user time differently.

Tabel 1 shows the list of platforms sorted according to performance. All Intel processor platforms used f2c and gcc. Note also that the FUJITSU VPP500 was running on a file system mounted from a VPP2610, which adds significantly to the wall clock, thus the flop rate increases to 300 Mflops if we disregard all the system time.

The vector architectures with high speed memory and/or multi pipes outperform the others. The flop rate does not increase with increasing problem size except on vector architectures. The SGI R8000 has the best performance of the cache based systems probably because the hole program has a size similar to the second level cache.

After a procurement process involving various selection criteria, DMI chose the NEC-SX4 to be the new supercomputer. The system offered the highest flexibility and the best performance on DMI's main applications. DMI supplied the HIRLAM reference version 2.4 as a benchmark together with the PARLAM [2] for benchmarking and NEC demonstrated close to 8.6 Gflops on 12 processors on the HIRLAM reference system on a relatively small problem size. The same program achieves 11.14 Gflops today after improvements both in the compiler and the hardware.

The current NEC system at DMI has 12 processors and 4GB of high speed main memory (SRAM) and 8GB of XMU (extended memory unit), divided into swap space, disk cache and file systems. The system will be upgraded to 16 processors in the spring of 1997. Each processor has a clock cycle of 8 ns. and 2 vector units performing up to 8 floating point operations each clock cycle thus the peak performance is two Gflops. Each vector unit has a two way path to memory, either *two loads* or *one load and one store*. The scalar peak performance is 250 Mflops. All floating point operations take place in 64 bit and there is hardware support of 64 bit integers as well.

1.2. Setting up the System

DMI received the second NEC SX4 in Europe, thus experience with this new NEC system was quite limited. The large number of options for configuring the file system and the batch system is somewhat confusing for the inexperienced NEC system manager.

The block size (cluster size) is the most important parameter in the file system configuration. If sufficient XMU cache is available, a 128 kb blocksize gives the same performance as a 1 MB blocksize, but the wasted disk space on typical source files is 96%. A small file system with a 4 kb blocksize is used for these files. The temporary file system (/tmp) is also a 4 kb blocksize file system, but this is in the XMU. The batch jobs run by default on a 3 GB 128 kb blocksize file system in the XMU.

There is also system support for the creation of virtual subsystems, called a *resource sharing groups* on the NEC and they may be reconfigured at any time. All login sessions run in the default group (0). Other groups should be accessed through batch. A special resource sharing group is used for the operational schedule. The resource sharing group feature is more powerful than the traditional UNIX share facility.

A user friendly system should allow parallelisation on the shell level and inside C and FORTRAN. The overhead in a fork on a shared memory multiprocessor system is small compared to execution of a batch or system command on another node with its own memory. The most frequently used shell level parallelisation is compilation in parallel spawned from make. Running *make* on a multiprocessor system is several times faster. Compiling with 8 tasks is much faster than using 1 task. Exactly how much depends on the load on the system. The large cache in the extended memory ensures that the system time is kept to a minimum. Load balance problems from make are normally not a problem, because there is a logical task for each object file.

1.3. Selecting the FORTRAN Compiler

The FORTRAN compiler is currently not available as a cross compiler, but is suitable as such, because it has a number of code and optimization list formats giving the user the possibility to optimize the code without measurements.

The HIRLAM forecast model is a traditional FORTRAN 77 program using a few FORTRAN 90 extensions such as include statements, namelists and automatic arrays. These extensions are all supported by the NEC FORTRAN 77 compiler with the exception of the routines with automatic arrays need to be compiled with a flag that invokes an preprocessor/optimizer called *fopp*. The FORTRAN 90 compiler is a new product on the NEC and is currently less efficient.

The FORTRAN 77 compiler has a number of options, which allow the user to select the word size and floating point format. The system supports 32 bit IEEE but with a performance degradation of 30% relative to 64 bits. The IEEE format is a new feature on the SX4 relative to the SX3, thus the likelihood of obtaining incorrect results was considered to be too high. Having the possibility to exchange data between platforms in IEEE is not important, because all data are packed in machine independent formats like *GRIB* anyway. In order to save memory, 32 bit IBM format is the best, but the accuracy is better in 64 bit, thus the CRAY format was chosen. The CRAY format enforces all word lengths to be 8 bytes long and double precision 16 bytes long.

The CRAY compatibility of the NEC SX4 FORTRAN compiler is not complete, but the 64 bit word length is all that the HIRLAM system needs.

1.4. The Implementation of the HIRLAM System on the NEC SX4

The DMI implementation of the HIRLAM system uses BUFR and GRIB as input and produces GRIB as output. All timings in this section is taken from the current operational GRV grid [3]. Each operational cycle consists of the following steps:

- **Creation of analysis observation file (AOF) from BUFR**
  This program runs on 1 processor on 34 seconds, which is half the time of the CONVEX 3880, but similar to that of an IBM 590. The program could be optimized in a machine independent manner to complete in less than 10 seconds with changes in the BUFR decoding software.
- **Analysis**
  This program is basically the ECMWF OI analysis with the memory manager replaced by automatic arrays. The program is running 5 times faster than the CONVEX 3880 on 8 processors. The low speedup is due to the fact that the program still includes a great deal of I/O, though all direct access files has been replaced with memory I/O. There are a number a cache levels for I/O on the NEC system. The FORTRAN runtime library uses main memory as cache, the operating system uses both main memory and extended memory as cache. The first cache level is controlled by a unit number dependent environment variable, thus the user specifies n times a typical record length for each unit for maximum performance. Based on the file information generated by the system we managed to reduce the system time by a factor of 3.5 by adjusting environment variables. The Gflop ratio of the analysis is not impressive yet (2 Gflop), but the execution time is close to 1 minute on 8 processors with 10 seconds of system time.
- **Initialization**
  The initialization is a part of the forecast model, but is still executed in a separate step. The program complete in 6 seconds on 10 processors.
- **Forecast**
  The forecast is running with a speedup of more than 15 on 10 processors compared to the CONVEX 3880 for the same problem size. This includes all I/O which means encoding 9500 GRIB records and decoding up to 2000 GRIB records. An operational 48 hour forecast with a 4 minute time step completes in two and a half minutes on 10 processors.

Thus the hole cycle completes in 4 minutes and 15 seconds.

1.5. Choosing the Parallelisation Strategy for the HIRLAM Forecast Model

The forecast is the most important part of the HIRLAM system, because it is executed more frequently than the other parts and runs for a longer period of time than the analysis. In addition many research activities running on the NEC system use this model.

A traditional parallelisation strategy is considered to be superior to a message passing strategy on the NEC architecture for several reasons:

- **The program scales well with an increasing number of processors.**
  The entire physics runs perfectly in parallel without the use of message passing. The difficult part of the forecast model to parallelize, *the adiabatic*, executes at more than 1.1 Gflops on each processor up to 12 processors. It is likely that the program will continue to scale, but we have yet to gain access to more than 12 processors.
- **Less memory requirements.**
  A message passing strategy with area decomposition requires overlapping data and all static data is allocated in each task. This is too expensive in practical programming on the NEC system due to the fact that main memory is expensive. It is possible to declare data to be stored in the slow memory, but this is considered to be too system dependent.
- **I/O is more straightforward.**
  The time needed to optimize the I/O for distributed memory programming in HIRLAM is not available at DMI.
- **Researchers should not spent time on message passing.**
  The current state of HIRLAM message passing models does not allow the researcher to add new fields without taking message passing into account, This has to do with an impractical declaration of the global data, which leads to explicit communication for each multilevel variable. The administration routines in the model have to be reorganized in order to hide the message passing for the researcher. Moreover the possibility of addressing globally is important in the development phase.
- **Utilization of the system.**
  In order to achieve maximum usage of the system, frequent adjustment of the number of physical (concurrent) tasks is needed. This would lead to a very poor performance with a message passing programming model. In the current parallelisation strategy we have

a global variable defining the number of logical tasks in each loop. The variable is updated every timestep from the execution time from the previous timestep. The current implementation is not very intelligent, only the wall clock is used to optimize the number of processors to use. However, this is sufficient for a program that run through the same subroutines time after time.

The NEC has a hardware monitor, thus the Mflops and vector length etc. are all available in runtime, thus the program can get insight in its own performance. A dramatic drop in performance shows that a high priority program has started and the number of logical tasks should be reduced. The system will run most efficient if the number of logical tasks is equal to the number of physical tasks. However, the program might become too sensible to the load on the system and the amount of work in each task.

The following example from execution of the HIRLAM forecast model shows the available information:

```
    ******  Program Information  ******
 Real Time (sec)          :         182.920101
 User Time (sec)          :        2160.992570
 Sys  Time (sec)          :           3.260046
 Vector Time (sec)        :        1861.376158
 Inst. Count              :        45970069890.
 V. Inst. Count           :        21244072593.
 V. Element Count         :      5091442027082.
 FLOP Count               :      2586429471443.
 MOPS                     :        2367.508383
 MFLOPS                   :        1196.871062
 MOPS   (concurrent)      :       28142.258738
 MFLOPS (concurrent)      :       14227.047870
 VLEN                     :         239.664123
 V. Op. Ratio (%)         :          99.516709
 Memory Size (MB)         :        2174.000000
 Max Concurrent Proc.     :                12.
  Conc. Time(>= 1)(sec):         181.796638
  Conc. Time(>= 2)(sec):         180.693171
  Conc. Time(>= 3)(sec):         180.549243
  Conc. Time(>= 4)(sec):         180.534364
  Conc. Time(>= 5)(sec):         180.506807
  Conc. Time(>= 6)(sec):         180.502878
  Conc. Time(>= 7)(sec):         180.497760
  Conc. Time(>= 8)(sec):         180.492618
  Conc. Time(>= 9)(sec):         180.485299
  Conc. Time(>=10)(sec):         180.453007
```

```
 Conc. Time(>=11)(sec):        179.429533
 Conc. Time(>=12)(sec):        175.421897
Event Busy Count      :                0.
Event Wait (sec)      :          0.000000
Lock Busy Count       :                0.
Lock Wait (sec)       :          0.000000
Barrier Busy Count    :                0.
Barrier Wait (sec)    :          0.000000
```

This information is similar to that of the hardware performance monitor on CRAY, but the information above is actually generated by the main program in the FORTRAN runtime library, which is calling the users main program.

There is no standard NEC routine for automatic adjustment of the number of physical tasks, but it is not difficult to write one. The most difficult problem is to distinguish between serial and parallel work. If the program is partly serial, task *one* would have more work than the other. That would also occur if the number of physical task is one less than the number of logical tasks. With a check of the user time spent in each task, serial work can be separated from parallel work, thus the routine for adjustment of the number of logical tasks does *not* need to know anything about the users program structure. There is a problem with this approach, which occur if a part of the user program is compiled for a fixed limited number of processors greater than one.

1.6. Optimization of the HIRLAM Forecast Model

The current operational HIRLAM forecast model at DMI uses

- **Rotated latitude/longitude grid**
- **Eta as vertical coordinate**
- **Eulerian advection**
- **Semi implicit time integration**
- **Fourth order horizontal diffusion**
- **Physics each third dynamic timestep**
- **Kuo scheme**
- **Local vertical diffusion scheme**

The NEC SX4 needs a certain amount of work in each task to compensate for the overhead in the creation of the tasks. NEC and DMI have developed a parallelisation over latitudes of the explicit dynamics and the explicit and implicit corrections from semi implicit time scheme. In addition horizontal diffusion is done in parallel over levels.

When optimizing for the NEC SX4 one should bear in mind, that vectorization pays off for loop length as short as 6 to 8 and that the vector-scalar ratio on the NEC SX4 is of the order 50 for multiply add constructs.

- **Boundary relaxation**
  Too much indirect addressing gave a bad vector performance. An extra loop level and a table for the loop boundaries removed all indirect addressing with a speedup of more than 10 although the loops length are relative short.
- **Physics**
  The HIRLAM physics is written for vectorization with parallelisation on a high level. In order to save memory and to reduce loop overhead a compiler directive **vdir shortloop* was inserted before each horizontal loops. This means that the length of each loop should be limited to the vector register length, which is 256. In practice 255 is used for reduction of memory bank conflicts. This method gains nearly 10% in speedd. The load balance in the physics on a grid that spawn over the tropics and one of the poles is poor if all tasks were distributed to the processors from south to north. An almost perfect load balance is achieved by distributing the tasks one by one, because the number of logical tasks is much larger than the number of physical tasks.
- **GRIB output**
  The efficiency of the GRIB encoding is important, but does not give many Mflops. The most efficient way to encode data into GRIB format on the NEC SX4 is to create a template record. The parameter, level type, level and data section are substituted into the template. This is done in parallel and each task is running with full vectorization. The total overhead of encoding 10000 records including write out is less than 6 seconds including the writeout.

1.7. Scalability

The scalability of VPP system is definitely an important issue, due to the competition from the MPP systems. Two experiments has been carried out to verify how the system scales with increasing number of processors in use. The number of gridpoints is kept constant in the first experiment, while the the work in each task is kept constant on each processor in the second experiment, thus the grid dimensions is smaller in the sequential run than in the parallel runs.

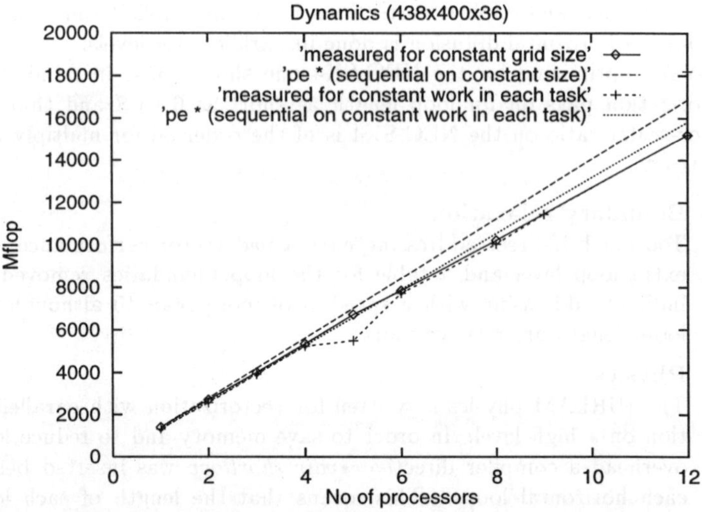

Figure 1: *Performance on HIRLAM forecast model.*

Figure 1 shows the performance in the dynamics and figure 2 shows the performance in the physics. The dynamics has the highest performance in both the sequential and parallel runs.

The diabatic part scales well although the time filter and time stepping of the surface fields are not parallelised. The two curves with linear speedup coincide with each other, because the the performance of the sequential runs are independent of the problem size.

Running the same problem with increasing number of processors is the most common way to measure how the system scales. Such a scaling is close to linear on a cache based scalar processor system, because their performance is more or less independent of the problem size. On a VPP system like a CRAY C90 and NEC SX4 we will see a drop in performance. This is, however, not the case for the physics in HIRLAM, because the maximum performance is reached on a relative short vector length (255).

The dynamics scales better when the work on each processor is kept constant, but the performance in the physics is similar to that of the constant grid experiment.

The reader may argue that the scaling in the constant work case is not fair, because the computers job is to perform a given task as fast as possible without scaling of the problem. The example is constructed for the purpose of understanding the system and to test the memory

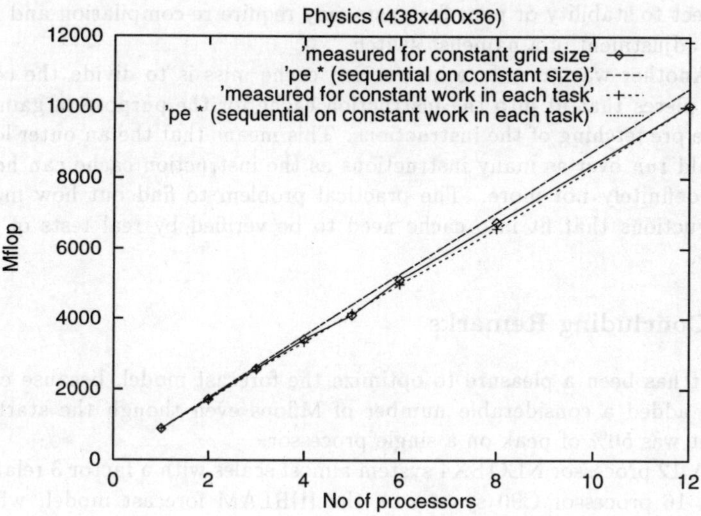

Figure 2: *Performance on HIRLAM forecast model.*

bandwidth, micro task overhead and instruction cache miss influence.

1.8. Further Optimizing and Tuning of the System

The adiabatic part runs at 15.2 Gflop on 12 processors and 1.3 Gflop on a single processor on the same problem, thus the memory bandwidth does not seem to be a limiting factor. In general we observe increased performance with increasing number of grid points, because the time spent on executing the code becomes less significant relative to the time spent on the floating point instructions. The code executed in each timestep can not fit into the instruction cache, thus the wait states between global memory and the scalar unit destroy the performance. This is consistent with table 1, which has a high frequency of instruction cache miss relative to floating point operations due to the small grid dimensions. The mean vector length is long enough for NEC, CRAY and CONVEX, but they are far from their flop rate on a large grid with the HIRLAM forecast.

One way to reduce the instruction cache miss frequency is to preprocess the source code as much as possible. The HIRLAM code is developed for research and contain a number of switches declared as *namelist members*, thus the compiler need to generate general code. However these switches are seldom changed and could be pre-processed in an operational forecast executable. The disadvantage is that debugging of a run with respect to stability or poor forecasts may require re-compilation and not just adjustment of a namelist switch.

Another way to reduce instruction cache miss is to divide the code into pieces that fit into the instruction cache for the purpose of gaining some pre fetching of the instructions. This means that the an outer loop should run over as many instructions as the instruction cache can hold, but definitely not more. The practical problem to find out how many instructions that fit into cache need to be verified by real tests of the code.

2. Concluding Remarks

It has been a pleasure to optimize the forecast model, because each step added a considerable number of Mflops even though the starting point was 50% of peak on a single processor.

A 12 processor NEC SX4 system almost scales with a factor 3 relative to a 16 processor C90 system on the HIRLAM forecast model, which is quite impressive compared to the price of such a system. The larger memory and especially the larger extended memory keep the system time at a minimum.

It is our experience that the NEC SX4 has a high availability and the the efficiency is high with the job mix at DMI. The system needed to be tuned after installation, which took months in the DMI case probably because the DMI system was one of the first installations in Europe.

Users need to pay attention to the performance of their programs, because the distance from 1 to 8 Gflop is surprisingly small. Poor load balance or a problem in the parallelisation may reduce the performance. Parallelisation requires discipline. The compiler can parallelize automatically, without parallelisation directives and with the right flags, but the code will most likely run slower than sequential code.

3. Acknowledgements

Thanks to Maryanne Kmit and Bjarne Amstrup for assistance in optimizing the HIRLAM system.

4. References

1. Gustafsson, N. (ed.) *HIRLAM 2 Final Report*, HIRLAM 2 Consortium, Norrkoeping,(1993).
2. Skålin, R.,Bjørge, D. and Lie, I. *Parallelization of HIRLAM for the Intel Paragon*, Manuscript, SINTEF. Trondheim (1994).

3. Sass, B. H. *The DMI Operational HIRLAM Forecast System, Version 2.3.*, DMI Technical Report 94-8, ISSN 0906-897X, Copenhagen (1994).

A PARALLEL VERSION OF THE HIRLAM FORECAST MODEL: STRATEGY AND RESULTS

KALLE EEROLA
Finnish Meteorological Institute, P.O.B. 503, 00101 Helsinki, Finland

DEBORAH SALMOND
SGI/Cray, 1530 Arlington Business Park, Theale,
Reading/Berks. RG7 4SB, United Kingdom

NILS GUSTAFSSON
Swedish Meteorological and Hydrological Institute,
S-60176 Norrköping, Sweden

JOSE A. GARCIA-MOYA
Spanish Meteorological Institute
Camino de las Moreras s/n, 28040 Madrid, Spain

PETER LÖNNBERG and SIMO JÄRVENOJA
Finnish Meteorological Institute, P.O.B. 503, 00101 Helsinki, Finland

ABSTRACT

Starting from a serial HIRLAM forecasting model a parallel version has been developed. The aim is to develop a code, which is easy to maintain and can be run efficiently on vector and parallel computers. Since the aim is to run the model operationally on a CRAY T3E system, the performance on this platform is important and the SHMEM routines are the main tool for data communications. A portable MPI-version also exists. The preliminary tests on CRAY T3E show the performance of almost 10% of the peak performance. Scaling to different number of processors shows almost linear speed-up from 20 to 64 processors. Especially, the data communication between processors is efficient. The areas, where improvements are foreseen, are single node performance and input/output.

1. Introduction

The HIRLAM forecasting system is the complete forecasting system developed in the international HIRLAM project. The system has been thoroughly described in [1] and the description is repeated here only if needed for the present discussion. The HIRLAM system is operational in several meteorological institutes, including the Finnish Meteorological Institute (FMI) and the Swedish Meteorological and Hydrological Institute (SMHI), and is run in both these institutes on vector computers. The motivation to parallelise the forecast model was that it will soon be possible at FMI and SMHI to use massively parallel CRAY T3E for numerical weather forecasting. The increase in computer power makes it possible to increase the horizontal and vertical resolution of the model. The aim is to go down to the scale of 10 kilometres in the horizontal.

The general principles behind the parallel code development are described in section 2. Section 3 describes the flow and distribution of data during different phases of the computations, which is the crucial point when developing parallel applications. Section 4 describes shortly the flow of computations to the extent, which is needed to understand the preliminary results on CRAY T3E, which are shown in section 5. Finally, section 6 gives the summary and some outlines to the future work.

2. General principles

The starting point for the parallel HIRLAM forecasting model was the latest serial version at the time the current project was initiated. In the beginning a set of principles was set up to be followed during the course of the work. The overall principle in all phases has been: "simple is beautiful". With this rule in mind we believed to be able to make an efficient, easy-to-read and easy-to-modify code, which would be a good basis for the new developments in the meteorological area.

We wanted to have only one code, which would be efficient both on vector machines and on parallel machines. It is very difficult to maintain several versions of the same code and vector computers will be still used in both experiments and development work, although operationally the HIRLAM model will be run on MPP systems at FMI and at SMHI. The new developments in the RISC processors have made it even possible to run the forecast model on efficient workstations.

When possible we have used general mathematical libraries, for instance BLAS, since for many platforms the computer vendors have made highly optimized versions of these routines. As an example of the efficiency the matrix multiplication routine SGEMM in the BLAS 3 library is running in a typical HIRLAM application on CRAY C90 with one processor with a speed of 872 Mflops, while the corresponding home-made routine reaches only the speed of 639 Mflops.

We decided to use dynamical memory allocation (automatic arrays) for the arrays in the code. This feature, which is a standard feature in FORTRAN 90, but exists already in some FORTRAN 77 compilers, makes the need to re-compile the code for different grid sizes and numbers of processors unnecessary, but at the same time the memory is not wasted for the maximum size arrays. There have been doubts about the efficiency of automatic arrays on CRAY T3D [2] when using a large number of processors, but this has not been observed on T3E. In general the FORTRAN 77 style has been used throughout the code.

The dynamics of the HIRLAM reference model has been coded using one-dimensional arrays in the horizontal to allow very long vectors, which normally makes the code efficient on vector machines. In the parallel code we decided to use two-dimensional arrays in the horizontal, which suits better to the logic of the data distribution strategy. This meant that we had to re-code a large part of the dynamics. The positive side-effect, however, was that we, at the same time, made the code more tidy and easy-to-read.

The target of the project is to make an efficient version of the HIRLAM forecasting model to be run operationally on CRAY T3E system. For this reason we decided to use the SHMEM routines as the primary tool for data communications between processors. SHMEM is by far the most efficient way of data communications on CRAY T3D and T3E systems, but the disadvantage is that it is not portable to other platforms. For this reason we also coded a version based on MPI message passing model.

All the communications are hidden into a small set of subroutines separated from the other parts of the code. A difference in our strategy compared to some others e.g. [3] is that in our solution these communication routines call directly either SHMEM routines or MPI routines. The selection of the memory model is made in the compiling phase by using a preprocessor with pre-processor directives and `#ifdef ... #endif` structures in the code. Our solution avoids any overhead caused by still another layer of subroutines. The preprocessor and `#ifdef` structures can be used to select also other machine-dependent parts or, as in our case, to select between a gridpoint or a spectral model.

The last but an important point in our strategy is that we should not destroy the readability or generality of the code while optimizing. This concerns especially the single node optimization.

3. Data distribution strategy

The data decomposition structure is often a crucial point for distributed memory machines, because the data exchange between processors is much more expensive in elapsed time than the computations.

In this paper we focus our attention only on the Eulerian gridpoint version of the HIRLAM model, although also semi-Lagrangian and spectral versions exist. The relevant features of the model when considering the data decomposition strategies are discussed below.

The HIRLAM model uses second-order finite differences in the Arakawa C-grid and it contains a full physical parametrization package. The time integration method is semi-implicit. The horizontal diffusion scheme is either explicit linear fourth order scheme or implicit linear scheme. The adiabatic non-linear normal mode initialisation scheme is built in the same code.

A schematic figure of the data decomposition in different phases of computations is shown in Fig. 1. For the explicit part of the dynamics (phase 1) the data dependencies are local in the horizontal and global in the vertical. So a natural data strategy is a two-dimensional decomposition into subdomains and every processor takes care of one and only one subdomain. Because of second order finite differences, a halo zone of one row and column is needed, where the data exchange between neighbouring processors is necessary. An exception is the surface pressure, for which a halo zone of two rows/columns is needed in the east/north direction because of the Arakawa C-grid. This is taken separately into account. In case of explicit fourth order diffusion the fourth-order operator is calculated in two phases calculating first the Laplacian, then swapping data in the halo zones and calculating the Laplacian again. So a halo zone of one row/column is enough also here.

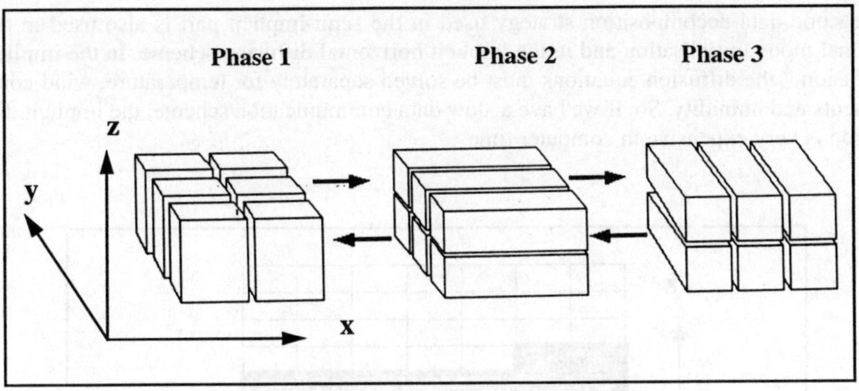

Fig 1. Data distribution during the explicit dynamics (Phase 1), during the FFT (Phase 2) and during the Gaussian elimination (Phase 3)

In the code the halo zone is treated as boundary points and the logic of the code is very similar to the original serial code except that in every subdomain the parallel code works only in local dimensions. When running on a shared memory machine we have only one subdomain and the calculations are done for the whole area.

In the semi-implicit time integration scheme we must solve a three-dimensional Helmholtz equation in every timestep. As described in detail in [1] this is done by decoupling the three-dimensional Helmholtz equation into a set of two-dimensional equations, which are solved using a direct method by applying FFT sine-transformation in x-direction and Gaussian elimination in y-direction. So we need two other data decomposition strategies, one for FFT calculations, which has all the points on a given row in x-direction in the same processor (phase 2 in Fig. 1), and the other for Gaussian elimination, which has all the points in a certain column in y-direction in the same processor (phase 3 in Fig. 1).

To get a good load balance between processors in phase 2 and phase 3 an equal number of rows/columns, if possible (the difference can be at most one row/column), is given to every processor. Fig. 2 illustrates the decomposition of data in phase 2. The X-axis is perpendicular to the paper. The first processor takes the first NLEV*NLAT/NPROC rows (with the exception that some processors must take one extra row), the next processor takes the next NLEV*NLAT/NPROC rows and so on. NLEV is the number of levels, NLAT the number of columns and NPROC the number of processors. The important thing is that the rows owned by a given processor need not be at the same level or in the same column. A similar thing is done in phase 3 with NLON instead of NLAT.

The same data decomposition strategy used in the semi-implicit part is also used in the normal mode initialisation and in the implicit horizontal diffusion scheme. In the implicit diffusion [4] the diffusion equations must be solved separately for temperature, wind components and humidity. So, if we have a slow data communication scheme, the implicit diffusion is very expensive in computer time.

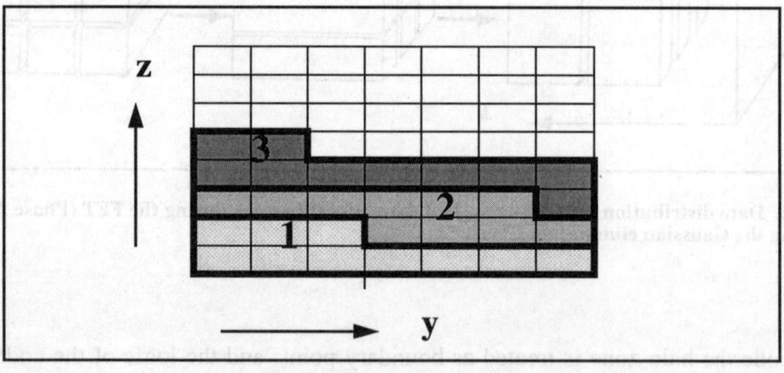

Fig 2. Distribution of rows to different processors during the FFT calculations (Phase 2)

In the physics the data dependencies are only in the vertical and the same data decomposition as in the explicit part of the dynamics is used. However, because of the Arakawa C-grid de-staggering before and staggering after the physics are needed. The calculations in the halo zone are not needed. Load imbalance can occur, because of dissimilar meteorological conditions in different subdomains, but at the moment we have not done anything to compensate for this.

4. Flow of computations

Before showing the preliminary results we will give a short description of the flow of computations. Only the essential steps needed to understand the results in the following section are described.

The essential steps are:

1. Read the input parameters containing the number of processors in x- and y-direction and the dimensions of the grid.
2. Calculate the data decomposition strategies for phases 1, 2 and 3.
3. Read the input fields and distribute to each processor its own data. At the moment the reading and encoding of GRIB takes place on one processor.
4. If new boundaries are needed, read the boundary data as in step 3.
5. Calculate the explicit part of dynamics and exchange the data in halo zones with other processors when needed.
6. In case of explicit horizontal diffusion perform the fourth order diffusion by calculating Laplacian twice and exchanging data in the halo zones between the calculations.
7. Perform the semi-implicit calculations:
   - data transposition from phase 1 to phase 2
   - FFT calculations in the x-direction
   - data transposition from phase 2 to phase 3
   - Gaussian elimination in the y-direction
   - data transposition from phase 3 to phase 2
   - inverse FFT calculations in the x-direction
   - data transposition from phase 2 to phase 1
8. Perform the physics computations using the phase 1 data decomposition.
9. If required do the implicit horizontal diffusion using the same data transpositions as in the semi-implicit phase. Note that all the data transpositions must be done for temperature, wind components, humidity and possible scalar fields.
10. Write out the model state and possibly the post-processed fields at given intervals
11. Repeat the phases 4 - 10 for every timestep.

5. Performance on CRAY T3E

In the test runs we have used the Eulerian version of the model with 258 x 200 x 31 gridpoints. The diffusion scheme is the linear explicit fourth order one and the physics contains Kuo-type of convection, large scale precipitation, radiation and non-local vertical diffusion schemes developed in the HIRLAM project, and the surface processes. They all are described in [1]. The physics is calculated in every timestep. The performance numbers are obtained by comparing the results from identical runs (with same code and same data) running on CRAY C90 and CRAY T3E.

In a 24-hour forecast run having the output of model state every 6 hours the parallel code with 16 processors ran with the speed of 56 Mflops/processor on T3E, which is 9.3 % of the theoretical peak performance. Only a limited amount of single node optimization has been done so far.

The scalability of the model is studied by running it with different number of processors from 20 to 64 and different number of processors in x- and y-direction. The speed-up compared to 20 processors is shown in Fig. 3. The speed-up compared to the ideal linear one seems to be good, almost linear: the speed-up being over three with 64 processors compared to 20. There is no sign of the scalability becoming worse, when coming close to 64 processors. The speed-up depends, however, on the number of processors in x- and y-direction. For instance, with 42 processors the execution time differs about 5% depending if we have 6 x 7 or 7 x 6 processors in x- and y- direction, respectively.

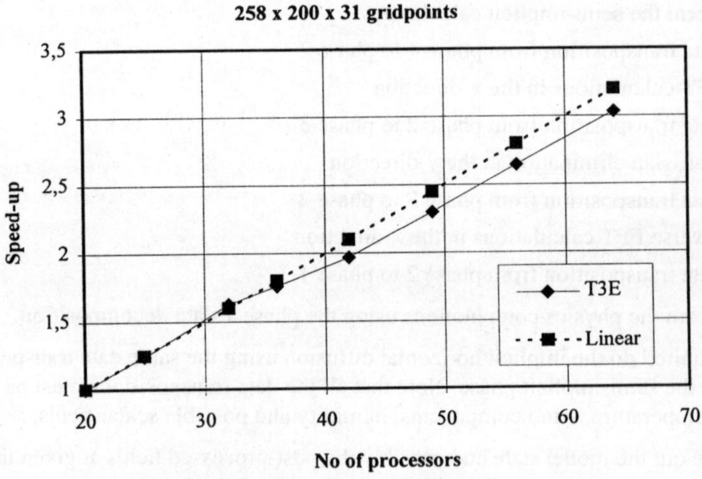

Fig 3. The speed-up of execution time depending on the number of processors compared to 20 processors for the 258 x 200 x 31 grid

For a smaller model, having only a quarter of gridpoints (130 x 100 x 31) of the previous one the scalability, as seen in Fig. 4, is not so good. The speed-up from 20 to 64 processors is only 2.6 and it seems to decrease, when the number of processors come close to 64. This supports the general opinion, that small problems are not at their best on massively parallel computers with many processors.

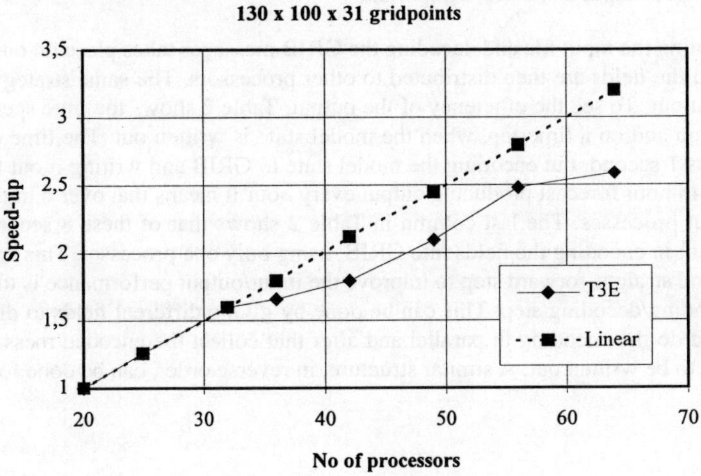

Fig 4. As Fig. 3 but for the smaller, 130 x 100 x 31 grid.

The percentage of time spent in different parts of the code on CRAY C90 with one processor and on T3E with 64 processors is shown in Table 1. The explicit part of dynamics takes about the same percentage of the total time on both computers. An interesting feature is that the semi-implicit part with all its data transpositions is not much more expensive on T3E. This suggests that we have developed a good data transposition method and the SHMEM routines are very efficient. By far the physics is the most expensive part and it is worth concentrating the optimizing work on this part of the code.

TABLE 1. Relative execution times of different parts during one timestep

	% on C94 (1 CPU)	% on T3E (64 CPUs)
explicit dynamics	14	15
hor. diffusion	4	6
semi-implicit part	12	14
physics	68	59
boundary relaxation	1	3
rest	1	3

In an operational environment the output from the numerical models is nowadays required every hour, i.e. a 48 hour forecast produces 48 output files. When going to higher resolution the boundary fields must also be provided every three hours or even every hour. The efficiency of input/output is therefore important.

Currently, reading the input file and decoding the GRIB messages takes place on one processor only and the fields are then distributed to other processors. The same strategy takes place in the output. To see the efficiency of the output, Table 2 shows the time spent on a normal timestep and on a timestep, when the model state is written out. The time of normal timestep is 1 second, but encoding the model state to GRIB and writing it out takes 8 seconds. In a 48 hour forecast producing output every hour it means that over 6 minutes is spent in output processes. The last column in Table 2 shows that of these 8 seconds 5.3 seconds is spent in encoding the fields into GRIB, using only one processor. This suggests that the first and straight-forward step to improve the input/output performance is to parallelise the encoding/decoding step. This can be done by giving different fields to different processors and do the encoding in parallel and after that collect the encoded messages to one processor to be written out. A similar structure, in reverse order, can be done for input fields.

TABLE 2. Total time and time spent in I/O phase in seconds

	normal timestep	encode GRIB + write	only encode GRIB
OUTPUT	0.0	8.0	5.3
TOTAL	1.0	9.0	6.3

6. Summary

A parallel version of the HIRLAM forecasting model has been developed. This model will replace the serial versions currently used both at FMI and SHMI in 1997, when both institutes will have access to CRAY T3E systems. The starting point has been the serial version of the model, but large parts of the dynamics have been re-coded to achieve a more up-to-date version as concerns the coding style and readability of the code.

The primary tool in data communications has been the SHMEM routines, which are available on CRAY T3E systems. For portability reasons MPI message passing system can also be used. The new code can efficiently be run both on vector machines and distributed memory platforms. In the test runs on T3E the performance of 9.3 % sustained performance of the peak performance was achieved with modest work on the single node optimization. The test runs also show that we have built a very efficient data transposition method and that SHMEM routines on CRAY T3E are very efficient.

The future work will concentrate on improving the single node performance, especially in the physics part. Attention will be also paid on the load balance in the physics. This becomes more important, when the number of processors will increase. Also improvements are needed in input/output performance before the model is ready for operational use.

7. References

1. E. Källen (ed.), HIRLAM Documentation Manual, System 2.5, 1996, Available from SMHI, S60176 Norrköping, Sweden.

2. A. Dickinson, P. Burton, J. Parker, R. Baxter. Implementation and Initial Results from a Parallel Version of the Meteorological Office Atmosphere Prediction Model, in Coming of Age, Hoffman and Kreitz (ed.) (World Scientific, Singapore, 1995).

3. D, Bjørge and R. Skålin, PARLAM, the parallel HIRLAM version at DNMI, Research Report No. 27, 1995, Norwegian Meteorological Institute.

4. W. H. Raymond, High Order Low Pass Implicit Tangent Filters for Use in Finite Area Calculation. Mon. Wea. Rev., *116* (1988), 2132-2124.

REMO - Implementation of a Parallel Version of a Regional Climate Model

Martin Kücken
Potsdam Institute for Climate Impact Research
Department Climate
P.o. Box 601203
14412 Potsdam, Germany

and

Ulrich Schättler
German Weather Service
Department for Research and Development
P.o. Box 100465
63004 Offenbach, Germany

ABSTRACT

REMO is a regional climate model developed by the Max Planck Institute for Meteorology and the German Climate Computing Centre in Hamburg based on the "Europa" and "Deutschland" models of the German Weather Service. A parallel version of REMO has been implemented on the IBM RS/6000 SP computer system of the Potsdam Institute for Climate Impact Research including parallel input/output processing, the explicit Eulerian time-step, the semi-implicit corrections, the normal-mode initialization and the physical parametrizations of the German Weather Service. The implementation utilizes Fortran 90 and the Message Passing Interface. The parallelization strategy used is a 2D domain decomposition. This report describes the parallelization strategy, the parallel I/O organization, the influence of different domain decomposition approaches for static and dynamic load imbalances and first numerical results.

1. Introduction

REMO is a meso-α-scale grid-point model for numerical climate prediction derived from the EM/DM ("Europa-Modell", "Deutschland-Modell") of the German Weather Service, a model for numerical weather forecasting. For the parallelization we therefore have to solve almost the same problems as in parallelization approaches used in numerical weather prediction systems.

Above all these problems are:
- Organization of parallel input/output processing.
- Development of effective parallel implementations for the explicit Eulerian time-step, the semi-implicit corrections and the physical parametrizations based on a data parallel domain decomposition approach.
- Avoidance of load imbalances.

The original serial version of REMO is running on a shared memory vector architecture (Cray-C90) and is written in Fortran 77 code. This version of REMO is already parallel in the sense that it works under the Cray macrotasking control. The parallel version for distributed memory computers has been developed and optimized on an IBM RS/6000 SP (69 Thin nodes (66 Mhz POWER 2) with 128 MB RAM, 3 Wide nodes (66 MHz POWER 2) with 1 GB RAM, High-Performance Switch (40 MB per seconds, 39 usec latency), TB2 Adapter). The implementation was carried out by using Fortran 90 coupled with the Message Passing Interface (MPI) to create a portable parallel version which is also suitable for execution on other parallel platforms. Fortran 90 was chosen to ensure an efficient dynamic memory management.

2. Parallelization of REMO

2.1. The Parallel Programming Model

The parallelization strategy based on the parallelization approach used for the EM/DM, the 2D domain decomposition. Every processor sets up its partial problem by specifying the local data structures and initializing organizational variables. In principle every subdomain can be considered as a small region running its own model. The parallelization of the different parts of the adiabatic model for the EM/DM is described in [1]. Only the organization of the I/O management is an exceptional case. There are processor groups with different task specifications. These processors are working code parallel.

For the practical implementation we chose the MPI as a portable message-passing library. All MPI - and I/O - processes are written in C and are concentrated in separate subroutines. The clear dividing line between these subroutines and hardware independent subroutines provides an easy way to manage different hardware architecture features.

For processor communications we use nonblocking send and blocking receive MPI subroutines. Nonblocking send means that the I/O operation is only initialized and there is no need to wait for it to end. This causes less waiting time, the processor can continue its computation. Of course, the sending array is not reuseable until the end of the send operation. In practice this problem is solved by creating barriers for all processors after common communication points.

The time measurement was realized by using subroutines of the Fortran 90 library, namely rtc -- a subroutine for wall clock time (also real time) -- and etime -- a subroutine for user elapsed time (also CPU time). The implementation of the communication subsystem library used was the User Space (US) communication subsystem (the hardware is configured with its high-performance switch feature).

2.2. *Parallel I/O Organization*

In REMO a huge amount of data is necessary for input as well as for output. For a 24 hour prediction in our actual model area (193 x 121 gridpoints, 20 vertical levels) the transfer of about 28 MB input data and 48 MB output data has to be realized. Additionally there is the choice to write restart files (180 MB). The prediction period reaches up to five years. These facts underline the importance of a good parallel I/O organization to ensure a reasonable speed-up of the whole model.

Every processor of the IBM RS/6000 SP has its own operating system and is able to perform I/O operations. In operational use of REMO it is necessary to distribute the initial data fields and the boundary data fields produced by a global climate model to all processors corresponding to their subdomains. Furthermore it is necessary to write result output files, in our case typically every six hours.

Our model input files are given as binary packed records in the GRIB (gridded binary) format. That means that a processor cannot get its data from disk space in a direct way. At first it is necessary to analyse the structure of the GRIB file and later to decode this GRIB file to build up input arrays. Therefore, only a single node (input node, generally node zero) reads the whole GRIB file to its local memory. This node analyses the GRIB structure and sends single, but complete, GRIB fields to the other nodes which then are able to unpack the fields. After decoding every node distributes the data to all other nodes according to their subdomains. Then it returns a message to the input node that it is ready for receiving new unpacked arrays. If such a message is not present for the input node it also starts decoding GRIB fields and distributing them to the subdomains. But for the input node the distribution of GRIB data to the other nodes has a higher priority. After decoding of the whole GRIB file every node has completed its own input. The tasks of the input node and the other nodes are represented in Figure 2.1.

input node   other nodes

GRIB file input
|
Analyse GRIB structure
|
Send GRIB array ----------------------------------- > Receive GRIB array
| |
Unpack GRIB array   Unpack GRIB array
| |
Send subdomains <-----------------------------------> Send subdomains
| |
Receive own subdomains   Receive own subdomains

Figure 2.1. Tasks of the input node and the other nodes

In the case of a small number of nodes this approach ensures a balanced distribution of user elapsed time, but for a higher number of nodes the load of the processors becomes more imbalanced. Figure 2.2. shows the distribution of user elapsed time for 4 processor nodes and for 32 processor nodes in the case of the input of the GRIB code initial data (including 120 GRIB data arrays).

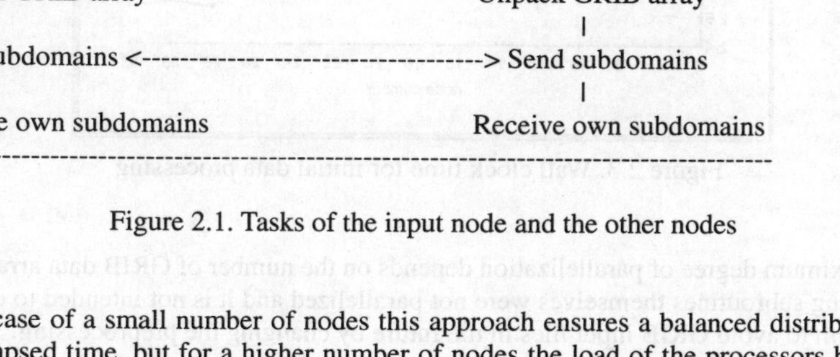

Figure 2.2. Distribution of user elapsed time for initial data input processing

When using a higher number of nodes a reason for a higher degree of imbalancing is the increasing impossibility of distributing all GRIB data arrays even to all processors.

Figure 2.3. shows the wall clock time for different numbers of processors in the case of initial data processing (6,7 MB input data).

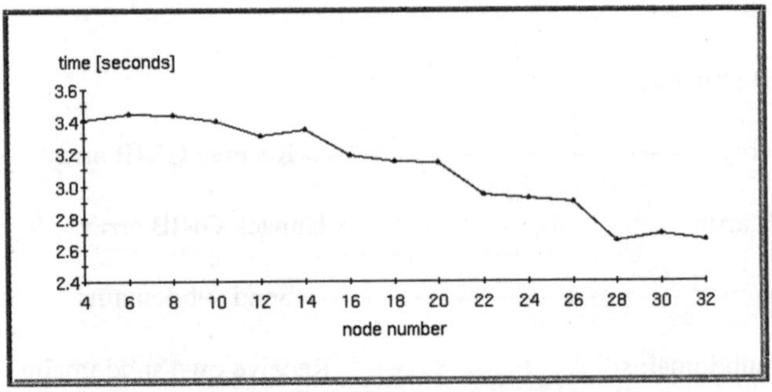

Figure 2.3. Wall clock time for initial data processing

The maximum degree of parallelization depends on the number of GRIB data arrays. The unpacking subroutines themselves were not parallelized and it is not intended to do so. It is our aim to avoid GRIB input files in the future by changing the preprocessing.

The model output consists of the calculated fields and of the restart data.
Three output possibilities exist for these files:
- Output of binary files using a C subroutine.
- Output of binary, unformatted files using a Fortran subroutine.
- Output of binary GRIB files.

On the IBM RS6000/SP we prefer writing output files with C subroutines. It is the most efficient output method in UNIX operating systems and very simple to manage. No processor communication is necessary. Every node performs its own output and can go on working.
The problem then is that the output disk space does not contain the complete model area files but only node dependent subdomains of these arrays. Therefore, after the writing of all output files every node creates a new child process (fork subroutine) and this child process then starts a programm (execlp subroutine) for collecting the different parts of an array. Later these arrays will be dumped on a Tapelibrary 3494 with six drives. This sequence of output processing is not simply a transfer of work to post processing. The child process works together with the parent process in time sharing mode. That guarantees a better use of system resources, e.g. by overlapping of processor time and I/O time. The Tapelibrary dumps and compresses created files without also making any demands of processor ressources.

The writing of output files with Fortran subroutines is done in the same way. The difference between binary Fortran files and binary C files is a different internal file structure.

The output as GRIB files ensures compatibility with Cray users. This output is realized in a similar way to the input of GRIB files, but in reverse mode. The GRIB code output, like the GRIB code input is unsuitable for use by parallel filesystems. But our current I/O approach supports also later use of parallel file systems without extensive modifications.

2.3. Dependence of load imbalances concerning different subdomain decomposition approaches

All area dimensioning parameters are defined in a separate control file. Therefore, these parameters can be changed in a very easy way without recompiling of any model source code. Before the model starts it is necessary to set different POE (partition manager) environment variables. These variables enable the specifications of such options as an input or output host file, methods of node allocation and processor count.

After the model starts every node is able to compute its own subdomain. Assuming a grid with N processors, where $N = NX * NY$, it is possible by variation of NX (number of processors in east-west direction) and NY (number of processors in north-south direction) to ensure an optimal result for different model areas concerning the computing time. It was our aim to investigate whether this way is suitable to obtain an acceptable load balance between all nodes without extensive source code modifications and enhancement of communication time. Two groups of load imbalances exist:

- Static load imbalances

The impossibility to create in every case absolutely identical subdomains for all single node causes load imbalances in the explicit Eulerian time-step and the semi-implicit corrections. Furthermore, the surface data are concentrated over the continents only which causes an irregular grid point distribution (see Figure 2.4.).

- Dynamic load imbalances

Many other sources for load imbalances are located in the physical part of REMO, above all in the computation of radiation and convection.

Our approach is a static one, it is not very well suited to handle dynamic load imbalances. The first test variant is a square domain decomposition (e.g. $NX = 4$, $NY = 4$) , the second test variant is a rectangular domain decomposition over complete latitudes (e.g. $NX = 1$, $NY = 16$). Mainly our model area covers the area of South America and the Southern Atlantic (193 longitudes,121 latitudes, see Figure 2.4.).

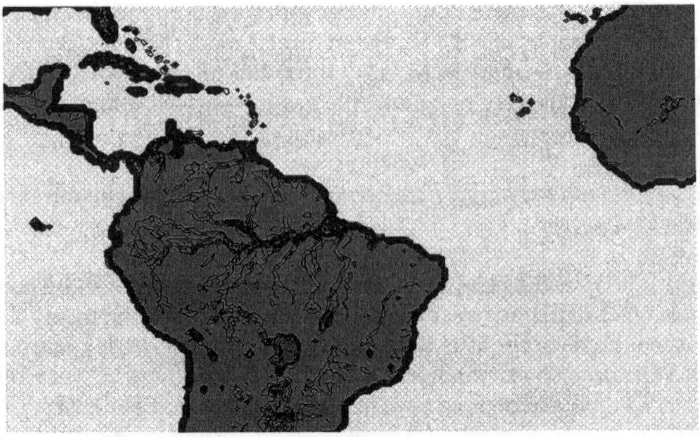

Figure 2.4. Investigated model area

The results shown in Figure 2.5. illustrate the influence of these decompositions on the distribution of the user elapsed time for soil parametrization and for Eulerian integration.

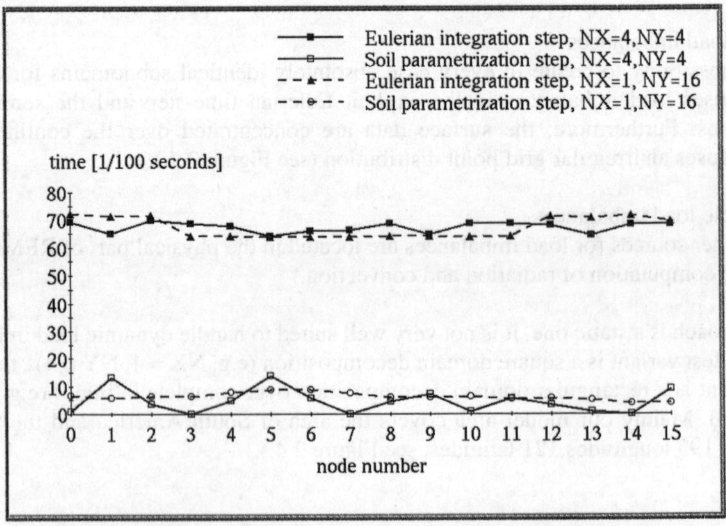

Figure 2.5. Distribution of user elapsed time for Eulerian integration and soil parametrization

In the case of soil parametrization the square domain decomposition provides a very bad balance between different processors. This result was expected (see soil distribution in Figure 2.4.). Splitting up subdomains by distributing only latitudes to different nodes results in a much better balancing of elapsed time.

These facts do not apply to the Eulerian time step. The elapsed time distribution between processors is much more homogeneous generally. An explanation for the differences are the facts that the subdomains are not absolute identical and that the boundary lines of the total domain are treated differently in the algorithms than the grid points in the interior of the domain.

Because soil parametrization is less time consuming it follows that in the case of rectangular decomposition increasing time of Eulerian integration and the decreasing time of soil parametrization cancel each other out.

Finally a square decomposition provides better results than a rectangular decomposition. Figure 2.7. illustrates this fact. So we prefer a processor distribution which approximates a square decomposition. In this way only a best possible subdomain decomposition for the Eulerian integration scheme decides the issue.

Figure 2.7. and Table 2.6. show the scalability of a full prediction step of REMO in the case of square decomposition and in the case of rectangular decomposition by use of up to 32 processors. The measured prediction step contains the Eulerian integration scheme, semi-implicite corrections, soil parametrization, radiation and convection. It is not representative for all prediction steps, because, e.g. the radiation is computed only every eighteenth time step.

From Figure 2.7. it can be seen that there is a superlinear speedup going from 4 to 12 processors. The reason for this is the memory requirement of the model.
The serial REMO version needs a memory storage of more than 500 Megabytes. The IBM RS/6000 SP (thin) nodes only have 128 MBytes available, so if running on 4 nodes the operating system uses disk space to extend its memory (paging). This paging space makes it possible to run models with memory demands which are higher than real processor memory, but the price is a lot of additional I/O operations and an increase of wall clock time. Use of paging space is a very efficient method to maximize the work done by several tasks for given computer resources, but for time-critical applications it should be avoided generally. Using of smaller arrays for a higher number of processors gives also a better cache memory performance.

Figure 2.7. also shows a better scalability of the square subdomain decomposition. A reason for this is the reduced increase of communication time in relation to a rectangular decomposition. Finally the performance of the rectangular approach is limited by the number of latitudes.

Balancing of radiation and convection is still an unsolved problem. Convection varies spatially and temporally in an unpredictable manner. In our current parallelization approach there is no possibility to balance these variabilities. An attempt to balance the computation of daylight points of radiation represents the parallelization approach over latitudes. To decide the efficiency it is necessary to compute at least a one day prediction period (288 time steps). Every load imbalance is equivalent to an efficiency loss, the slowest processor determines the computation speed.

The load imbalances are adding over time steps, because at the end of every time step there is the need for synchronization points for all processors e.g. to compute semi-implicit corrections. On the other hand every attempt to minimize dynamic load imbalances can cause performance loss, too. Future work will investigate a load balance scheme which works with a global data base on a wide node and dynamic distribution of subdomains to the other processors (thin nodes), perhaps connected with another integration scheme, namely the finite element scheme. We will test these ideas to determine the cost of more communications and the influence of better numerical accuracy on prediction results.

Table 2.6. Measured wall clock times [seconds] of a full prediction step

Number of processors	rectangular decomposition	square decomposition
4	365.2	377.2
8	122.1	126.2
12	66.0	50.8
16	48.6	37.1
20	45.0	31.8
24	39.2	27.3
28	35.5	24.0
32	37.3	19.8

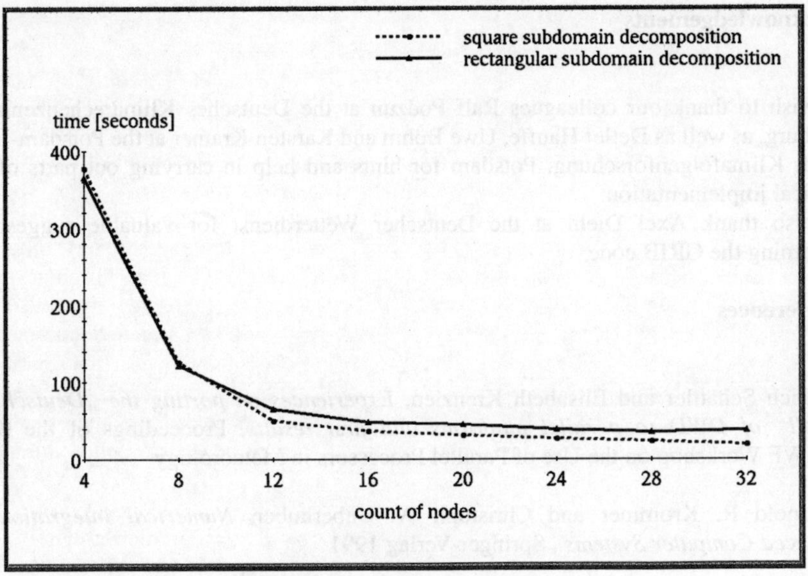

Figure 2.7. Scalability of a full prediction step

3. Conclusions

For a good performance of a parallel program a good performance of I/O operations is absolute essential. A factor for efficient input/output processing is the suitability of I/O files for parallel file processing algorithms. GRIB files are not very suitable for parallel I/O processing. Therefore we try to avoid the use of them as far as possible.

A still unsolved problem is the problem of load imbalances. With a static domain decomposition and fixed local data for every processor node it is impossible to balance unpredictable workloads. Static load imbalances are not so important and are better to manage. Future work will study the possibility to solve all balancing problems with only one approach, the approach of dynamical distributed subdomains to the nodes corresponding to their workload. This approach could also be a way to manage irregular grids.

4. Acknowledgements

We wish to thank our colleagues Ralf Podzun at the Deutsches Klimarechenzentrum, Hamburg, as well as Detlef Hauffe, Uwe Böhm and Karsten Kramer at the Potsdam-Institut für Klimafolgenforschung, Potsdam for hints and help in carrying out parts of the practical implementation.
We also thank Axel Diehl at the Deutscher Wetterdienst for valuable suggestions concerning the GRIB code.

5. References

1. Ulrich Schättler and Elisabeth Krenzien, *Experiences on porting the „Deutschland Modell" of DWD to parallel platforms and first results*, Proceedings of the Sixth ECMWF Workshop on the Use of Parallel Processors in Meteorology

2. Arnold R. Krommer and Christoph W. Ueberhuber, *Numerical Integration on Advanced Computer Systems* , Springer-Verlag 1991

3. J. Drake and I. Foster, *Introduction to the special issue on parallel computing in climate and weather modeling*, Parallel Computing 21 (1995)

4. Ian T. Foster and Brian R. Toonen, *Load-Balancing Algorithms for Climate Models,* Argonne National Laboratory, Argonne, IL 60439

SHARED AND DISTRIBUTED MEMORY IMPLEMENTATIONS OF THE CANADIAN MC2 MODEL

M. DESGAGNE, S. J. THOMAS, R. BENOIT, M. VALIN
Recherche en prévision numérique,
Environnement Canada, Dorval, Québec

A. V. MALEVSKY
Centre de recherche en calcul appliqué
5160, boul. Décarie, Montréal, Québec, Canada, H3X 2H9

ABSTRACT

The Mesoscale Compressible Community (MC2) model is an extension of a fully compressible limited area model developed by Monique Tanguay, Andre Robert and Rene Laprise in the mid-1980's. The model employs a three-time-level semi-implicit, semi-Lagrangian time discretisation with modified terrain following Gal-Chen coordinates. The semi-implicit scheme results in an elliptic boundary value problem with first-order derivative terms in the vertical direction and thus the resulting discretised system of equations has a large sparse non-symmetric coefficient matrix. The original alternating direction implicit (ADI) elliptic solver has been replaced with a flexible Generalised Minimum Residual (GMRES) Krylov method with variable preconditioning. We report on the performance of successive over relaxation (SOR) and vertical line relaxation (ADI) as parallel preconditioners to improve the convergence rate of GMRES. Parallelisation of the model for a distributed-memory computing environment has been based on the MSG toolkit developed for the P_SPARSLIB package of parallel iterative solvers. Performance of the model on the Cray T3E MPP and NEC SX-4/32 SMP is presented.

1. Introduction

The Mesoscale Compressible Community (MC2) model relies on a semi-implicit time discretisation to integrate the compressible governing equations for atmospheric flow in a limited area domain [2]. The model extends the earlier work of Tanguay *et al.* [24] with the addiction of terrain-following Gal-Chen coordinates in the vertical direction [12]. When a semi-implicit scheme is applied to gravity and acoustic waves, it effectively reduces their phase speeds and thus removes time step limitations due to numerical stability considerations. In a semi-implicit formulation of the compressible equations, algebraic elimination of the momentum and thermodynamic equations leads to an elliptic boundary value problem for a pressure perturbation. In a projected coordinate system, the resulting elliptic operator contains vertical buoyancy terms with first derivatives and thus the coefficient matrix of the discretised elliptic problem is nonsymmetric. The magnitude of these terms is governed by the Brunt–Väisällä frequency N and the speed of sound C_S. To build an efficient distributed-memory implementation of the MC2 model, the original alternating direction implicit (ADI) iterative solver in the sequential MC2 code has been replaced with a flexible variant of the preconditioned GMRES algorithm of Saad and Schultz [20, 17]. We report on the performance of successive over relaxation (SOR) and vertical line relaxation (ADI) as parallel preconditioners to improve the convergence rate of GMRES.

Numerical time step limitations associated with the explicit treatment of advection dominated problems can be further relaxed by employing variants of the method of characteristics. Such methods are now widely used in the atmospheric sciences where they are known as semi-Lagrangian schemes. Staniforth and Côté [23] present an overview of their use in atmospheric models. The MC2 model employs a centered (or off-centered) three-time-level semi-Lagrangian scheme and an efficient parallel implementation of this algorithm is crucial to achieving good performance [26]. We will describe a single program multiple data (SPMD) implementation of the MC2 model for distributed-memory, message-passing architectures and we will also discuss recently proposed domain decomposition type algorithms for semi-Lagrangian advection.

2. Model Formulation

The Euler equations can be written in several different equivalent forms since the equation of state for an ideal gas $p = \rho RT$ relates pressure, temperature and density. By introducing a metric projection term at reference latitude ϕ_0, $S = m^2$, $m = (1 + \sin\phi_0)/(1 + \sin\phi)$ and Coriolis parameter $f = 2\Omega\sin\phi$, the compressible governing equations for a polar stereographic projection become

$$\begin{aligned}
\frac{DU}{Dt} &= fV - K\frac{\partial S}{\partial X} - RT\frac{\partial q}{\partial X} + F_X \\
\frac{DV}{Dt} &= -fU - K\frac{\partial S}{\partial Y} - RT\frac{\partial q}{\partial Y} + F_Y \\
\frac{Dw}{Dt} &= -g - RT\frac{\partial q}{\partial z} + F_z \\
(1-\alpha)\frac{Dq}{Dt} &= -S\left[\frac{\partial U}{\partial X} + \frac{\partial V}{\partial Y}\right] - \frac{\partial w}{\partial z} + \frac{F_T}{T} \\
\frac{DT}{Dt} &= \alpha T\frac{Dq}{Dt} + F_T \\
\frac{DM}{Dt} &= F_M \\
\frac{DC}{Dt} &= F_C
\end{aligned} \tag{1}$$

where R and C_p are the gas constant and heat capacity for dry air at constant pressure and g is the gravitational acceleration. In the above equations, the dimensionless variable $q = \ln(p/p_0)$ has been introduced, where p_0 is a constant reference pressure. We also adopt the notation $\alpha = R/C_p$, so that the potential temperature is $\theta = Te^{-\alpha q}$. It follows that the thermodynamic relation is equivalent to $D\theta/Dt = 0$, implying that the potential temperature obeys a conservation law along particle trajectories. U, V and w are the wind images in projected $\mathbf{X} = (X, Y, z)$ coordinates. M and C represent water vapour and liquid water content. $K = (U^2 + V^2)/2$ is the pseudo kinetic energy per unit mass, F_T represents a heat source or sink affecting the temperature T and (F_X, F_Y, F_z) are sources or sinks affecting momentum. F_M and F_C are sources of sinks for moisture and liquid water.

In a conformal projection, cartesian coordinates are scaled by m and rotated according to the transformation (see Haltiner and Williams [8])

$$\begin{bmatrix} dX \\ dY \end{bmatrix} = m \begin{bmatrix} -\sin\lambda & -\cos\lambda \\ \cos\lambda & -\sin\lambda \end{bmatrix} \begin{bmatrix} dx \\ dy \end{bmatrix}$$

Wind images (U, V) are related to local wind components $\mathbf{u} = (u, v)$ via

$$\begin{bmatrix} U \\ V \end{bmatrix} = \frac{1}{m} \begin{bmatrix} -\sin\lambda & -\cos\lambda \\ \cos\lambda & -\sin\lambda \end{bmatrix} \begin{bmatrix} u \\ v \end{bmatrix}$$

and the Lagrangian derivative is

$$\frac{D}{Dt} = \frac{\partial}{\partial t} + S\left[U\frac{\partial}{\partial X} + V\frac{\partial}{\partial Y}\right] + w\frac{\partial}{\partial z}.$$

2.1 Vertical Coordinate System

The above equations form the basis for the limited area model described in Tanguay *et al* [24]. The only difference between these equations and the MC2 model formulation are the introduction of a terrain following vertical coordinate system based on Gal-Chen and Sommerville [6]. Here the cartesian coordinate z is mapped into a vertical coordinate Z by way of the stretching transformation

$$\zeta(X, Y, z) = H\left[\frac{z - h_0(X, Y)}{H - h_0(X, Y)}\right]$$

where $h_0(X, Y)$ is the height of topography and ζ is the Gal-Chen height. The vertical coordinate is a monotonically increasing function $Z(z)$, with the bottom and top of the atmosphere corresponding to the constant surfaces $Z = 0$ and $Z = H$. By applying the chain rule for differentiation, the following relations can be derived [10]

$$\left[\frac{\partial\zeta}{\partial X}\right]_z = -\frac{H - \zeta}{H - h_0}\frac{\partial h_0}{\partial X} = \frac{g_1}{g_0}, \qquad \left[\frac{\partial\zeta}{\partial Y}\right]_z = -\frac{H - \zeta}{H - h_0}\frac{\partial h_0}{\partial Y} = \frac{g_2}{g_0}$$

and

$$\frac{\partial\zeta}{\partial z} = \frac{H}{H - h_0} = \frac{1}{g_0}$$

where

$$g_0 = \frac{H - h_0}{H}, \qquad g_1 = -\frac{H - \zeta}{H}\left[\frac{\partial h_0}{\partial X}\right]_z, \qquad g_2 = -\frac{H - \zeta}{H}\left[\frac{\partial h_0}{\partial Y}\right]_z$$

Here the notation $[\cdot]_z$ implies differentiation for a fixed z. Variable resolution in the vertical direction is introduced via the stretching transformation $\zeta(z)$. Transformation metrics are derived as follows,

$$\left[\frac{\partial Z}{\partial X}\right]_z = \left[\frac{\partial Z}{\partial\zeta}\right]\left[\frac{\partial\zeta}{\partial X}\right]_z, \quad \left[\frac{\partial Z}{\partial Y}\right]_z = \left[\frac{\partial Z}{\partial\zeta}\right]\left[\frac{\partial\zeta}{\partial Y}\right]_z, \quad \frac{\partial Z}{\partial z} = \frac{\partial Z}{\partial\zeta}\frac{\partial\zeta}{\partial z}$$

Since Z is a function of ζ, the horizontal derivatives of Z are zero for constant ζ. A vertical coordinate is defined simply by specifying discrete values of Z with respect to ζ, resulting in a thickness ratio $\partial Z/\partial\zeta \approx \Delta Z/\Delta\zeta$ between model levels ΔZ and Gal-Chen coordinates $\Delta\zeta$. In model ΔZ coordinates, the above relations become

$$\left[\frac{\partial Z}{\partial X}\right]_z = \frac{G_1}{G_0}, \qquad \left[\frac{\partial Z}{\partial Y}\right]_z = \frac{G_2}{G_0}, \qquad \frac{\partial Z}{\partial z} = \frac{1}{G_0}$$

where G_i are the metric factors

$$G_0 = g_0\frac{\partial\zeta}{\partial Z}, \qquad G_1 = g_1, \qquad G_2 = g_2$$

The vertical discretisation of the MC2 model uses uniform spacing in the model coordinate ΔZ. Uniform spacing is also specified in the horizontal direction for ΔX and ΔY.

2.2 Compressible Governing Equations

In a vertically stretched, terrain-following coordinate system, the material or Lagrangian derivative is

$$\frac{D}{Dt} = \frac{\partial}{\partial t} + S\left[U\frac{\partial}{\partial X} + V\frac{\partial}{\partial Y}\right] + W\frac{\partial}{\partial Z}. \tag{2}$$

where

$$W = \frac{\partial Z}{\partial t} = S\left[U\frac{\partial Z}{\partial X} + V\frac{\partial Z}{\partial Y}\right] + w\frac{\partial Z}{\partial z}$$

and the fully compressible governing equations can now be written in the following form

$$\begin{aligned}
\frac{DU}{Dt} &= fV - K\frac{\partial S}{\partial X} - RT\left[\frac{\partial q}{\partial X} + \frac{G_1}{G_0}\frac{\partial q}{\partial Z}\right] + F_X \\
\frac{DV}{Dt} &= -fU - K\frac{\partial S}{\partial Y} - RT\left[\frac{\partial q}{\partial Y} + \frac{G_2}{G_0}\frac{\partial q}{\partial Z}\right] + F_Y \\
\frac{Dw}{Dt} &= -g - \frac{RT}{G_0}\frac{\partial q}{\partial Z} + F_Z \\
(1-\alpha)\frac{Dq}{Dt} &= S(\,F_1U + F_2V\,) - S\left[\frac{\partial U}{\partial X} + \frac{\partial V}{\partial Y} - \frac{1}{G_0}\frac{\partial G_0 W}{\partial Z}\right] + \frac{F_T}{T} \\
w &= WG_0 - S(\,G_1U + G_2V\,) \\
\frac{DT}{Dt} &= \alpha T\frac{Dq}{Dt} + F_T \\
\frac{DM}{Dt} &= F_M \\
\frac{DC}{Dt} &= F_C
\end{aligned} \tag{3}$$

where

$$F_1 = \frac{1}{g_0 H}\left[\frac{\partial h_0}{\partial X}\right]_z, \qquad F_2 = \frac{1}{g_0 H}\left[\frac{\partial h_0}{\partial Y}\right]_z$$

A more detailed derivation of the above MC2 model governing equations can be found in Laprise [10]. In the next section we describe the semi-implicit, semi-Lagrangian time discretisation of the compressible Euler/Navier-Stokes equations represented by (3).

3. Time Discretisation

3.1 Trajectories

A semi-Lagrangian time discretisation scheme approximates the material derivative (2) of a scalar field $\psi(\mathbf{X}, t)$, along trajectories defined by

$$\frac{dX}{dt} = S\,U(\mathbf{X}, t), \quad \frac{dY}{dt} = S\,V(\mathbf{X}, t), \quad \frac{dZ}{dt} = W(\mathbf{X}, t) \tag{4}$$

A three-time-level approximation of $D\psi/Dt$ on a trajectory originating at the upstream departure point $(\mathbf{X}^-, t_{n-1})$ and terminating at a grid point $(\mathbf{X}^+, t_{n+1})$ is

$$\frac{D\psi}{Dt} \approx \frac{\psi(\mathbf{X}^+, t_{n+1}) - \psi(\mathbf{X}^-, t_{n-1})}{2\Delta t}$$

Interpolation is used to compute $\psi(\mathbf{X}^-, t_{n-1})$ at the upstream point. In essence, the method depends on the accurate backward integration of (4) to obtain displacements $\alpha = (\alpha, \beta, \gamma)$ where $\mathbf{X}^- = \mathbf{X}^+ - 2\alpha$ and $\mathbf{X}^0 = \mathbf{X}^+ - \alpha$ are the departure point and midpoint. To integrate (4), Robert [15] combined a first-order estimate of the displacements with two iterations of a second-order midpoint Runge-Kutta method. Since the velocity is interpolated between grid points, this approach is very costly in 3D. Smolarkiewicz and Margolin [22] and Thomas *et al.* [27] demonstrate that Adams predictor-corrector type methods similar to the one proposed by Malevsky [11] provide the same accuracy and are far more efficient. In this case, the velocity is obtained by linear extrapolation

$$\begin{aligned} U(\mathbf{X}^+, t_{n+1}) &= S\,[\,2\,U(\mathbf{X}^+, t_n) - U(\mathbf{X}^+, t_{n-1})\,] \\ V(\mathbf{X}^+, t_{n+1}) &= S\,[\,2\,V(\mathbf{X}^+, t_n) - V(\mathbf{X}^+, t_{n-1})\,] \\ W(\mathbf{X}^+, t_{n+1}) &= 2\,W(\mathbf{X}^+, t_n) - W(\mathbf{X}^+, t_{n-1}) \end{aligned} \tag{5}$$

followed by one iteration of a midpoint corrector to obtain displacements

$$\begin{aligned} X^0 &= X^+ - \Delta t\, S\, U\,(\mathbf{X}^+ - \Delta t\, U(\mathbf{X}^+, t_{n+1}), t_n) \\ Y^0 &= Y^+ - \Delta t\, S\, V\,(\mathbf{X}^+ - \Delta t\, V(\mathbf{X}^+, t_{n+1}), t_n) \\ Z^0 &= Z^+ - \Delta t\, W\,(\mathbf{X}^+ - \Delta t\, W(\mathbf{X}^+, t_{n+1}), t_n) \end{aligned} \tag{6}$$

3.2 Interpolation

Semi-Lagrangian advection schemes can be interpreted as Eulerian finite difference schemes that have been shifted to the upstream grid cell containing the foot of a characteristic curve. Interpolation of a field ψ within this cell includes all high order cross terms in the Taylor series expansion of ψ about the departure point. Consider a regular cartesian grid with uniform spacing Δx. At the upstream grid cell $[x_{i-1}, x_i]$ containing the departure point, the piecewise cubic polynomial interpolating ψ is

$$p(x) = \psi_{i-1}(1-\hat{x}) + \psi_i \hat{x} + \beta_{i-1}\psi_{xxi-1} + \beta_i \psi_{xxi} \tag{7}$$

where $\psi_{xxi} = \psi_{i+1} - 2\psi_i + \psi_{i-1}$, $\hat{x} = (x - x_{i-1})/\Delta x$ and

$$\beta_{i-1} = -\frac{1}{6}\hat{x}(1-\hat{x})(2-\hat{x}), \qquad \beta_i = -\frac{1}{6}\hat{x}(1-\hat{x})(1+\hat{x})$$

Computational costs can be significantly reduced by neglecting higher-order terms which do not affect the global truncation error. For example, Ritchie *et al.* [14] construct 'truncated' Lagrange polynomials by dropping $\mathcal{O}(\Delta x^2 \Delta y^2)$ terms. A truncated form of the Newton polynomial is described in Thomas *et al* [27]. When combined with a predictor-corrector ODE solver, this resulted in a 50% improvement in efficiency over the original semi-Lagrangian algorithms implemented in MC2.

3.3 Semi-Implicit Scheme

A semi-implicit treatment of the linear terms appearing in the compressible governing equations implies that a scalar field $\psi(\mathbf{X}, t)$ is averaged in time along a particle trajectory according to

$$\bar{\psi}^t = \frac{\psi(\mathbf{X}^+, t_{n+1}) + \psi(\mathbf{X}^-, t_{n-1})}{2} \tag{8}$$

The effect of time averaging is to reduce the phase speed of the gravity waves and sound waves associated with these terms. A linear stability analysis of the semi-implicit scheme applied to the Euler equations is presented in the appendix of Tanguay *et al* [24]. To minimize distortions due to topographic forcing, nonlinear right-hand side terms R_i appearing in the time-discretised system of equations are averaged in space according to

$$\bar{\psi}^x = \frac{(1+\varepsilon)\,\psi(\mathbf{X}^+, t_n) + (1-\varepsilon)\,\psi(\mathbf{X}^-, t_n)}{2} \tag{9}$$

where $\varepsilon \ll 1$ is a small uncentering parameter (Tanguay *et al.* [25]). Moreover, the time average (8) of linear terms is replaced by the weighting

$$\bar{\psi}^t = \frac{(1+\varepsilon)\,\psi(\mathbf{X}^+, t_{n+1}) + (1-\varepsilon)\,\psi(\mathbf{X}^-, t_{n-1})}{2} \tag{10}$$

The combination of these space and time averages is sometimes referred to as an Eulerian treatment of mountains [13]. The overall effect is to reduce the spurious resonant response of a semi-Lagrangian scheme to stationary forcing in the Eulerian frame of reference.

3.4 Time-Discretised Equations

In order to apply both semi-implicit and semi-Lagrangian time integration schemes, thermodynamic fields are decomposed into a basic state and perturbation, $T = T^*+T'$ and $q = q^*+q'$. The basic state is an isothermal atmosphere in hydrostatic equilibrium,

$$\begin{aligned} T^* &= \text{constant} \\ q^*(z) &= q_0^* - \frac{gz}{RT^*}, \qquad \text{with} \qquad q_0^* = \text{constant}. \end{aligned}$$

The linear terms responsible for acoustic and gravity waves are isolated by writing equations (3) as follows. Terms containing the metric factor g_0/G_0 have been added to both the vertical momentum and continuity equations.

$$\begin{aligned} \frac{DU}{Dt} + RT^* \frac{\partial q'}{\partial X} &= R_1 \\ \frac{DV}{Dt} + RT^* \frac{\partial q'}{\partial Y} &= R_2 \\ \frac{Dw}{Dt} + RT^* \frac{g_0}{G_0} \frac{\partial q'}{\partial Z} - g\frac{T'}{T^*} &= R_3 \\ (1-\alpha)\left[\frac{Dq'}{Dt} - \frac{gw}{RT^*}\right] + S\left[\frac{\partial U}{\partial X} + \frac{\partial V}{\partial Y}\right] + \frac{g_0}{G_0}\frac{\partial G_0 W}{\partial Z} &= R_4 \\ w - G_0 W &= R_5 \\ \frac{DT'}{Dt} - \alpha T^* \left[\frac{Dq'}{Dt} - \frac{gw}{RT^*}\right] &= R_6 \\ \frac{DM}{Dt} &= R_7 \\ \frac{DC}{Dt} &= R_8 \end{aligned} \tag{11}$$

where the right-hand sides R_i are defined by

$$\begin{aligned} R_1 &= fV - K\frac{\partial S}{\partial X} - RT'\frac{\partial q'}{\partial X} - RT\frac{G_1}{G_0}\frac{\partial q'}{\partial Z} + F_X \\ R_2 &= -fU - K\frac{\partial S}{\partial Y} - RT'\frac{\partial q'}{\partial Y} - RT\frac{G_2}{G_0}\frac{\partial q'}{\partial Z} + F_Y \\ R_3 &= RT^*\frac{g_0}{G_0}\frac{\partial q'}{\partial Z} - \frac{RT}{G_0}\frac{\partial q'}{\partial Z} + F_z \\ R_4 &= S(\,F_1 + F_2 V\,) - \frac{(1-g_0)}{G_0}\frac{\partial G_0 W}{\partial Z} + \frac{F_T}{T} \\ R_5 &= -S(\,G_1 U + G_2 V\,) \\ R_6 &= \frac{\alpha T'}{1-\alpha}\left\{S(F_1+F_2V) - S\left[\frac{\partial U}{\partial X} + \frac{\partial V}{\partial Y}\right] - \frac{1}{G_0}\frac{\partial G_0 W}{\partial Z} + \frac{F_T}{T}\right\} + F_T \\ R_7 &= F_M \\ R_8 &= F_C \end{aligned} \tag{12}$$

In a centered three-time-level or leapfrog discretisation, the time-average (8) is applied and the right-hand side terms R_i appearing in (12) are evaluated at the trajectory midpoint $(\mathbf{X}^0,\, t_n)$. The resulting time discretised equations are then given by

$$\begin{aligned}
U + \Delta t RT^* \frac{\partial q'}{\partial X} &= Q_1 \\
V + \Delta t RT^* \frac{\partial q'}{\partial Y} &= Q_2 \\
w + \Delta t RT^* \frac{g_0}{G_0}\frac{\partial q'}{\partial Z} - \Delta t\, g\, \frac{T'}{T^*} &= Q_3 \\
(1-\alpha)\left[q' - \Delta t \frac{gw}{RT^*} \right] + \Delta t\, S \left[\frac{\partial U}{\partial X} + \frac{\partial V}{\partial Y} \right] + \Delta t \frac{g_0}{G_0}\frac{\partial G_0 W}{\partial Z} &= Q_4 \\
\Delta t(\, w - G_0 W\,) &= Q_5 \\
T' - \alpha T^* q' + \Delta t \frac{\alpha g}{R}\, w &= Q_6 \\
M &= Q_7 \\
C &= Q_8
\end{aligned} \tag{13}$$

with $Q_i = P_i^- + 2\Delta t\, R_i^0$. In an off-centered scheme, the space (9) and time (10) averages are applied. Therefore Δt is replaced by $(1+\varepsilon)\Delta t$ in (13) and $(1-\varepsilon)\Delta t$ in (14) below. The right-hand side terms in (13) then become $Q_i = P_i^- + 2\Delta t\, \bar{R}_i^x$. An additional benefit of such an off-centered scheme is a reduction in the number of upstream interpolations [25]. Instead of interpolating seven fields at both the midpoint and departure point, interpolation is now only required at the departure point in the semi-Lagrangian advection scheme. In both cases, right-hand side terms P_i are given by

$$\begin{aligned}
P_1 &= U - \Delta t RT^* \frac{\partial q'}{\partial X} \\
P_2 &= V - \Delta t RT^* \frac{\partial q'}{\partial Y} \\
P_3 &= w - \Delta t RT^* \frac{g_0}{G_0}\frac{\partial q'}{\partial Z} + \Delta t\, g\, \frac{T'}{T^*} \\
P_4 &= (1-\alpha)\left[q' + \Delta t \frac{gw}{RT^*} \right] - \Delta t\, S \left[\frac{\partial U}{\partial X} + \frac{\partial V}{\partial Y} \right] - \Delta t \frac{g_0}{G_0}\frac{\partial G_0 W}{\partial Z} \\
P_5 &= -\Delta t(\, w - G_0 W\,) \\
P_6 &= T' - \alpha T^* q' - \Delta t \frac{\alpha g}{R}\, w \\
P_7 &= M \\
P_8 &= C
\end{aligned} \tag{14}$$

A semi-implicit, semi-Lagrangian time step can be summarized as follows. First, components of the right-hand side terms Q_i are interpolated at the departure point. By taking the horizontal divergence of the momentum equations in (13), algebraic elimination with the vertical momentum and continuity equations leads to an elliptic partial differential equation (PDE) for q'. Solving for q', the remaining fields are updated via back substitution.

3.5 Elliptic Boundary Value Problem

To derive an elliptic PDE for the log pressure perturbation q', we first take the horizontal divergence of momentum

$$S\left[\frac{\partial U}{\partial X}+\frac{\partial V}{\partial Y}\right]+\Delta t RT^{*}S\nabla^{2}q'=S\left[\frac{\partial Q_1}{\partial X}+\frac{\partial Q_2}{\partial Y}\right]$$

then multiply by Δt and substract from the continuity equation to give

$$\left[(1-\alpha)-\Delta t^{2}RT^{*}S\nabla^{2}\right]q'+\Delta t\left[\frac{g_0}{G_0}\frac{\partial G_0 W}{\partial Z}-(1-\alpha)\frac{gw}{RT^{*}}\right]=A_4 \qquad (15)$$

$$A_4=Q_4-\Delta t S\left[\frac{\partial Q_1}{\partial X}+\frac{\partial Q_2}{\partial Y}\right]$$

Next, multiplying the thermodynamic equation by $g\,\Delta t/T^{*}$ results in

$$C_1 w+\Delta t RT^{*}D_1(q')=A_5 \qquad (16)$$

$$C_1=1+\Delta t^{2}\frac{\alpha g^{2}}{RT^{*}},\quad D_1(q')=\frac{g_0}{G_0}\frac{\partial q'}{\partial Z}-\frac{\alpha g}{RT^{*}}q',\quad A_5=Q_3+\Delta t\frac{g}{T^{*}}Q_6$$

To eliminate w from the system of equations, $w=G_0W+Q_5/\Delta t$ is inserted into equations (15) and (16),

$$C_1G_0W+\Delta t RT^{*}D_1(q')=A_1$$

$$\left[(1-\alpha)-\Delta t^{2}RT^{*}S\nabla^{2}\right]q'+\Delta t\,D_2(G_0W)=A_3$$

where we have defined

$$A_1=A_5-\frac{Q_5}{\Delta t}C_1,\quad A_3=A_4+Q_5(1-\alpha)\frac{g}{RT^{*}},$$

$$D_2(G_0W)=\frac{g_0}{G_0}\frac{\partial G_0W}{\partial Z}-(1-\alpha)\frac{gG_0W}{RT^{*}}$$

Finally, eliminating G_0W results in an elliptic PDE for q'.

$$C_1\left[(1-\alpha)-\Delta t^{2}RT^{*}S\nabla^{2}\right]q'-\Delta t^{2}RT^{*}D_2\left[D_1(q')\right]=A_2$$

with right-hand side

$$A_2=C_1A_3-\Delta t\,D_2(A_1).$$

The current implementation of the MC2 model is based on a nesting strategy within a limited area domain. Lateral boundary conditions are derived from a global or driving atmospheric model. Inflow conditions on the velocity (U,V,w) are specified at the boundary and these are implemented within the semi-Lagrangian scheme. In order to solve a well-posed elliptic boundary value problem for the log pressure perturbation q', a Neumann condition $\partial q'/\partial \mathbf{n} = 0$ is imposed. Once q' is known at t_{n+1}, the remaining fields are computed by back substitution

$$
\begin{aligned}
W &= \frac{1}{C_1 G_0}\,[\,A_1 - \Delta t R T^* D_1(q')\,] \\
w &= G_0 W + \frac{Q_5}{\Delta t} \\
T' &= Q_6 + \alpha T^* q' - \Delta t \frac{\alpha g}{R} w \\
U &= Q_1 - \Delta t R T^* \frac{\partial q'}{\partial X} \\
V &= Q_2 - \Delta t R T^* \frac{\partial q'}{\partial Y} \\
M &= Q_7 \\
C &= Q_8
\end{aligned}
\tag{17}
$$

A Davies type relaxation scheme is applied at the end of a time step to combine the above solution with the driving model solution at the boundaries. This scheme requires an attenuation function [2] to minimize spurious reflections of waves at the boundary. A Robert time filter is then applied to damp the computational solution introduced by the centered (or off-centered) three-time-level discretisation [1].

4. Space Discretisation

The MC2 model is based on finite-difference approximations of space derivatives, where fields are staggered on an Arakawa 'C' type grid [16]. The standard convention in atmospheric modeling is to denote fields located at grid cell centers by q_{ijk} and T_{ijk}. Staggered wind images are denoted $U_{i-1/2,j,k}$, in the X-direction, $V_{i,j-1/2,k}$, in the Y-direction, and $W_{i,j,k-1/2}$, in the Z-direction. For example, the horizontal divergence D_{ijk} is located at grid cell centers and is computed according to

$$
D_{ijk} = \bar{S}^{XY}\left[\frac{U_{i+1/2,j,k} - U_{i-1/2,j,k}}{\Delta X} + \frac{V_{i,j+1/2,k} - V_{i,j-1/2,k}}{\Delta Y}\right]
$$

where

$$
\bar{S}^{XY} = \frac{1}{4}\,[\,S_{i,j} + S_{i+1,j} + S_{i,j+1} + S_{i+1,j+1}\,]
$$

Space derivatives are centered second order finite difference approximations. The first derivative of the pressure perturbation term q' is approximated by

$$
\frac{\partial q'}{\partial X} \approx [q'_X]_{i-1/2,j,k} = \frac{q'_{ijk} - q'_{i-1,j,k}}{\Delta X}
$$

The right-hand side terms R_i appearing in (12) are located at the same grid points as the associated fields. To compute R_1, the temperature T' must be averaged, since R_1 is located at the same grid points as U.

$$
-RT'\frac{\partial q'}{\partial X} \approx -R\left[\frac{T'_{ijk} + T'_{i-1,j,k}}{2}\right]\left[\frac{q'_{ijk} - q_{i-1,j,k}}{\Delta X}\right]
$$

Given the above space discretisation scheme, the right-hand side A_2 of the ellipitic boundary value problem is located on the same grid as the pressure perturbation q'.

5. Parallel Semi-Lagrangian Advection

Given a logical $P = P_X \times P_Y$ processor mesh with P_X processors in the X-direction and P_Y in the Y-direction, a $N_X \times N_Y \times N_Z$ grid is partitioned so that a subdomain contains $(N_X/P_X) \times (N_Y/P_Y) \times N_Z$ points. Such a distribution minimizes surface to volume effects for typical finite-difference and interpolation computations, since only local boundary data along the surface of this cube need be communicated to satisfy data dependencies. In a direct parallel implementation of the sequential semi-Lagrangian algorithm, the overlap is determined by C. The stencil (7) would therefore require a fixed $\lceil C+1 \rceil$ grid point wide overlap region. For a horizontal decomposition of the computational grid, the departure point could reside in any of the eight adjacent processors. In fact, for large enough C, a trajectory could in principle traverse several subdomains. For a limited area model such as MC2 it is safe to assume that the Courant number is $\mathcal{O}(1)$ and thus each processor must send and receive 4 messages (2 in the X direction and 2 in the Y direction), to update the values of ψ in the halo region including 'corners'. A fixed overlap for all processors can be pre-computed by estimating the maximum possible windspeed for the entire model integration. An alternative strategy would be to expand and contract the size of the halo region dynamically [5]. The former approach is currently implemented in MC2.

During development and testing of parallel semi-Lagrangian algorithms, it was found that parallel model outputs did not agree with the sequential results. The problem was traced to the semi-Lagrangian trajectory algorithms and local computation of upstream grid cell indices was identified as the source of discrepancies. The computation of these indices is very sensitive to floating point rounding errors. Within a processor subdomain, the range of local indices can be quite different than those computed over the global domain. Therefore, it is crucial that the location of the upstream grid cell index and coordinates of the trajectory departure point be computed from global information. Given the global coordinate of the starting index in a subdomain **gcld**, the X displacement **xd** and time step **dt**, a code fragment which computes the departure point coordinate **xdu** and grid cell index **ix** containing this point is

```
offi = gcld(1,myproc) - 1
xdu  = float(offi+i) - dt * xd
ix   = int(xdu)
capx = xdu - float(ix)
ix   = ix - offi
```

The MC2 model adopts the approach described in Robert [16], where the staggered wind images are interpolated to grid cell centers and displacements α are computed there. Displacements are interpolated onto the corresponding grid before components of the right-hand side terms Q_i are advected from their upstream positions at time t_{n-1} to an Eulerian grid point at time t_{n+1}.

6. Elliptic Solvers

The principle difficulty for an SPMD model of parallel computation are the global data dependencies present in the elliptic boundary value problem. Moreover, the elliptic operator contains first derivative terms associated with vertical buoyancy

$$D_2\left[D_1(q')\right] = \left(\frac{g_0}{G_0}\right)^2 \frac{\partial^2 q'}{\partial Z^2} - \left[\frac{N^2}{g} + \frac{g}{C_s^2}\right] \frac{g_0}{G_0} \frac{\partial q'}{\partial Z} + \frac{N^2}{C_s} q'$$

where

$$N^2 = \frac{g}{C_p T^*}, \qquad C_S^2 = \frac{RT^*}{(1-\alpha)}$$

and thus the resulting discretised system of equations has a large sparse nonsymmetric coefficient matrix when derivatives are approximated by centered second order finite differences. The magnitude of the corresponding matrix elements is determined by the Brunt-Väisällä frequency N, the speed of sound C_S, and the aspect ratio $\Delta X/\Delta Z$.

6.1 Preconditioned Krylov Subspace Methods

The GMRES algorithm was proposed by Saad and Schultz [20] for solving general sparse nonsymmetric linear systems. A flexible variant of the GMRES algorithm (FGMRES) developed by Saad [17] allows the preconditioner to vary from step to step, a feature that is quite useful in domain decomposition type algorithms or in any parallel computing implementation. The flexible variant also permits the introduction of a secondary iterative procedure as a preconditioner. In standard 'non-flexible' techniques, these inner solutions must be 'exact' or highly accurate within each subdomain. For FGMRES and other flexible techniques this does not have to be the case. Moreover, this allows the inner preconditioning steps to be completely asynchronous, a feature which may help minimize communication and synchronization costs in a parallel code. Consider a linear system of the form

$$Ax = b, \tag{18}$$

where A is a large sparse nonsymmetric real matrix of size $n \gg m$. The GMRES algorithm is a method which minimizes the L_2 norm of the residual vector $\|b - Ax\|_2$ by projecting the initial residual $r_0 = b - Ax_0$ onto the Krylov subspace

$$K_m = \text{Span}\{r_0, Ar_0, \ldots, A^{m-1}r_0\}$$

When 'right-preconditioning' is applied to the above linear system, a preconditioned linear system

$$(AM^{-1})(Mx) = b. \tag{19}$$

is solved implicitly instead of (18). FGMRES allows the right preconditioner M to be different at each step j. The algorithm described in Saad [17] is given below.

ALGORITHM **6.1 Flexible GMRES (FGMRES)**

1. **Start:** *Choose x_0 and a dimension m of the Krylov subspaces. Define an $(m+1)\times m$ matrix $\bar{H}_m$ and initialize all its entries $h_{i,j}$ to zero.*
2. **Arnoldi process:**
   (a) *Compute $r_0 = b - Ax_0$, $\beta = \|r_0\|_2$ and $v_1 = r_0/\beta$.*
   (b) *For $j = 1, ..., m$ do*
   - *Compute $z_j := M_j^{-1} v_j$*
   - *Compute $w := Az_j$*
   - *For $i = 1, \ldots, j$, do* $\begin{cases} h_{i,j} := (w, v_i) \\ w := w - h_{i,j} v_i \end{cases}$
   - *Compute $h_{j+1,j} = \|w\|_2$ and $v_{j+1} = w/h_{j+1,j}$.*

   (c) *Define $Z_m := [z_1, ..., z_m]$.*
3. **Form the approximate solution:** *Compute $x_m = x_0 + Z_m y_m$ where $y_m = \text{argmin}_y \|\beta e_1 - \bar{H}_m y\|_2$ and $e_1 = [1, 0, \ldots, 0]^T$.*
4. **Restart:** *If satisfied stop, else set $x_0 \leftarrow x_m$ and goto (ii).*

The Arnoldi loop constructs an orthogonal basis of the preconditioned Krylov subspace

$$K_m = \text{Span}\{v_1, AM_1^{-1}v_1, \ldots, AM_{m-1}^{-1}v_{m-1}\} \tag{20}$$

by a modified Gram-Schmidt process, where each new vector to be orthogonalized is generated from all previous basis vectors. When the preconditioner is constant, $M_j = M$ for $j = 1, \ldots, m$, the method is equivalent to the standard GMRES algorithm, right-preconditioned with M. The approximate solution x_m obtained from this modified algorithm minimizes the residual norm $\|b - Ax_m\|_2$ over $x_0 + \text{Span}\{Z_m\}$.

6.2 Parallel Preconditioners

Simple and yet effective parallel preconditioners to improve the convergence rate of a Krylov type iteration such as the FGMRES algorithm described above can be based on classical stationary methods such as the point Jacobi or Gauss-Seidel iterations [19]. These methods are derived from a splitting of the coefficient matrix $A = D - E - F$, where D is the diagonal of A, $-E$ is the strictly lower triagular part and $-F$ is the strictly upper triagular part of the matrix. Unless otherwise stated, the natural ordering of grid points is assumed. The Jacobi iteration is the simplest method and is also perfectly parallel since each grid point can updated in independently of all others,

$$Dx_{k+1} = (E + F)x_k + b$$

The Gauss-Seidel iteration results when updated values of the solution are used immediately within the iteration

$$(D-E)x_{k+1} = Fx_k + b$$

The left hand side contains a lower triangular matrix, implying a recursion and thus the algorithm does not vectorize. A well-known solution to this problem is to apply the Gauss-Seidel iteration separately to two or more sets of grid points. The resulting algorithms are known as multi-colour Gauss-Seidel iterations and this strategy can also be applied at the level of subdomains. The simplest two-colour paritioning of the nodes is also called a red-black ordering and results in the following block partitioned coefficient matrix

$$A = \begin{bmatrix} D_R & -F \\ -E & D_B \end{bmatrix}$$

where D_R and D_B are diagonal. The Gauss-Seidel iteration is then split into two independent systems of equations $D_R x_{k+1}^R = F x_k^B + b^R$ and $D_B x_{k+1}^B = E x_k^R + b^B$ which can be solved in parallel. Both the Jacobi and Gauss-Seidel iterations are linear fixed-point iterations of the form

$$x_{k+1} = M^{-1}Nx_k + M^{-1}b = (I - M^{-1}A)x_k + M^{-1}b \tag{21}$$

where $A = M - N$ is a matrix splitting. It follows that $M = D$, $N = A - D$ for the Jacobi iteration, whereas $M = D - E$, $N = A - D$ in Gauss-Seidel. Thus, (21) can be viewed as a 'preconditioned' fixed-point iteration for the system of equations $M^{-1}Ax = M^{-1}b$, where M is the preconditioner [19]. These same preconditioners can equally be applied to accelerate Krylov methods, where the matrix $I - M^{-1}A$ appearing in (21) is replaced by a matrix polynomial $P(M^{-1}A) \in K_m$ where K_m is the Krylov subspace appearing in (20).

When an over relaxation parameter ω is introduced to accelerate convergence, the Gauss-Seidel iteration becomes the method of successive over relaxation

$$(D - \omega E)x_{k+1} = [\,\omega F + (1-\omega)D\,]x_k + \omega b$$

with preconditioning matrix $M = \omega^{-1}(D - \omega E)$. There exists an extensive literature on finding the optimal ω for elliptic PDE's. Recent work by Saad [18] and DeLong and Ortega [4] indicates that $\mathcal{O}(10)$ 'inner' iterations of SOR with $1.5 \le \omega < 2.0$ results in a very effective preconditioner for GMRES. The optimal ω is of course problem dependent and can usually be found empirically. For the MC2 elliptic BVP, GMRES(10)-SOR(10) with $\omega = 1.8$ exhibits excellent convergence rates and these results are consistent with the those reported in [4]. The optimal ω may vary depending on the grid resolution and the degree to which the coefficient matrix is nonsymmetric. Thus, in the MC2 model, the value of ω can be selected at run time. A strictly local preconditioner is implemented by setting values of x_k to zero in the halo region at subdomain boundaries and applying the SOR iteration internally on each processor [3].

The alternating direction implicit (ADI) method of Peaceman and Rachford [19] was orginally proposed as a method for solving parabolic partial differential equations such as the heat conduction equation $u_t = \Delta u + r$ with time-dependent forcing term r. When this PDE is discretised by the method-of-lines, the resulting system of first-order ordinary differential equations with constant coefficient matrix $u' = Au + r$. In two dimensions, an ADI time integration scheme is based on the splitting $A = H + V$, where the individual matrices represent finite difference stencils in each coordinate direction,

$$(I - \frac{\Delta t}{2}H)u_{k+1/2} = (I + \frac{\Delta t}{2}V)u_k + \frac{\Delta t}{2}r \quad (22)$$

$$(I - \frac{\Delta t}{2}V)u_{k+1} = (I + \frac{\Delta t}{2}H)u_{k+1/2} + \frac{\Delta t}{2}r \quad (23)$$

An iteration for solving elliptic problems can be derived if the time step $\Delta t/2$ is replaced with an acceleration parameter ρ_k. By cycling through a sequence of ρ_k accross the spectrum of A, fast convergence rates may be achieved and the original MC2 solver was based on a stable generalisation of this algorithm to three dimensions. For constant ρ, matrices corresponding to the splitting $A = M - N$ are [19]

$$M = \frac{1}{2\rho}(H + \rho I)(V + \rho I), \qquad N = \frac{1}{2\rho}(H - \rho I)(V - \rho I).$$

Hill and Marshall [9] note that the discretised elliptic operator in a non-hydrostatic pressure solver will be dominated by the vertical terms when the aspect ratio $\Delta X/\Delta Z$ is large, implying that vertical motion is predominantly hydrostatic. Therefore, an effective preconditioning strategy is to invert the vertical components of the elliptic operator. Skamarock *et al* [21] extend this strategy to a semi-implicit Generalized Congugate Residual (GCR) Krylov type solver for non-hydrostatic analestic and compressible atmospheric models by using a vertical ADI line relaxation. In this case, one iteration of (23) is applied as the preconditioner with $u_{k+1/2} = u_k$ and $u_0 = 0$. By adopting this approach in MC2, the resulting solver is denoted GMRES(10)-ADIV(1). If the parallel data decomposition is strictly in the horizontal direction, such a preconditioner requires no communication and should result in good parallel performance.

6.3 Numerical Results

In order to compare the convergence rates of the preconditioners described above, the relative residual norm $\|r_k\|_2/\|r_0\|_2$ is plotted in Figures 2 and 3 for representative grid sizes defined on the hemispheric domain displayed in Figure 1. An acceptable convergence tolerance was determined to be $\varepsilon = 1.0 \times 10^{-4}$. For all simulations, the model lid is set to 25000m and 31 vertical levels are employed, corresponding to a vertical spacing of $\Delta Z = 806$m in model coordinates. Four different problem sizes were examined, with grid dimensions $119 \times 119 \times 31$, $239 \times 239 \times 31$, $321 \times 321 \times 31$ and $521 \times 359 \times 31$. Uniform horizontal resolution for these grids are 200km, 100km, 75km and 50km respectively. Physical parameterisations were turned off and in each case the model was run for 12 time steps at a maximum Courant number of $C = 0.64$.

Three different preconditioners were implemented for the GMRES accelerator; SOR(10), red-black SOR(10) and vertical ADI(1). Both SOR preconditioners were run with $\omega = 1.8$ as the over relaxation parameter, while $\rho = 1.0 \times 10^{-3}$ was chosen as the acceleration parameter for the single vertical ADI sweep. These are compared with original Richardson–ADI solver. It was found in all cases that the original ADI iteration exhibited the fastest convergence rate, although the trend at higher horizontal resolution was towards GMRES(10)-SOR(10) with a red-black ordering. Execution time is of course machine dependent. Despite the slower convergence rate of GMRES(10)-ADIV(1), this solver exhibited the best cross-platform performance since the preconditioner is very inexpensive. We observe that the vertical ADI preconditioner is most effective when the aspect ratio $\Delta X/\Delta Z$ is very large, as in the first two problems. Here the elliptic problem is dominated by vertical terms. Vertical ADI becomes less efficient as the grid becomes more isotropic. The single processor and parallel performance of the solvers is discussed in a later section.

7. Message-passing tools

To implement inter-processor communication, the Message-Passing Interface (MPI) standard was chosen both for portability and efficiency (Gropp *et al* [7]). Our primary goal was to keep the MC2 model code machine independent and thus adaptable to new architectures or new message passing paradigms such as one-sided get/put communication proposed for MPI–2. A substantial number of routines did not require any modification. However, some basic algorithms were not suitable for distributed memory computation and were replaced. Inter-processor communication is isolated in a limited number of subroutines and an intermediate layer has been created between the functional routines of the model and the MPI message passing library. These routines form an application interface known as the Message-passing tools for a Structured Grid (MSG). The MSG toolkit provides a simplified programming model defined in terms of boundary exchanges and global reductions such as a distributed dot product on a tensor-product grid. A node or grid point is defined by a vector of indices with one index per dimension. It is assumed that nodes are partitioned across a logical processor mesh in such a way that every node is assigned to one and only one processor. The computational domain (corresponding to a global array) is partitioned into subdomains (local arrays) residing in the memories of separate processors. The size and shape of the halo regions on each processor can then be computed by MSG from the overlap at subdomain boundaries and boundary conditions for the problem domain in each of the three coordinate directions.

The MSG toolkit must be supplied with the coordinates of overlapping regions, dimensions of the global and local arrays, and global coordinates of the subdomain corners within the global array. The MSG setup routines then automatically establish the processor connectivity. Global operations such as the sum of elements of a distributed array comprise a second type of inter-processor communication. These operations require synchronization of all processors and are directly implemented by most message passing libraries. A program performing a boundary exchange between the subdomains of a tensor-product grid typically incorporates the following sequence of calls.

```
call MSG_tsetup ( ... )  -- set-up tables
...
do iter = 1, niter
  ...
  call MSG_tbdx_send ( ..., ptrn, ... )
  -- first call opens a communication channel
     indicated by ptrn and sends boundary data
     to network; channel remains open
  ...
  ... calculations involving local data are
  ... performed while waiting for boundary data
  ...
  call MSG_tbdx_receive ( ..., ptrn, ... )
  -- boundary data is available at this stage
  ...
enddo
...
call MSG_tbdx_close ( ..., ptrn, ... )
-- channel closed
```

Different fields within the same model may be distributed in different ways. For example, the thickness of halo regions at the boundaries of the subdomains may differ depending on the order of a finite-difference scheme. In the case of a semi-Lagrangian algorithm, the Courant number C determines the amount of overlap. The MSG toolkit is capable of simultaneously establishing and maintaining several different communication patterns. The dimensions of an array and the halo size remain fixed for each communication pattern. The parameter `ptrn` passed to `MSG_tbdx_send` in the above code fragment identifies a unique pattern. On the first call to `MSG_tbdx_send`, a communication channel is opened between participating processors. Processors then exchange addresses of their send/receive buffers, along with the length and type of message. Subsequent calls to `MSG_bdx_send` with the same `ptrn` employ the corresponding pattern. An unnecessary 'handshake' between processors can be avoided, thus reducing communication latency.

8. Parallel Performance on the Cray T3E and NEC SX–4

8.1 Cray T3E

The Cray T3E is a next generation MPP computer and successor to the Cray T3D system from Cray Research. Unlike the Cray T3D, the T3E is a stand-alone massively parallel system which retains the $3D$ periodic torus mesh interconnect and contains up to 2048 processing elements (PEs). The architecture of the machine can be characterised as distributed shared-memory (DSM) where a global address space is supported through remote get/put type operations. A distributed memory programming model is provided by the Cray SHMEM one-sided communication library as well as implementations of the MPI and PVM message passing libraries. The T3E interconnect is three times faster than the T3D, with a peak inter-processor communication bandwidth of 1.8 Gbytes/sec. CRAY T3E-300 system PE's are based on the 300MHz DECchip 21164 (DEC Alpha EV5) from Digital Equipment Corporation, with a peak performance of 600 MFLOPS. Each PE has its own local memory ranging from 64 Mbytes to 2Gbytes of DRAM. The EV5 RISC microprocessor contains an on-chip 12288 word, 3-way set associative Scache which is augemented by a 4-64 Mbyte on-board cache. Prefetching is accelerated through a set of 6 stream buffers which can sustain 1 Gbyte/sec of bandwidth to DRAM. Remote memory access is supported by a set of 512 memory-mapped E-registers, which can also be used as a register file. E-register to EV5 bandwidth peaks at 600 MBytes/sec.

Execution times on a 64 PE Cray T3E system for the first three problem sizes described earlier in Section 6.4 are plotted in Figure 4. The timings have been broken down into total execution time with and without the communication time t_{comm} versus the number of PE's on a log-log plot. A slope close to -1 is observed, giving an indication of the excellent scaling properties of the Cray T3E for fixed problem sizes. Figure 5 is a plot of the total Mflop/sec execution rate achieved on these problems up to 64 PE's. These plots indicate that we were able sustain approximately 40 Mflops/sec per PE when the problem size per processor was at least $60 \times 40 \times 31$ points. Performance then drops off gradually from this level as the number of PE's increases.

8.2 NEC SX-4

The NEC SX-4 supercomputer consists of a maximum of 16×32 processor, shared memory symmetric multiprocessor nodes interconnected by a crossbar network. Each processor has a peak performance of 2 Gflops/sec or 64 Gflops peak performance per node. The SX-4 is based on 64 bit wide data paths to memory. The vector and scalar units support 32 and 64 bit operands. The scalar unit also supports 8, 16, and 128 bit operands. Each processor has hardware support for three different floating point data formats: IEEE 754, Cray, and IBM.

A single SX-4 CPU contains a vector unit and superscalar unit. The vector unit consists of eight vector pipeline VLSI chips and 32 vector registers. The eight chips are connected by crossbar and comprise 32 vector pipelines arranged as sets of eight add/shift, eight multiply, eight divide, and eight logical pipes. A set of eight pipelines serves a single vector instruction and pipelines can operate concurrently. The peak performance of a concurrent vector add and vector multiply is 2 Gflops/sec and typical scientific codes written in Fortran can sustain 1 Gflop/sec with sufficiently long vector lengths. The SX-4 provides support for traditional shared-memory macro and loop-level micro-tasking on a single node and now also provides both a shared and distributed memory implementation of MPI. Shared-memory MPI/SX is implemented using POSIX threads and macro-tasks, whereas the distributed memory MPI accross nodes is an implementation of the MPICH package from Argonne National Labs.

Version 3.4.3 of the MC2 model is compatible with both distributed and shared memory models of computation. Loop-level parallelism accross vertical layers of the atmosphere is achieved with micro-tasking. The micro-tasked version of the model with the original Richardson–ADI solver was able to sustain an execution rate of 6.3 Gflops on 12 SX-4 processors, completing 12 time steps in a wall-clock time of 40sec for the $511 \times 359 \times 31$ grid. This corresponds to a parallel speedup of $S_p = T(1)/T(N) = 6$. Single processor performance was rated at 694 Mflops for this run. When the $231 \times 231 \times 31$ problem is run with the GMRES(10)-ADIV(1) solver, the sustained single processor performance jumps to 944 Mflops/sec and the integration completes in 109secs. Thus, one is tempted to make a price/performance comparison between a single SX-4 processor and a Cray T3E with 32–64 PE's.

To estimate the CPU time required to complete a 48 hour forecast with the MC2 model, consider a single processor capable of sustaining 1 0 Gflops/sec and assume that 12 dedicated processors are available. For example, an SX-4/16 is currently in operation at the Canadian Meteorological Center. Given a parallel version of the model based on the GMRES(10)-ADIV(1) solver that can achieve close to a level of 100% parallel efficiency (either shared or distributed memory), it would then be possible to run the current configuration of the 48 hour Canadian regional forecast on a $245 \times 189 \times 31$ grid with full physics in less than six minutes.

9. Conclusions

Our goals in creating a distributed-memory implementation of the mesoscale compressible community model were simple, yet very ambitious since this is the first project of its kind within recherche en prévision numérique. We hoped to deliver the simplest possible standalone version of the MC2 model that could

1. Run on any shared or distributed memory platform.
2. Run as fast as possible on any number of vector or cache processors (clearly these are conflicting requirements !).
3. Provide full dynamic memory allocation on a per processor basis with configuration options available to the user.
4. Ensure bit reproducibility.

If we consider the issue of code validation, it is obvious that reproducibility of the sequential results by the parallel code would be very difficult to achieve when one considers the basic structure of the parallel elliptic solver. However, it can be achieved by either having the sequential solver execute in the same order as the parallel code over subdomains or by forcing the parallel solver to execute global reductions in a predefined sequential order. For the first implementation of MPP-MC2, neither strategy has been implemented but should be in future releases. It must be emphasized however that validation of the parallel code was achieved up to the solver by simply imposing a known solution for the elliptic BVP in both the sequential and parallel codes. Validation in this context simply means total bit reproducibility of the results at every time step.

Much work remains to be done and we envision several exciting and productive years of model development ahead of us. Among the major outstanding issues yet to be addressed by the MC2 development team and other modeling groups is the possible processor load imbalance associated with physical parameterisation packages. Parallel I/O remains a concern for very large problems, not only to partition the I/O related functions accross processors, but also how to manage and post process the Terabytes of data generated by these simulations. In terms of algorithm development, the investigation of scalable multi-level domain decomposition type preconditioners will be an important area of research. To support more basic research efforts at RPN, the model will be used for very high-resolution studies of atmospheric turbulence and the development of new physics packages for mesoscale modeling and forecasting efforts.

A new level of support will also be required for the MC2 user community, since an entirely new class of computers has been added to the MC2 répertoire which now ranges from workstations to very high-end supercomputers. Indeed, the MC2 model is already being considered as the primary benchmark in the next supercomputer procurement cycle at the Canadian Meteorological Center.

Acknowledgements

For encouragement and advice throughout our MPP-MC2 project, we are most grateful to Roar Skålin (UNST), Dag Bjørge (DNMI) and Ivar Lie (NDRE). Access to Cray T3E MPP machines was generously provided by John Tulley and Jimmy Scott of Cray Canada. In particular, we would like to thank Charles Grassl of Cray Research in Eagan for providing technical support and suggestions which allowed simplified and rapid porting of the MC2 code onto the Cray T3E platform. Toine Beckers and Christian Lantwin from NEC-ESS Germany kindly arranged for time on the NLR NEC SX-4/16 in Amsterdam. Dave Parks and Per Nyberg from HNSX in Houston kindly provided technical documents related to MPI/SX and performance tuning information.

References

[1] R. A. Asselin. Frequency filter for time integrations. *Mon. Wea. Rev.*, **100** (1971), pp. 487-490.

[2] R. Benoit, M. Desgagne, P. Pellerin, S. Pellerin, Y. Chartier and S. Desjardins. The Canadian MC2: A semi-Lagrangian, semi-implicit wide-band atmospheric model suited for fine-scale process studies and simulation. *Mon. Wea. Rev.*, submitted.

[3] D. Bjørge. Implementation of the semi-implicit scheme in a message passing version of HIRLAM. *Parallel Supercomputing in Atmospheric Science.* Proceedings of the 6th ECMWF Workshop on the use of Parallel Processors in Meteorology editors. G-R Hoffman and N. Kreitz, World Scientific, Singapore, 1995, 75-90.

[4] M. A. DeLong and J. M. Ortega. SOR as a preconditioner. *Appl. Numer. Math.*, **18** (1995), pp. 431-440.

[5] J. B. Drake, I. T. Foster, J. G. Michalakes, B. Toonen and P. H. Worley. Design and Performance of a Scalable Parallel Community Climate Model. *Parallel Computing*, **21** (1995) pp. 1571-1592.

[6] T. Gal-Chen and R. C. Sommerville. On the use of a coordinate transformation for the solution of the Navier-Stokes equations. *J. Comp. Phys*, **17** (1975), pp. 209-228.

[7] W. Gropp, E. Lusk and A. Skjellum. Using MPI: Portable Parallel Programming with the Message-Passing Interface, MIT Press, Cambridge, MA, 1994.

[8] G. J. Haltiner and R. T. Williams. Numerical Prediction and Dynamic Meteorology. 2nd ed. John Wiley, New York, 1980.

[9] C. Hill and J. Marshall. Application of a parallel Navier-Stokes model to ocean circulation. Proceedings from *Parallel Computational Fluid Dynamics*, Pasadena, CA, Elsevier, 1995, pp. 545-552

[10] R. Laprise. Formulation of the mesoscale compressible community (MC2) model. *Atmos. Ocean.* , in press (1996).

[11] A. V. Malevsky. Spline-characteristic method for simulation of convective turbulence. *J. Comp. Phys*, **123** (1996), pp. 466-475.

[12] J. P. Pinty, R. Benoit, E. Richard and R. Laprise. Simple tests of a semi-implicit semi-Lagrangian model on 2D mountain wave problems. *Mon. Wea. Rev.*, **123** (1995), pp. 3042-3058.

[13] H. Ritchie and M. Tanguay. A comparison of spatially averaged Eulerian and semi-Lagrangian treatments of mountains. *Mon. Wea. Rev.*, **124** (1996), pp. 167-181.

[14] H. Ritchie, C. Temperton, A. Simmons, M. Hortal, T. Davies, D. Dent, and M. Hamrud. Implementation of the semi-Lagrangian method in a high-resolution version of the ECMWF forecast model. *Mon. Wea. Rev.*, **123** (1995), pp. 489-514.

[15] A. Robert. A stable numerical integration scheme for the primitive meteorological equations. *Atmos. Ocean.*, **19** (1981), pp. 35-46.

[16] A. Robert. Bubble convection experiments with a semi-implicit formulation of the Euler equations. *J. Atmos. Sci.*, **50** (1993), pp. 1865-1873.

[17] Y. Saad. A flexible inner-outer preconditioned GMRES algorithm. *SIAM J. Sci. Stat. Comput*, **14** (1993), pp. 461-469.

[18] Y. Saad. Highly parallel preconditioners for general sparse matrices. In G. Golub, M. Luskin and A. Greenbaum, editors, *Recent Advances in Iterative Methods, IMA Volumes in Mathematics and its Applications*, **60** (1994), Spinger–Verlag, New York, pp. 165-199.

[19] Y. Saad. Iterative Methods for Sparse Linear Systems. PWS Publishing, Boston, MA, 1996, 447 pp.

[20] Y. Saad and M. Schultz. GMRES: A generalized minimal residual algorithm for solving nonsymmetric linear systems. *SIAM J. Sci. Stat. Comput*, **7** (1986), pp. 856-869.

[21] W. C. Skamarock, P. K. Smolarkiewicz and J. B. Klemp. Preconditioned conjugate-residual solvers for Helmholtz equations in nonhydrostatic models. *Mon. Wea. Rev.*, in press (1996).

[22] P. K. Smolarkiewicz and L. G. Margolin. On forward-in-time differencing for fluids: An Eulerian/semi-Lagrangian nonhydrostatic model for stratified flows. *Atmos. Ocean.* , in press (1996).

[23] A. Staniforth and J. Côté. Semi-Lagrangian integration schemes for atmospheric models - a review. *Mon. Wea. Rev.*, **119** (1991), pp. 2206-2223.

[24] M. Tanguay, A. Robert and R. Laprise. A semi-implicit semi-Lagrangian fully compressible regional forecast model. *Mon. Wea. Rev.*, **118** (1990), pp. 1970-1980.

[25] M. Tanguay, E. Yakimiw, H. Ritchie, and A. Robert. Advantages of spatial averaging in semi-implicit semi-Lagrangian schemes. *Mon. Wea. Rev.*, **120** (1992), pp. 113-123.

[26] S. J. Thomas and J. Côté. Massively parallel semi-Lagrangian advection. *J. Simulation, Practice and Theory*, **3** (1995), pp. 223-238.

[27] S. J. Thomas, R. Benoit, M. Desgagne and A. V. Malevsky. A new dynamics kernel for the MC2 model I: Semi-Lagrangian advection. *Q. J. R. Meteorol. Soc.*, submitted (1996).

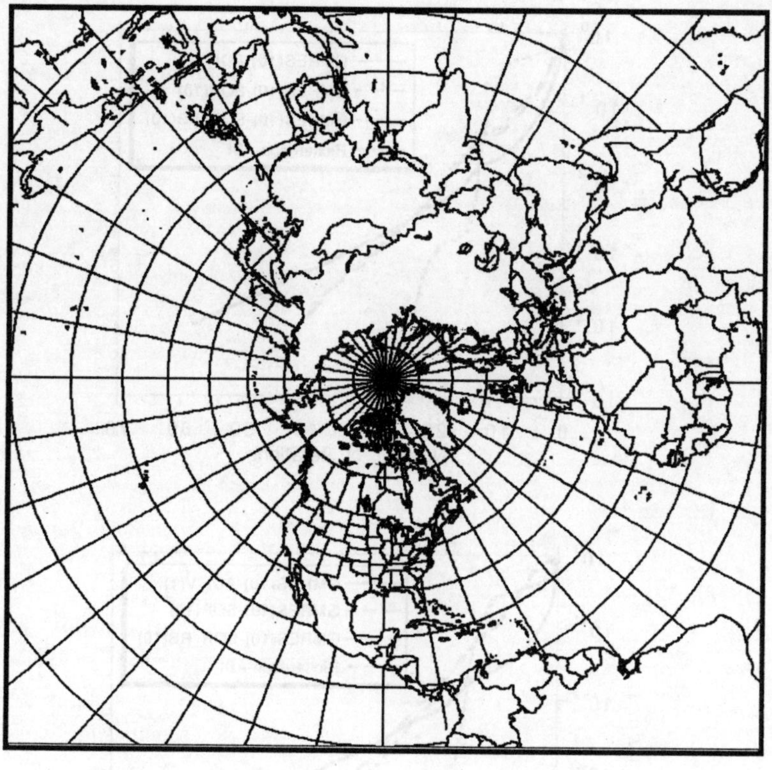

Figure 1: Hemispheric problem domain used in test cases.

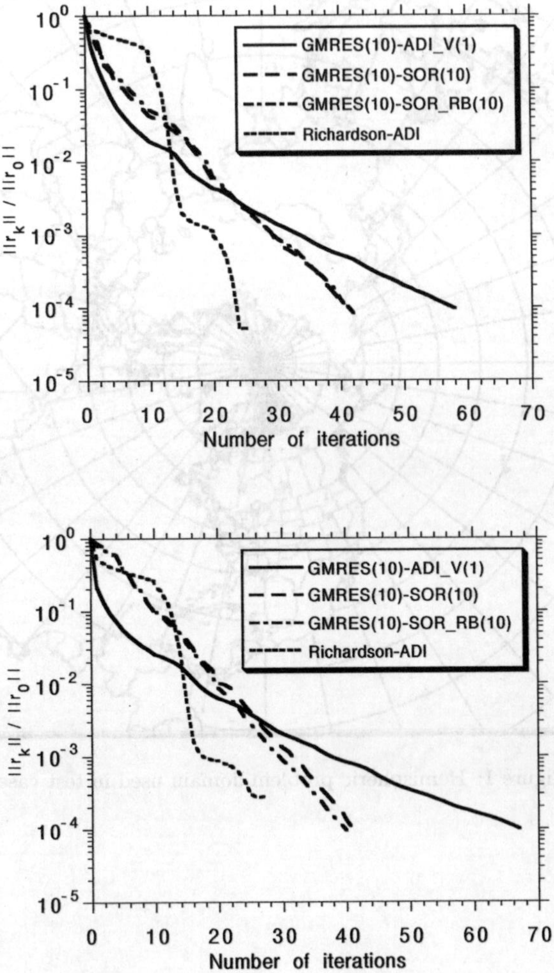

Figure 2: Iterative solver convergence rates. Top: $119\times119\times31$ grid. Bottom: $231\times231\times31$ grid.

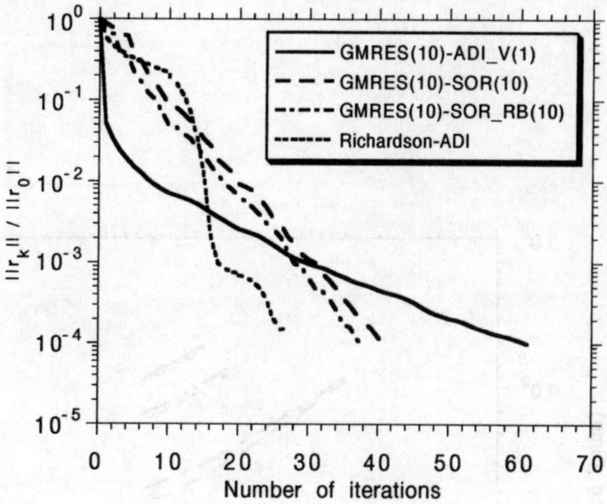

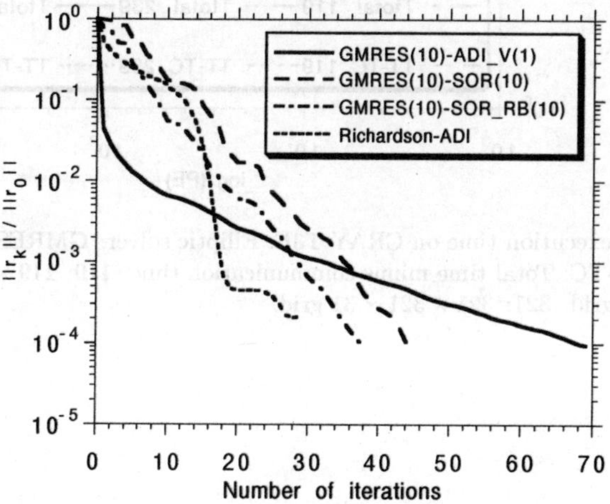

Figure 3: Iterative solver convergence rates. Top: $321 \times 321 \times 31$ grid. Bottom: $511 \times 359 \times 31$ grid.

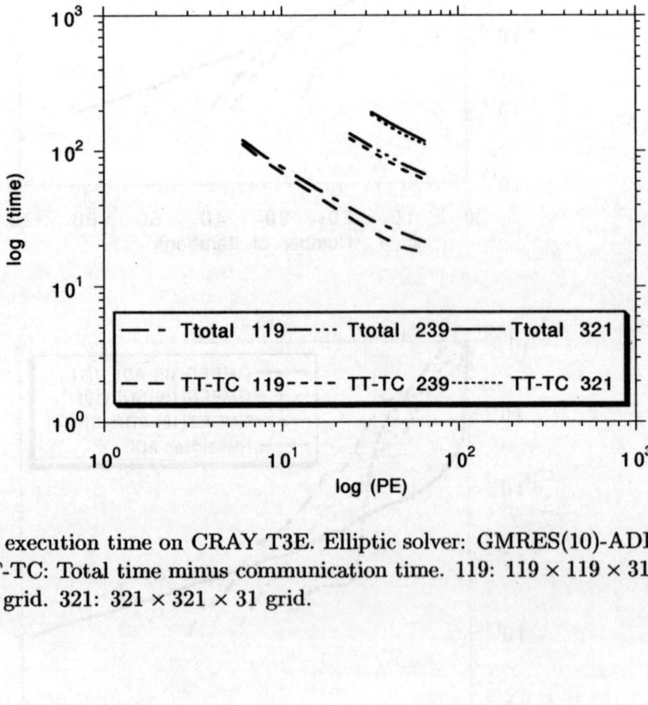

Figure 4: MC2 execution time on CRAY T3E. Elliptic solver: GMRES(10)-ADIV(1). TT: Total time. TT-TC: Total time minus communication time. 119: $119 \times 119 \times 31$ grid. 239: $119 \times 119 \times 31$ grid. 321: $321 \times 321 \times 31$ grid.

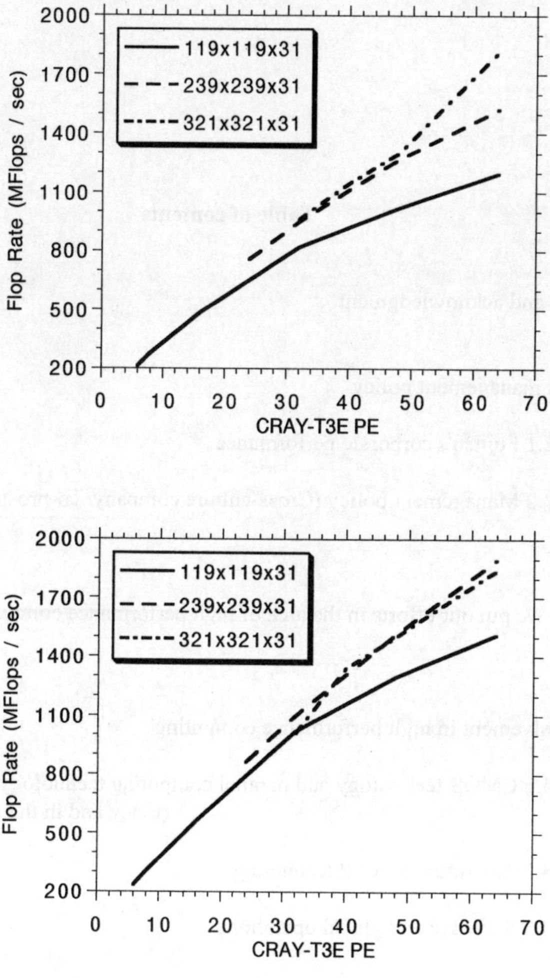

Figure 5: MC2 total Mflops execution rate on CRAY T3E. Top: GMRES(10)-ADIV(1). Bottom: GMRES(10)-SOR(10)

ECMWF-Workshop Speech by Chairman Yamamoto
Fujitsu's involvement in high performance computing

Table of contents

1. Address and acknowledgment

Honorable guests, Ladies and gentlemen.
I feel honored to have this opportunity to speak on behalf of the supercomputer vendors at this workshop which has such a high reputation for its engineering and for its distinguished members. I would like to thank Director Burridge and his staff for their kind courtesy. Today, I would like to share with you Fujitsu's activities in the supercomputer business, as well as our vision towards the 21st century.

I remember some time ago, when I read a book on supercomputers that there was a chapter about Dr. L.F. Richardson, a British scientist who, shortly after the First World War, tested numerical weather simulation. I was deeply impressed by his creative idea, not only because it was so long ago, but also because the numerical weather model he used was not substantially different from the model we have today.

It is exciting that, in this location, in the country where this concept was invented, one of our supercomputer products will play an important role in predicting the weather for Europe, thus contributing to people's lives as well as to the economy.

2. Fujitsu's management policy

2.1. Fujitsu's corporate performance

Before I come to the main subject of my speech, I would like to make some short remarks about our business situation at Fujitsu and our basic business strategies. There are about one-hundred-and-sixty-thousand employees in the Fujitsu group. Our revenues in fiscal 1995 reached $38 billion, with profits of $1.57 billion. Compared to companies in other industries in Japan, which continue to struggle for economic recovery, I have all reasons to be rather satisfied with these figures.

This year, revenues from semiconductor products have declined, broadly due to a worldwide slump in this field. However, supported by the strong demand for telecommunications and networks, in which Fujitsu has many years of engineering experience, as well as the continuing growth in our mainframe and PC business, I expect that we will be able to reach targets in line with last year's.

Added to this, our movement into developing countries, the expansion of worldwide markets and the efforts of our sales forces are making a great contribution to our business results. Last but not least, our long-term business restructuring efforts, as exemplified by changes in our personnel policy, are having a significant impact on improving our overall productivity.

2.2 Management policy (Cross-culture company/co-prosperity)

We are now living in a time where the world's economy should not be built just by exporting products designed and produced in Japan, creating further trade deficits with the rest of the world. Today, we can not neglect the national economies and employment issues of the countries we want to trade with. Among Japanese corporations, Fujitsu was relatively early to start expanding production-related corporate activities such as component procurement, and production testing, on an international scale, and we have steadily distributed equipment, financial resources, and personnel into all parts of the world. To continue our contribution to local regions and maintain our strong competitive position, we are now including research and development, and services in our promotion of globalization.

Currently, Fujitsu has 12 production facilities in Japan, and 21 abroad. More than 30% of our total revenue already comes from markets abroad, and we continue to grow into a cross-culture company that transcends geographical limits. While the main part of our revenue still comes from Japan, our economic direction is "co-prosperity". This means that our corporate direction is to achieve three targets equally: Firstly, to contribute to the performance of our customers. Secondly, to help develop the economy of our customers' countries. And thirdly, to improve our own company profit.

In this sense, in managing the company, I am paying just as much attention to the development of the worldwide economy as to the future direction of the information industry. I have never taken my eyes off the recovery of the American economy, the economic development of South-East Asia, especially China, or the movement towards European unification. Speaking about the United States, the development of new technology, with the IT industry and electronics at its core, and the dynamic economy that comes from that require that both Japan and the US rely on each other's high quality products, components and software, and their mutually dependent economies will become an inseparable structure.

Speaking on the other hand about the Asian countries, it is evident that the unprecedented speed of their development and their enormous populations make necessary large investments into the infrastructure for communication equipment etc.. As this is an area in which Fujitsu specializes, we make it our business not only to export our goods and products, but to transfer the technology which is required for creating high-tech corporations. Seeing the recent developments in China and the other Asian countries, and taking into consideration that their growth will continue, there is no doubt that, in the 21st century, Fujitsu will receive quite a large share of its business from this region.

In Europe, we have for many years enjoyed large-scale business cooperation with Siemens-Nixdorf and ICL, and with respect to the acquisition of leading-edge technology and the development of specialized software, it seems that our technological relationships should be further intensified. With the same intention, FECIT, the Fujitsu European Centre for Information Technology, was founded as an organization for broader exchange of European technology. We hope that FECIT will help Fujitsu to establish links with high-class European partners.

3. Why do we lay such importance on the Area of High-performance Computing?

I would now like to speak about Fujitsu's involvement in the area of science and engineering. Quite some time ago, when I was project manager for development of mainframes and supercomputers, I was for a long time involved in hardware, operating systems, and compiler technology. At that time, the strengthening of the research and development sector was extremely important for a country like Japan that was rather a poor country in terms of economic strength and available resources. It was assumed that computers would be a key technology for this process. The vanguard of today's engineering force was created by following the idea that it would be worthwhile to try and push our technology to the limits of what is possible. We had a strong conviction that research and technology would increase the wealth of our country, and at the same time open the way to a future in which many of the hardships of mankind, like environmental pollution, food and energy shortages, would be solved. Even today, the direction of our technology, as well as the concept that we are making a contribution to society, has remained fundamentally unchanged.

Fujitsu uses the phrase "What mankind can dream, technology can achieve" as a corporate slogan. As a matter of fact, most of the technological dreams of mankind have already been realized. Many of Fujitsu's engineers have the tendency to focus on realizing our technological potential rather than concerning themselves with the potential for new business. From the management point of view this is not a very desirable tendency, and we would of course prefer our engineers to develop products that bring in revenue which is higher than our development costs. Still, we secretly expect that it is this group of people that will find challenging new business areas, or create hit products. I think that this approach to research and technological development is one of Fujitsu's strong points.

In the US, where science and engineering are most developed, start-up HPC vendors have in the last few years developed products based on unique ideas and positioned them as weapons in the marketplace. Still, with markets that were not fully developed, the overall impression has been that many of these companies have quickly disappeared, had to abandon their business for management-related reasons, or have come under the umbrella of other corporations. Recently, there has been a thinning out of demand for supercomputers from such traditional mainstay areas as national defense and government, however, it would not be wise to expect too much business from government customers. There should be sufficiently strong demand for HPC products from the private industry.

It is true that strong competition from the never-ceasing flow of new, start-up rivals, the performance increases of engineering workstations, and of course the long-standing competition from Japanese rivals, as well as user demand for high product quality, high performance, and high-quality services, certainly make this a high-pressure field. However, to the management of American HPC companies, who are facing the same difficulties, I would like to say "This area is a tough market, but many people are eagerly waiting for our products, so let's try our best!". As a result of severe competition, users receive better and better products and can use more and more advanced technology which will -in turn- create new markets.

Just as long-term challenges are important for the engineering area, I believe that it is important from the management point of view that we keep on with our long-term challenge in this, admittedly, rather small-scale, HPC market. Although many of the traditional features of Japanese business culture are now being eroded, we are keeping the habit of setting up long-term business targets, and I believe that such long-term targets are important for the HPC area as well.

4. Our involvement in high performance computing

Exactly 2 years ago, we set up a dedicated HPC Group, which consists of a specialized group of 300 people including marketing personnel, hardware, Operating System, and language developers, as well as engineers for the development and porting of parallel applications. These people have many years of experience, are familiar with the latest technology, and have deep knowledge in and enthusiasm for the high-performance computing area. This is right in line with our policy slogan "What mankind can dream, technology can achieve".

In actual fact, we are now offering 2 product lines to respond to the needs of the market. One of these lines consists of parallel machines based on vector technology like the VPP700. And we had the pleasure of receiving an order for this kind of machine from ECMWF. These machines will offer superior cost-performance to future supercomputer users. The other of these product lines consists of high-performance scalar parallel machines for market segments in which vector machines are not an adequate solution. This year, we have introduced our AP3000 series; and I expect that, with this machine, we will be able to serve new HPC areas beyond the traditional research and engineering segment.

4.1 CMOS technology and parallel computing technology (Today and in the 21st century)

Now, I would like to speak about the practical implementation of our product development. There are two recent trends in processor technology. One is the decrease in the price per MIPS, the other is the movement towards multi-processing. Let me start with an explanation for the first trend. The factor that triggered the decrease in the price per MIPS is the invention of CMOS semiconductor technology in the 1970's. High packaging density and low power consumption, as well as advantages in manufacturing techniques, meant that this invention made a big contribution to the miniaturization of computers. There used to be the disadvantage that small-sized packaging and low power consumption went hand in hand with slow processing speeds but, now that we have overcome this problem, the use of CMOS technology is spreading rapidly. The trend went from personal computers to mainframes and is rapidly moving to supercomputers as well. In our VPP series, CMOS technology was employed throughout the systems, which made it possible to achieve a dramatic decrease in the price per FLOPS.

As we enter the 21st century, it is difficult to see how any new technology will replace the current technology, and thus we intend to continue developing our future supercomputer products based on CMOS technology.

It is expected that the second trend, multiprocessing, is an approach that will become necessary in the long run, as speed improvements based on semiconductor technology will, sooner or later, reach their limits. Around 1990, the keyword "grand challenge" was used everywhere, and there was demand for TERAFLOPS machines. However, there were limits to the number of processors that could be connected, and it became clear that it was impossible to produce such a system with shared-memory technology. We were very early to dream about developing a TERAFLOPS system with a distributed memory machine that would use a crossbar switch. Still, we did not have the technology at that time that would allow us to produce such a high-speed machine at acceptable costs.

A cooperative research project with the National Aerospace Laboratory of Japan to solve problems of computational fluid dynamics for a numerical wind tunnel led to the development of a distributed memory type parallel vector system. The results of this work enabled us to provide the market with the distributed-memory multiprocessing HPC system, the VPP500. The VPP300, as well as the VPP700, which we provided to ECMWF, replaced the technology of this machine with CMOS technology. By 1998, ECMWF plans to expand the VPP700 to its maximum configuration, and I believe that being able to offer a machine with such high performance levels and yet so small and with such low power consumption is a great success for our technology.

4.2 The future of VPP technology

In areas outside meteorology, there is also demand for supercomputers which exceed TERAFLOPS performance for solving various so-called “grand challenge” problems. We will continue to improve the single-processor performance of our machines, try to achieve higher speeds for the crossbar switches by which multiple processors are connected, and seek to advance our technology for achieving higher speeds and even greater miniaturization.

It is to be expected that future advances in supercomputing will not be restricted to the development of faster processors: High-speed networks will also play an important role. For applications like remote real-time visualization, real-time environmental simulation, shared use of supercomputers in real time, shared use of supercomputing centers etc., data transfer rates of several GIGA Bits per second and processing speeds in the TERAFLOPS range will be necessary. We are targeting the development of the high-speed network and processing technology that will be appropriate for these tasks. In 1996, we have the technology for 600 MEGA Bits per second, and our research efforts are proceeding to plan to make connections with 2.4 GIGA Bits per second available by 1999. We expect that there will soon be research projects which will find new ways of using supercomputers, ATM networks and so forth, that are based on this technology.

4.3 - Software and global operations

The enormous advances in the area of engineering workstations as well as the extreme improvements in cost-performance mean that part of the high-end HPC market environment is changing rapidly. Besides that, the competitive situation in the market for supercomputers is also changing depending on whether there is application software that efficiently uses newly-employed hardware architectures.

Through FECIT, we try to pursue not only the development and improvement of European software but of the software for other regions as well. Considering the broad usage of technology with supercomputers, the speed, technology and resource within the company is inadequate. Therefore, we need to gather skills and knowledge not only from Europe, but from all over the world, in order to speed up our development, and meet our customer demands.

Fujitsu has received a lot of positive and frank feedback about our high-quality product developments and services from our international user community. With the HPC Group, we have set up a business unit that can work quickly and flexibly. Therefore you can continue to expect the best from our products as well as from our services. Furthermore, there are high hopes not only from our customers, but also from inside our company, that HPC technology, in its role as leading-edge computer technology, will have a great impact on product development besides that of supercomputers; Which is my hope as well.

And in conclusion, I have much pleasure in announcing today, that as a further commitment to European research in weather forecasting, we will provide an additional VPP300 system to ECMWF, in order for you to run full 4D-VAR for the Re-Analysis Cycle at state of the art resolution. We are confident that the scientists at ECMWF and the member states will be able to achieve major new scientific results in the area of seasonal forecasting, and the provision of a consistent set of re-analyzed data using the advanced techniques on the VPP300 system.

I am sure that the close cooperation between ECMWF and Fujitsu will be further extended by this additional support that Fujitsu is giving to the Meteorological community.

Thank you for your kind attention.

Crisis or mutation in high-performance computing

Paul Caseau

Electricité de France
2, rue Louis Murat - 75008 Paris, France

Abstract . When we look at the concepts prevailing around 1990 - the "Grand Challenges" and the race to Teraflops - we may have a feeling of failure. The growth of the market promoted servers and workstations - rather than parallel computers.

These facts are reflecting the emergence of a new paradigm in numerical computation, that we may call "virtual reality". The mutation towards this new model has been resisted, when solving industrial problems, by the owners of physical experiments. It should have succeded on more global problems (like climate prediction) but has suffered from the formidable difficulties linked with complexity, and from a general lack of organisation. Nevertheless, a culture of parallelism, of collaboration and coupling between processors, is slowly emerging around the motto "the network is the computer". Parallelism will lead to scalability and, as the industrial problems are still there, scalability will lead to escalation.

During the last few years, several worrying events have happened concerning high-performance computing evolution, such as : the failure of most manufacturers of parallel computers (KSR - TMC - MEIKO - SSI - ACRI), the slackening of ECL technology, CRAY Research difficulties and the collapse of CRAY Computer. These failures are likely to question the R&D policy first, then the effectiveness of the "Grand Challenge" concept as well as its associate "race for Teraflops", and finally the HPC future.

It's no use denying the fact there has been a crisis, which has taken two aspects, apart from the difficulties met by manufacturers.

First the increase in the number of computers has been unfavorable to the most powerful systems. This situation can be figured on a pyramid. The HPC concept should have corresponded to the rise of the apex (sketch A, the height of the pyramid increases). But the evolution has been quite different : it doesn't even correspond to a homothetic pyramid (sketch B) but to a widening of the base (sketch C, meaning an increase of servers and stations).

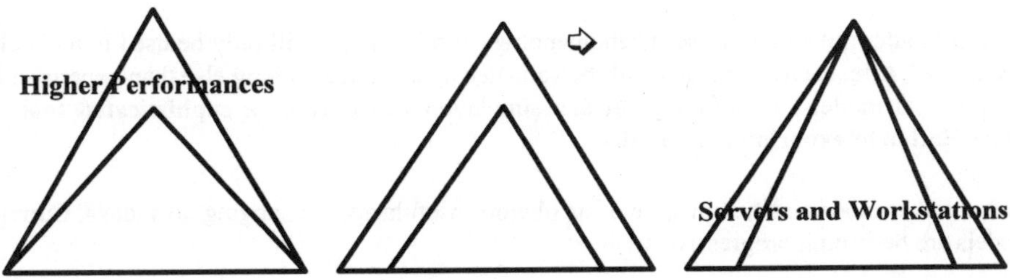

Moreover, the small number of victory records referring to "Grand Challenges" should be noted. It's in the climatological field that the most obvious improvements have happened (in particular the ocean-atmosphere coupling), independently of Teraflop computing : the steady progress of computers (a 30 factor every 10 years) has replaced the breaks over several decades.

Concerning the direct simulation of turbulence or the resolution of Navier-Stokes equations around a whole airplane, improvements are real but slow. The situation is identical in the fields of combustion or molecular dynamics.

The above mentioned facts are enough for a crisis diagnosis. The purpose of this paper is to examine the diagnosis more thoroughly : it will lead us to speak of mutation rather than crisis.

Computer simulation is becoming a new tool. This tool will be important enough to generate strong resistances : **its future place can't be completely clarified.**

1 What was computer simulation, what it has become

Today as well as yesterday, a computer model has always been composed of a "corpus" of equations, generally formalized a long time ago (for example Maxwell, Boltzmann or Navier-Stokes equations). It's also composed of "phenomenological relations" referring to medium properties or, more generally, to interactions between different phenomena which are not taken into account in basic equations. These relations proceed directly from experimentation.

The proportion of both groups of equations can vary tremendously. Their existence reveals the dual origin of computer simulation, which stems from theoretical as well as experimental models. And this dual origin leads the users to proceed with modesty.

When a computer model comprises very few (or even no) phenomenological relations, it will be used by physicists to check the validity of the starting equations, for example with a key-experiment. The simulation is then a **sophisticated tool of falsification** as defined by Karl Popper [1]. And nothing more.

If the computer model comprises phenomenological relations, it will only be used in its "field of validity". After each extension it will be validated again, since nothing else than experiments (on physical models) is unfailing. In fact simulation will only be a **sophisticated tool of interpolation of experimental results**.

Such a rather modest status, in spite of its obvious usefulness, is changing nowadays. Current models are becoming progressively :

- non-phenomenological : the importance of phenomenological relations is becoming small enough to witness a change,

- non-deterministic : the coefficients used in the model (medium properties or boundary conditions) and the observed quantities are varying “randomly”,
- nearly experimental : with the observation of results (which are no longer a priori predictable) models are becoming similar to common physical models.

We are going to develop these three points :

1 As soon as computing power has increased, computer models have got rid of their original defects : approximations due to 2D or 3D representations have nearly disappeared and 3D evolutionary models are (almost) becoming the only reference. As for the “phenomenological” part, it represents something else : it is linked with the fact that each time we choose our scale for representing reality, what is known about lower scales must be summed up either in a theoretical model (a conventional example is the coefficient of viscosity or diffusion) or in a phenomenological model.

Computer ability to describe multi-scale phenomena has then brought an opening in this classical process. More precisely :

- in the numerous cases (turbulence is the most famous example) when all scales interact with each other, phenomenological approximations are made useless with complete simulation,
- when the separation between the different scales is clear, the need for a phenomenological model remains but simulation enables us, while moving back the non-represented field, to rely on a level of knowledge which is more general than the level of the initial problem, and therefore to benefit from much larger experimental verifications.

2 With the decline of the representation scale, a phenomenon which the first simulations managed more or less to hide is appearing : model coefficients vary randomly. What is expected is the average result (with an assessable dispersion) of a large number of simulations whose data are varying.

But there is something more serious. It has been revealed through the systematic study of non-linear phenomena (turbulence, meteorology, climatology) : even when there is no identifiable data variations, equations exhibit chaotic behavior and results can no longer be represented without a thought on variability, gaps, possibility of sudden variations.

3 All this points to the fact that computer models can no longer come up with predictable results, as it has been the case up to now with the procedures of falsification or interpolation. What they are producing now is a “**virtual reality**” which can become as difficult to understand and observe as physical reality so that the researcher’s or engineer’s behavior becomes similar to the experimenter’s. He can’t foresee what is going to happen : **the exploitation of the model will enable him to discover elements which the elaboration of the model would not foresee.**

Consequently it will be necessary to look for steadiness and order in chaos if we want to progress in terms of understanding and predictability.

This step forward from conventional computer models to virtual reality represents the mutation which is taking place at the moment. It goes far beyond the problem of "Grand Challenges". It raises justifiable questions and comes up against many oppositions.

2 Facing the new paradigm : doubts, difficulties and oppositions

Justified questions...

As we mentioned earlier, the new paradigm of simulation can be considered as a challenge to common sense. What is the use of building new models if their parameters can't be precisely defined ? Is it worth simulating chaotic phenomena if simulation is as difficult to analyze as the initial phenomena ?

When we reach a high level of complexity, physical phenomena may be modeled thanks to simulation progress : and for all that can they be predictable ? Wouldn't it be better to limit our ambitions and ask the computer the questions whose answers have already been more or less given by experimentation ?

These questions are likely to be linked with the mutation experienced by computer simulation. They have nevertheless been asked a long time before computers started to play the part they have at the moment. In a passage from his book "*The Poverty of Historicism*" (1957), Karl Popper underlines the difference between "prophecy" and "technological prediction".

Technological prediction concerns the world we have built and which can be modified - and simplified - until prediction becomes possible.

Prophecy, on the other hand, concerns the phenomena which can't be influenced, so that we are faced with such a situation as to predict everything or to predict nothing at all.

Karl Popper's thesis concerns mainly human sciences but one of his examples belongs to meteorology. He indicates that scientific success is largely related to the simplified world created by technology and industry : beyond this world there is a kind of wall of complexity. Since problems can't be isolated and limited, they can only be referred to under common propositions without any predictive value.

This questioning must be taken very seriously, as there is a definite correspondence of the notions of "technological prediction" with conventional modeling, and of "prophecy" with virtual reality. Thus, when industrial simplifications are not possible precautions must be increased and a "naive" exploitation of models can't be the solution.

The experimenter must define questions whose answer is stable : whether the characteristics of a steady state are concerned, or the length of a transient state, or the presence of a bifurcation. He must also link the variability or the uncertainty of the results with the uncertainty of the initial and boundary conditions.

However, if virtual reality is synonymous with rigor, it also offers large advantages. The first is the ability to experiment (even in situations where this is physically impossible) and the acceleration of time. The second is the formalism of computer modeling. The answers are open to criticism and as a consequence more reliable than the results of an exhausting (and expensive) campaign of measurements which nobody will ever bother to start again.

But all these advantages call into question methods and situations which have stood the test of time. That's why, when we speak of interrogations or reserves, it is not enough : oppositions must also be mentioned.

Those who deal with "conventional problems" refuse, at the moment, to change for a new organization...

It is hardly necessary to insist on the fact that the industrial and scientific establishment refuses to change for another paradigm. Finding support in a simplified version of Popper' criticism, it express its doubt and hostility for any attempt to get rid of any testing facility and to go straight to a prediction which wouldn't be an interpolation.

It is worth adding than that to build a model which corresponds to the experiment-simulation logic, a **Gigaflop computer is enough most of the time**.

It has often been emphasized that the advanced work station and the memory-sharing server are part of the "natural sociology" of researchers and engineers : a laboratory, a team, a computer. But an equally important fact has not been particularly noticed : one of the qualities of these computers is that they neither question the conventional model nor hinder the testing facility where the team is gathering.

If oppositions are important as far as problems and common organization are concerned, what can be said about problems concerning "prophecy" (in Popper's language) for which the new paradigm is in fact the only possible solution ?

... And for new problems an effective organization is lacking

These problems are most of the time nationwide or worldwide. They are obviously multidisciplinary and need wide collaborations and a long-term effort (ten years or more) to be solved. They generally require a very sophisticated data management and quality assurance to prevent splitting. But there is nothing new in this : it simply points to the difficulties linked with complexity.

The trouble is all difficulties happen at the same time, and what should help to cope with them - that is to say a strong leadership, a flawless mastery of the system - is generally what is missing. For, with a few exceptions, there is no responsibility whomsoever : challenges offered to humanity are for nobody in particular. Dangers and profits are too difficult to attribute to give a living chance to the necessary organization.

That's why problems dealing with global hazards (continental pollution, regional flooding...) make little use of simulation on Teraflop computers. The few exceptions, such as climate forecasts as the most remarkable, must be noticed. Nevertheless, it is clear that it is only the existence of a previous organization that has permitted its development : this organization was based on National Meteorological Offices and the corresponding worldwide network.

Another example is given by the ASCI (*) program : after the ban of nuclear tests, the U.S. have launched this program for simulating what would no longer be physically experimented. ASCI wholly corresponds to the paradigm of virtual reality : it can be in the near future one of the great systems which will enable us to evaluate difficulties and successes of virtual reality.

To sum up what has just been said, in 1990 the focus on some "Grand Challenges" has been a **dual mistake**, paid for by computer industry and the world of HPC ever since.

- The basis and the body of the pyramid were not ready (and they are not ready yet) to change the paradigm in simulation. Everything led to an evolution independent of any efforts on the biggest computers, and to the choice of specifications - "the same but much cheaper" - which moved away from the evolution of the "apex" and the race to Teraflop computers.

- Problems which correspond obviously to the new model do exist and their importance will grow. But the obstacle of complexity, which helps to benefit from virtual reality, makes the necessary organization difficult.

Teraflop power is not necessarily the most important obstacle. As one of the managers of the ASCI project pointed out : "what will be really difficult will be to make people work together on such a large scale project".

Shall we conclude then, as some have dared to, that we have had **"a Teraflop before it's time"** ?

3 And yet mutation is on the way

It would be wrong to conclude, following the presentation of the above arguments, that the HPC case can be abandoned for better days. The situation is actually changing in a rather positive though paradoxical way.

(*) Accelerated Strategic Computing Initiative

The **culture of parallelism is spreading** : it is spreading **right at the base of the pyramid.** Work stations or server networks have planned to "defeat the CRAY computers". But beyond this target there is a more general concept : "the network is the computer". Whether right or wrong this concept will lead everyone on the right path.

As a matter of fact the stumbling blocks on the way to parallelism is in the management of DMS (Distributed Memory System). The effort must then be on the algorithmic development and the application software. This effort is the same and improvements are the same whether working on a work station network or a T3E or any other distributed memory system.

Three of these improvements can be mentioned :

- programming models are appearing such as BSP (Bulk Synchronous Parallelism). Each programmer is asked to think of coupling, and to dissociate communication and synchronization. Any algorithm which can accept these rules leads to "scalability" : the station network is the incentive, but Teraflop is clearly the outcome.
- the methods of domain decomposition, with or without overlapping have appeared (or rather re-appeared) with parallelism. Researchers who have experienced the triumph of previous direct resolution methods find again much of the know-how which existed in the 70s... In that field too, the size and power of the hardware are pointless : **scalability is there**.
- the methods of probabilistic mechanics (FORM and SORM) have tackled the problem of random data even long before Teraflop computing. They naturally lead to the coupling of several codes and to specialization of tasks : computation - differentiation and sensitivity - optimization.

In these three examples, the size of the problem is pointless. The efficiency of the network can be very moderate (if the cost is very low). The basic achievement lies within application software, data and process distribution and organization of the dialog between the different phases of the same algorithm.

Taking advantage of this culture of parallelism, which is progressively spreading, the new paradigm of simulation is going to show its efficiency. From now on, we can describe some of the most significant phases of this general mutation. We will start, on purpose, with the base of the pyramid to finish with the applications which will bring into action several Teraflops.

1 People are going to investigate random phenomena in simple situations through Monte-Carlo methods. This can be done on station networks (specially if the problem is about computing power rather than data access). After this introduction to random phenomena, assessment of sensitivity by differentiation will come immediately.

2 Computation using several scales is going to develop, in particular in molecular dynamics and materials theory. These computations are generally rather easy to program, and can replace simple experiments. They may be run on station networks or medium-size SMP processors. The most complex would of course reach the Teraflop.

3 The simulation of turbulence and combustion finds its place right above the previous field as far as complexity is concerned. The simulation of a complete "industrial" situation is not at stake first : the purpose will be to tackle simple cases corresponding to academic experimentation. These computations extend from SMP network to Teraflop computers.

4 The previous simulations (2 and 3) dealt only with "theoretical" cases. But they can reach a higher level of complexity, and tackle industrial problems : they will then replace current test benches.

We have already insisted on obstacles and negative reaction. We must also emphasize the fact that (as shown in the grid below) "problems are still there" and that the advantages of computer simulation in cost and time is absolutely decisive. We must also note that with industrial problems we need Teraflop computers (even if SMP processor networks can achieve great things) and that difficulties of organization and data management are becoming serious.

Aeronautics and turbine machine : some data

Computations	**Today**	**Wanted**
COMPR./TURBINES	3 canaux, 1 000 000 points 700 Mo, 150 h	10 canaux, 24 000 000 points 20 Go, Néq*2 ~ 6 000 h
L.E.S.	scale 1/15, 100 000 meshes 150 000 time steps , 30 h	scale 1 ou 3D, 1 000 000 meshes Néq*2 ~ 600 h
SUPERSTATOREACT	300 000 meshes, 400 Mo 150 h	dxyz, Néq*2, 7 500 000 meshes 10 Go, ~ 15 000 h
PEUL	30 000 meshes, 14 000 particles CH4 17 espèces, 71 réactions 9 000 pas, 40 h/itération E/L	200 000 meshes, 90 000 particles Kéro. 80 espèces, 250 réactions 9 000 pas, 1 000 h/itération E/L

5 We are now reaching "Grand Challenges", those at least which are part of the new paradigm of simulation. Both climate forecasting and ASCI project are working in virtual reality. Difficulties but also successes have been mentioned. A power of 1 to 10 Teraflops (depending on cases) is necessary to achieve complete credibility.

Some more industrial problems can be added to this category when the two following characteristics are present :

- the obligation to turn to virtual reality because experiments are either impossible or much too costly to be performed,
- a level of complexity which means that the methodology of cases 3 and 4 must be abandonned, that possibilities and scenarios of actions become flourishing, and that data uncertainty bring forth the problem of predictability.

Two examples linked to power production and transmission can be mentioned : the analysis of severe accidents in nuclear power plants and the analysis of continental interconnected power grids.

4 Conclusion

By now it must be clear enough that the HPC is more the object of a mutation than the victim of a crisis.

Its slightly mythical key-tool, the 1 Teraflop Distributed Memory Computer is linked with a new paradigm of simulation : the use of virtual reality instead of an experimentation.

This model is disturbing. Created to tackle complex problems, it has to face all difficulties and hazards which are generated by the fact of dealing with such complexity. Created to replace impossible or economically unaffordable experimentation, it seem to be a threat for the balance between experiment and modeling which we have been used to.

These difficulties, which can hardly be overcome directly, will certainly be overcome thanks to tools and culture. The motto "The network is the computer" has spread from tops to bottom on the pyramid of organizations. To distribute and couple procedures, to "build parallel" rather than centralized is becoming a reflex. And it is thanks to scalability that the desire to climb up the scale, that is to say to process more and more complex problems, will become natural.

As a consequence, when the difficulties of organizing and financing some important projects based on the prediction of the future - for which teraflop is compulsory - are overcome, the paradigm of simulation in virtual reality will be validated as well as the necessary tools. They will spread because culture of parallelism will have shown them the way.

[1] Karl Popper : The logic of scientific discovery.

VECTOR AND CACHE PERFORMANCE OF OCCAM

R. W. FORD and D. F. SNELLING
Centre for Novel Computing,
Department of Computer Science,
The University of Manchester, U.K.

ABSTRACT

This paper compares the performance of a recent version of the OCCAM Ocean Model and a more vector oriented version on two architectures, a parallel cache based SMP (an SGI Challenge) and a parallel vector based machine (a Fujitsu VPP 700). OCCAM was originally developed to run efficiently on cache based parallel platforms, with little attention paid to its vector performance. However, with the continuing price/performance levels of the latest parallel vector architectures and their continued use in meteorology, an efficient vector version of OCCAM may prove useful. Given the potential for blocking vector codes to improve cache performance and the recent implementation of streaming to improve the vector performance of caches, it may be possible to obtain a single implementation of OCCAM, which is efficient on both types of architecture. Preliminary results show the potential for such an approach.

1 Introduction

The Ocean Circulation and Climate Advanced Modelling (OCCAM) code [2, 3], has been developed by the OCCAM project [1] to enable them to study the climate, investigate global ocean characteristics, and use the code as a test bed to improve ocean model physics. The model is an extended version of the Modular Ocean Model Array (MOMA) [4] code which is based on the Modular Ocean Model (MOM) [5]. MOM is, in turn, derived from Bryan's Ocean General Circulation Model [6]. The world is represented as a 3-D latitude-longitude-depth grid and the time dependent equations [6] are discretised using a central finite difference scheme, stepped forward in time with an explicit leapfrog method.

Previous global ocean codes have been designed for vector architectures, with arrays dimensioned to make latitude and longitude contiguous in memory and the inner loops of the code over latitude and longitude. In contrast, the OCCAM code has been designed for cache based parallel architectures, with depths contiguous in memory and the codes inner loops over depths. Unfortunately, whilst each code formulation performs well on the architecture for which it is designed, each performs poorly on the alternative architecture. This dichotomy is predictable as latitude and longitude are typically very large, giving rise to long vector lengths and therefore good vector performance. Such a formulation gives little data re-use for caches to exploit. Similarly, depths are typically small, giving a small working set of data and thereby allowing good cache re-use. Such a formulation gives very small vector lengths and therefore poor performance on vector architectures. As both the atmosphere and oceans exhibit a much greater number of latitude-longitude points, with respect to depth, this problem pervades meteorology.

Whilst the OCCAM code performs well on cache based architectures, the price/performance levels of the latest parallel vector architectures, coupled with their continued use in meteorology (and therefore availability), make an efficient vector version of OCCAM potentially useful. However, the OCCAM group only want to develop and maintain a single code, therefore our code modifications must also maintain or improve cache performance. Given the potential for blocking vector codes to improve cache performance [8] (effectively reducing the working set size until it fits into cache) and the recent implementation of streaming to improve the vector performance of caches, it may be possible to obtain a single implementation of OCCAM, which is efficient on both types of architecture.

Section 2 explains the OCCAM code structure, section 3 details our strategy to improve vector performance and section 4 describes the parallel machines used. Section 5 discusses the OCCAM code resolution and porting details and section 6 details any code restructuring and presents our results. Finally, section 7 gives our conclusions and discusses any future work.

2 OCCAM Structure

The (fast moving) surface gravity waves (barotropic mode) were found to restrict the model time-step, therefore they are treated separately from the rest of the solution (baroclinic mode), allowing a longer time step for the latter. The current model uses a free surface scheme for the barotropic mode. A two grid method [7] is used to avoid grid convergence problems at the North Pole. The smaller grid covers the North Atlantic and is rotated by 90 degrees so that its North Pole is over land. The larger grid covers the the rest of the worlds oceans. The main time-stepping section of the OCCAM code is shown in Figure 1. In this parallel implementation each processor is assigned a region and each processor makes a list of the sea points inside this region. A list of sea points allows regions to be arbitrarily shaped (to obtain good load balance) and ensures calculations only need take place at sea points i.e: there is no redundant computation over land.

The finite difference discretisation in the horizontal direction uses a nine point stencil. To support the resulting communication, each processor's domain is separated into a 'core' region, surrounded by a ring of 'inner halo' points, which in turn are

Figure 1: Code structure

```
do main time step
  do l=1,number of points held
    find i and j
    call baroclinic(i,j)
  end do ! l
  Exchange halo information for all levels
  do barotropic time step
    do l=1,number of points held
      find i and j
      call barotropic (i,j)
    end do ! l
    Exchange halo information for surface points
  end do ! barotropic time step
end do ! main time step
```

surrounded by a ring of 'outer halo' points. The core region represents points that are not required by neighbouring processors, the inner halo points are points required by neighbouring processors, and the outer halo points are copies of neighbouring processors points that are required by this processor. At the end of a time-step, the updated inner halo points are sent to the appropriate neighbouring processors.

The explicit separation of core points from inner halo points allows overlap of communication and computation. The list of sea points first contains all core points then inner halo points. Therefore, as remote halo data need only be received for inner halo points, communication of halo data and computation of core points for the next iteration can be overlapped.

The data structures of the OCCAM code have been organised so that depths are contiguous in memory and the inner loops are over depths i.e: the world is mapped to a 3-D depth-latitude-longitude grid. This formulation gives good cache use as depths are typically small (see Section 5.1) and therefore give a small working set when calculating all depth related data per grid point.

3 Cache and Vector Performance

3.1 Structure

OCCAM is written with depths in the inner loop, therefore, at each grid point all required computation is performed for the depths associated with that grid point. As the depths are small this gives a small working set of data which can fit into cache. To obtain good performance on a vector architecture the loops must be reordered, as vector performance cannot be obtained over depths. To maintain cache performance the code can be blocked so that a small chunk of latitude-longitude points and all associated depths are computed at a time. Note that blocking to the level of every single latitude-longitude point is logically equivalent to performing depths in the inner loop, although in practice various scalar optimisations (e.g: loop unrolling) would be inhibited. However, it is possible that cache performance may actually be improved by such a method, as cache use may improve with a larger working set. There are a number of problems with this approach, a potential increase in memory use, the use of indirection, and redundant computation. These are discussed in the next section.

3.2 Overheads

If loops are re-ordered, memory use may increase, as any temporary arrays passing depth data in-between loops must be expanded to the blocking value (note, for vector machines this could be the total number of latitude-longitude points, depending on the trade-off between memory use and vector length).

The latitude-longitude points are stored as lists of elements and are not contiguous in memory - they are stored in the second and third dimension. To obtain good vector performance, either the indirection must be removed, or the vector architecture must efficiently support indirect addressing. There are good reasons for keeping indirection. Indirection allows any shaped region to be held by a processor. This means that no redundant calculation is required over land points.

Table 1 shows the total number of grid points for each grid, the number of land and sea points on the grid and the percentage of sea points. Notice that significant amounts of redundant calculation would be performed if computation were done at all points. The typical solution is to subdivide ('shrink wrap') the grid to minimise redundant computation. However, a list of points removes all redundant computation and also allows varying numbers of grid points per processor. This is useful for load balancing as, deeper parts of the ocean require more computation than more shallow parts of the ocean. Therefore, in this paper, any vector performance loss through indirection is accepted.

In the current formulation, a map of depths for each grid point is stored and the inner loop depth calculations are performed from the sea surface to the grid points depth, therefore, there is also no redundant 'in the dirt' calculation performed. If the loops are reordered and the inner loop is over latitude and longitude then it appears no longer possible to do this (without a test for every level and every grid point), as different grid points may have different depths. The simplest scheme is to allow redundant 'in the dirt' computation for all depths. Therefore the depth for each ocean point is considered to be the maximum ocean depth. This is the scheme used for these preliminary results.

Table 1 shows the number of points that are considered to be the sea, assuming that all sea points equal the maximum depth (sea box grid pts); then gives the actual

Table 1: Grid Information

	grid 1	grid 2
grid	1440x577	438x434
land pts	329191	108946
sea pts	501689	81146
% sea	60	43
sea box grid pts	18060804	2921256
actual grid pts	14606996	1899383
% sea grid	81	65

number of sea and depth points. The percentage gives an estimate of the amount of redundant computation that would be performed in the simple case.

This scheme may be acceptable for vector architectures but the extra computation will reduce the performance of cache based machines. As stated in Section 1 the aim of this work is to improve vector performance and maintain cache performance. Redundant computation can be removed by storing the list of 'core' points and subsequently 'halo' points in depth order. If an index of the number of points for each depth is kept the latitude-longitude loop can run from 1 to the index. This technique removes all redundant computation and the index and list can be sorted once at the beginning of the model run.

If this method does not benefit the cache based architecture, as various scalar optimisations may be hindered by the loop ordering, a further optimisation would be to add a pre-processing macro which ordered the latitude-longitude and depth double loops, depth in the inner loop for the cache based machine and latitude-longitude in the inner loop for the vector machine, see below. This is not possible in the original code as the latitude-longitude loop is not distributed.

```
#ifdef cache
  DO L=1,LAT-LONG
    I=IINDEX(L)
    J=JINDEX(L)
    DO K=1,DEPTH(I,J)
#else ! vector
  DO K=1,MAXDEPTH
    DO L=1,LAT-LONG(K)
      I=IINDEX(L)
      J=JINDEX(L)
#endif
```

4 Parallel machines

The two machines used in this work are representative of parallel cache and parallel vector architectures. They are an SGI Challenge, owned by the Centre for Novel Computing (CNC) and a Fujitsu VPP700, owned by the Fujitsu European Centre for Information Technology (FECIT). The SGI Challenge is a cache based SMP. Processors are connected to a shared memory through a bus. Caches are kept coherent through hardware snooping on the bus. The CNC machine has four 100MHz MIPS R4400 processors with a 16KByte level one data cache, 1MByte level two cache and 0.5GBytes of memory. The operating system is IRIX 5.3, and the Fortran compiler is the standard compiler with that operating system version. The compiler options were -r8 -i4 and -O2, linked with standard PVM 3.3.8.

The Fujitsu VPP700 has four 2.2Gflop/s processors and 8Gbytes of main memory. Unlike traditional vector architectures it is made from CMOS technology. Memory is distributed and processors are connected through a crossbar. The Fortran compiler is version 1.1. The compiler options were -Ad (double precision) -Wv,-Of -Of (full vector and scalar optimisation) -Wl,-P (Parallel object) -J (vector object) -dn (link statics), linked with version 10 of Fujitsu's in house PVM, VPP PVM V10.

5 Model details

5.1 Resolution

The OCCAM distribution comes with three test models. Each of these is a full quarter degree model giving 1440x577 and 438x434 latitude-longitude grid points for the two grids and a maximum of 36 depths. Notice how few depths there are compared with latitude and longitude points. The first model runs for six time-steps, the second model runs for 1.5 days and produces a restart data file. The third model runs for six time-steps from the restart file. All results presented here use the first (six time-step) model.

The full model resolution requires more than 4GBytes of memory which is much too large to run on the SGI Challenge, as it only has 0.5GBytes of memory. To reduce the memory requirements, a smaller model was produced by masking out the majority of the oceans and leaving a small strip of water where the two grids connect. This was achieved using a program kindly written by Andrew Coward from the OCCAM group at the Southampton Oceanography Centre.

5.2 Porting

The OCCAM distribution comes with three preprocessor options *inline*, *debug* and *pvm/shmem*. *Inline* inlines the inner subroutine to reduce the subroutine call overhead. *Debug* gives various levels of debug reporting from zero to five (this was set to the minimum value for all timing runs). Finally, *pvm/shmem* switches between the two communication mechanisms.

Figure 2: Split Loop to Remove Communication

```
do inner time loop
  do l=1,inner points held
    find i and j
    call barotropic (i,j) ! code inlined
  end do ! l
  recv halo data
  do l=inner+1,all points
    find i and j
    call barotropic (i,j) ! code inlined
  end do ! l
end do ! inner time loop
```

There were few problems with porting the code. Some of the PVM integer sizes were incorrectly defined and some PVM calls had incorrect number of arguments. When these were fixed the code ran to completion. The code gives summary model information at the end of each time-step. The model run used has six time-steps. These outputs for the SGI and Fujitsu are shown below.

```
SGI:                              Fujitsu:
ke= 1.047833209347e-03            ke= 1.047833209347E-03
dtemp= 3.128371141458e-08         dtemp= 3.128371141443E-08
dsalt= 1.351995812502e-12         dsalt= 1.351995812492E-12
```

To verify code modifications all results were checked to see if they gave exactly the same results as above. Note that the SGI and Fujitsu give slightly different results. This requires further investigation with the OCCAM group. All results are from 3 processor runs where one processor is the master and the other two are assigned to the two separate grids. This is the minimum number of processors required to run the model. The results presented are timings from the larger of the two grids.

6 Code Restructuring and Results

6.1 Barotropic

The barotropic section of the code only requires a single value for all depths at a latitude-longitude point. The weighted contributions for each depth are added in later. This means that this section of code is potentially highly amenable to vectorisation, as there are no inner loops over depths and the data structures have latitude and longitude points contiguous in memory. The default code formulation

Table 2: Barotropic Results

	Fujitsu		SGI	
	all	frees	all	frees
asis	24.8s	-	36.5s	-
inline	22.9s	-	34.7s	-
split loop	15.1s	9.0s 0V	32.7s	19.8s
promote if	7.3s	1.07s 100V	33.9s	19.9s
no latency	8.3s	1.01s	-	-

has the loop body in a subroutine. Fortunately, there is a preprocessor option that inlines this loop body. As the latitude-longitude points are kept as a list, a directive must be added to the loop to inform the compiler that it is safe to vectorise the loop. However the code will not vectorise efficiently as there is an 'if' test inside the loop that receives halo data after all 'core' points have been calculated (see Section 2). To remove this the loop was split into two segments, see Figure 2. The code still does not vectorise efficiently due to another 'if' test inside the loop. However, this test is loop invariant and can therefore be removed from the loop. The results of these modifications for the two machines on the larger grid are given in Table 2. As expected, inlining the code benefits both machines, as the calling overhead is removed and the compiler has more opportunity to perform cache optimisation. Splitting the loop also benefits both machines again presumably because removing the 'if' test allows better scalar optimisation. This benefits the Fujitsu more than the SGI Challenge. Promoting the loop invariant 'if' allows the baroclinic section to be vectorised, more than halving the total baroclinic execution time (including communication) and making the computation (frees) part run nearly nine times faster. Therefore, in the vectorised case, communication costs dominate the overall time. The SGI Challenge performs slightly worse, however the performance is still slightly better than the original inlined code. On both machines the difference between 'all' and 'frees' is the time spent waiting for the communicated data. The final 'no latency' result is a test to see the effect on vector performance of splitting the loop. The halo data is received before any computation is done and the two loops are merged. The vector performance is only slightly improved and the extra second of latency hiding (overlapping of computation and communication) is lost adding an extra second to the total execution time.

6.2 Baroclinic

The baroclinic section of the code is split into five subroutines calls, see Figure 3, each of which contain a number of loops over depths. As described in section 3.2, our strategy is to reorder the loops so that latitude-longitude are the inner loops. As in the

Figure 3: Baroclinic Code Structure

```
do l=1,no. of points held
  find i and j
  call setvbc
  call ppmix
  call clinic
  call tracer
  call cfl
end do ! l
```

Table 3: Baroclinic Results

	Fujitsu		SGI	
	all	ppmix	all	ppmix
asis	5.86s	-	33.9s	2.5s
remove comms	5.97	-	33.1s	2.25s
separate	5.22s	0.19s 100V	30.8s	3.18s
reorder ppmix	5.07s	0.049s 100V	31.5s	4.56s

previous Section, the first stage of this process is to split the latitude-longitude loop into two loops to remove the communication therein. As table 3 shows, this causes the total time on the Fujitsu to increase slightly. The SGI performs slightly better, however both changes are minor. The second stage is to split each of the five baroclinic calls into separate latitude-longitude loops. To achieve this all temporary arrays used to pass information between between calls must be expanded. The OCCAM code keeps approximately 18 full arrays and 20 surface arrays. Expanding temporary arrays adds two full and six surface arrays. Note that, as mentioned in Section 3.2, blocking to improve cache performance reduces the size of these arrays by the blocking factor.

As a preliminary test of our method we chose to concentrate on the loops in the *ppmix* call. The next stage is to distribute the latitude-longitude loop to each scalar portion and loop in *ppmix*; this did not require any further array expansion. Table 3 shows the overall time decreasing slightly on both architectures. As a result of this change the *ppmix* call can now be timed on the Fujitsu; previous timing overheads meant this was not possible. All depth loops have been vectorised at this point. The *ppmix* time on the SGI increases. This is most likely due to a loss of data re-use between *ppmix* and the other routines. Finally, the (perfectly nested), latitude-longitude and depth loops were reordered. To achieve this the bounds of the depths loop were changed from a depth index (giving the depth at that grid point) to the maximum depth. Clearly this means that there is now redundant 'in the dirt' computation being performed. The Fujitsu benefits greatly from this re-ordering, with ppmix running nearly four times faster. However this is at the cost of extra computation which shows up in increased time to calculate *ppmix* on the SGI. Note, the overall time is still less than the original, but this is most likely due to fluctuations in the communication time.

7 Conclusions and future work

In conclusion, the barotropic portion of the model can be easily modified to improve its vector performance, without adversely affecting the performance on a cache based machine. Modifications to the baroclinic routines show that an improvement in vector performance can be achieved by re-ordering the latitude-longitude and depth loops, at the expense of performing redundant computation and increasing working set size, thereby reducing cache performance. However, we have discussed how blocking can reduce the working set size and potentially provide improved cache re-use compared with the current formulation. We have also discussed how re-ordering the lists of grid points, or pre-processing the loop ordering, can remove any redundant computation. Therefore, the technique described in this paper offers the real possibility of a single version of OCCAM which performs well on both cache based and vector based parallel architectures.

Future work will re-order all baroclinic routines, test the potential for blocking the data, remove redundant calculation by re-ordering the latitude-longitude lists and examine the communication overheads. These modifications will be tested on the full model resolution and an increased number of processors.

8 Acknowledgements

The authors would like to thank Fujitsu Systems Europe (FSE) for funding this work and Fujitsu European Centre for Information Technology (FECIT) for allowing access to their VPP700. The authors would also like to thank the OCCAM group at the Southampton Oceanography Centre for allowing us access to their code and, in particular, Andrew Coward who gave valuable advice.

References

[1] http://www.soc.soton.ac.uk/JRD/OCCAM/welcome.html

[2] C.S. Gwilliam, A.C. Coward, B.A. de Cuevas, D.J. Webb and E. Rourke, *The OCCAM Global Ocean Model: current status and future plans within the UK Oceanographic Community* Technical Report, Southampton Oceanography Centre, Southampton, U.K.

[3] C.S. Gwilliam, *The OCCAM Global Ocean Model*, Proceedings of the Sixth ECMWF Workshop on the Use of Parallel Processors in Meteorology, Editors Geerd-R Hoffman and Norbert Kreitz, World Scientific, 1995.

[4] D. J. Webb, *An Ocean Model Code for Array Processor Computers*, Internal Document No. 324, Institute of Oceanographic Sciences, Deacon Laboratory, 1993

[5] R. C. Pacanowski, K. Dixon and A. Rosati, *The GFDL Modular Ocean Model Users Guide*, Technical Report No. 2, GFDL Ocean Group, GFDL/NOAA, Princeton University, Princeton, N.J. 08542, U.S.A, 1990

[6] K. Bryan, *A Numerical Method for the Study of the Circulation of the World Ocean*, J. Comp. Phys., Vol 4:347-376, 1969.

[7] A.C. Coward, P.D. Killworth and J.R. Blundell, *Tests of a Two-Grid World Ocean Model*, J. Geophys. Res., Vol 99:22725-22735, 1994.

[8] D. Dent, L. Isaken, G. Mozdzynski, G. Robinson, F. Wollenweber and M. O'Keefe, *The IFS Model Performance Measurements*, Proceedings of the Sixth ECMWF Workshop on the Use of Parallel Processors in Meteorology, Editors Geerd-R Hoffman and Norbert Kreitz, World Scientific, 1995.

A General Programming Model for Developing Scalable Ocean Circulation Applications*

Aaron C. Sawdey and Matthew T. O'Keefe
Laboratory for Computational Science and Engineering
4-174 EE/CS Building
University of Minnesota
Minneapolis, MN 55455
(612) 625-6306
okeefe@lcse.umn.edu

Wesley B. Jones
Silicon Graphics Computer Systems
2011 N. Shoreline Blvd. MS 580
Mountain View, CA 94043

January 6, 1997

Abstract

In this paper we describe our efforts in porting a global ocean model—the Miami Isopycnic Coordinate Ocean Model or MICOM—to clusters of symmetric multiprocessors (SMPs). This work extends our previous efforts in porting this same application to the massively parallel Cray T3D

*This research was supported by the Office of Naval Research under grant no. N00014-94-1-0846, by the National Science Foundation under grant no. CDA-9414015 and grant no. ASC-9523480, by Silicon Graphics Inc. through an equipment grant and the "Distributed Virtual Memory Research Project," a joint research project between Silicon Graphics and the U.S. Army Research Laboratory supported through the DoD High Performance Computing Modernization Program.

machine. Our programming environment provides a "defensive" yet general programming model that can efficiently work across PVP, MPP, SMP cluster and DSM (Distributed Shared Memory) hardware. SC-MICOM, the SMP cluster version of our ocean code achieves good scalability both within and across the SMP architecture and has achieved up to 120 Mflops (C90 equivalent) per processor on the new SGI Origin2000 machine with a single 195 MHz R10000 processor and almost 90 Mflops per processor on a 30-processor Origin machine. This code simultaneously exploits locality and parallelism to improve single node performance while tolerating potentially large network latencies and slow main memory systems.

1 Introduction

With the recent introduction of the Cray T3E, SGI Origin2000, HP Exemplar and the Fujitsu VPP500, the performance of highly parallel computers is beginning to rapidly outstrip the performance of traditional shared-memory vector machines. For some users in the scientific computing community, such as the European Centre for Medium Range Weather Forecasting (ECMWF) [5] and NRL's Layered Ocean Model (NLOM) team [9] [10], parallel processing has become routine. These groups and other pioneering users in the community have been willing to parallelize their applications and endure rapid changes in the underlying parallel technology. The lessons learned in these developments must now be distilled into better tools (including languages, compilers, source-to-source translators, libraries, debuggers, operating systems, and pre- and post-processing software), better codes (through a consistent set of code design principles), and mechanisms to distribute this technology throughout the high performance computing user community. Much work remains before the larger climate, weather, and oceanographic applications community can benefit from parallelism.

1.1 The Fortran-P Programming Model

Our team has focused on developing a programming model and supporting tools that make parallel programming for climate, weather and oceanographic models and other regular grid applications. This focus arose out of the extremely demanding requirements for this class of applications:

- Atmospheric codes, especially operational weather models, typically must run within a fixed time limit or the numerical results are uninteresting
- Oceanographic codes, though less likely to have operational time constraints, are notorious for requiring huge amounts of both memory and processor time.

As Bleck [4] has pointed out:

> Though basin-scale ocean circulation simulations can resolve wind-driven horizontal circulation cells (gyres) and thermally-driven vertical-meridional overturning cells with course-mesh models at least qualitatively, obtaining a quantitatively correct picture of oceanic current systems requires models capable of resolving not only the gyre-scale motion but also flow instabilities which manifest themselves on scales of only a few tens of kilometers. Thus, computing requirements in ocean modeling have always been and will continue to be high.

In addition to weather and oceanographic applications, we have collaborated with Paul Woodward and his team to develop parallel versions of PPM, a compressible gas dynamics code used in the study of turbulence, convection in stars, and a variety of high Mach number flows [1]. This effort focused on source-to-source translation of a subset of Fortran 77—known as Fortran-P—that allows programmers to cleanly express numerical algorithms based on regular grids. The goal of the Fortran-P model is to allow programmers to write serial Fortran code which can then be translated to parallel code[1]: a domain decomposition is applied so that the regular grid is partitioned across processors in the parallel machine. Aggressive use of halo regions allows many loop nests to execute before communication or synchronization are required [1]. We were successful in developing a Fortran-P translator and applied various versions of it to help parallelize several codes, including PPM [1, 13] and ARPS[6, 7], a non-hydrostatic mesoscale weather prediction model developed by K. Droegemeier at the University of Oklahoma.

The Fortran-P source-to-source translator was targeted to generate CM Fortran, a Fortran 90 dialect with directives to indicate data layout in distributed memory machines. An early precursor of HPF (High Performance Fortran), CM Fortran was successful in allowing programmers to quickly write codes in data parallel style and execute them on the CM-5. However, users who demanded high efficiency often found that the high-level expression of parallelism inherent in CM Fortran made it difficult to express (to the compiler) optimizations that would significantly increase performance, often by factors of 2 to 10 [1, 3, 9]. In addition, very low performance sometimes resulted from using certain source code expressions (loop structures versus array syntax patterns) that were equivalent to others that were highly efficient; it was generally hard to develop any intuition about what would perform well and one depended on experimentation and discussions with other users to determine the programming idioms necessary to attain good performance.

---

[1] This parallel code could be data parallel (such as HPF), message-passing, threads, or perhaps even F−− [14].

2 The Massively Parallel MICOM: MP-MICOM

Driven by a desire to gain more portability and control over performance achieved, and to explore how Fortran-P would map to other parallel machines, we began to explore message-passing implementations for Fortran-P-based applications late in 1993. At around that time we began a collaboration with two groups: one was the U. of Miami ocean modeling team, led by Rainer Bleck, and the other was with Cray Research through analyst Robert Numrich. The goal of the Miami collaboration was to modify (initially manually, later semi-automatically) the source code of the Miami Isopycnic Coordinate Ocean Model (MICOM) so that it would fit the constraints of the Fortran-P model. Cray was introducing the Cray T3D in the Fall of 1993 and we targeted this machine using Numrich's SHMEM (get/put) one-sided message passing. Numrich, the developer of the SHMEM interface, was looking for users to try out his model and we were willing: we found it to be both elegant and efficient relative to other message-passing implementations, such as PVM, available at that time.

However, we had a problem: we had no direct access to the Cray T3D during the Fall of 1993 which meant that we could not get started on the MICOM port to the T3D until early 1994, unless we could find a way to simulate Numrich's SHMEM API on some other hardware. Conveniently, we had a 20-processor R4400-based SGI Challenge machine available to us for pre- and post-processing simulation datasets. We developed a SHMEM library for the SGI Challenge using IRIX mechanisms for mapping together portions of share group processes' address spaces so that they could transparently share data with other processes in the group across the SHMEM interface [2, 8].

In the case of MICOM, halo regions around each subdomain mapped to a processor were shared using this mechanism. Using the SHMEM library and software routines for parallel input and output and dataset preparation developed at Minnesota, we successfully parallelized MICOM and performed some very high-resolution North Atlantic circulation calculations with the Cray T3D [4, 24]. We call this first parallel version the ocean code MP-MICOM (short for MultiProcessor or Massively Parallel, depending on your architectural bias). On these machines, both with relatively small secondary caches, we achieved about 10 Mflops per processor (Alpha 21064, MIPS R4400) [24].

3 The SMP Cluster MICOM: SC-MICOM

At this point, we could have gone back directly to our original mission of developing parallelizing translators for regular grid applications but we felt that, although our MICOM ports to the Cray T3D and SGI Challenge machine were successful, these architectures were not fully general in that they had only a single level of cache locality and parallelism. On the horizon we could see future parallel machines that would consist of clusters of symmetric multiprocessors—some vendors, such as HP/Convex (Exemplar) and Silicon Graphics[2] (Power Challenge Array) were already beginning to launch products—and that our original approach would not provide the efficiency and scalability that applications demanded on these architectures[3]. We wanted to determine the parallel source code constructs we should generate to be most efficient for this new, general class of parallel machines.

In addition, with processor secondary cache capacities growing rapidly (they are now at 4 MBytes and next-generation microprocessors will likely have 8- to 16-MBytes of secondary cache) relative to efficient grid subdomain sizes, we saw that the day would soon come when we could easily fit all of a single subdomain from a domain-decomposed grid into secondary cache, gaining a tremendous performance boost and putting microprocessors within striking distance of the fastest vector processors.

So we began an effort to introduce another level of domain decomposition within the "outer" level of domain decomposition across SMPs. For our SMP cluster version of MICOM—which we have dubbed SC-MICOM—this "inner" decomposition, shown in Figure 1, would provide parallelism that multiple processors within the SMP could efficiently load-balance and execute, would also provide a tuning mechanism to precisely control the subdomain size and aspect ratio to optimize cache performance, and opened up significant opportunities for overlapping communications with computation between SMP boxes [11]. We were, however, very careful to isolate the parallel control code from the physics and dynamics routines so that the API for the modeler and physics package developers was kept simple and understandable.

In SC-MICOM, each SMP is assigned a rectangular subsection of the global problem domain; in Figure 1 four subsections (SMP 1-4) have been assigned to the four SMP machines in this configuration. Within each SMP, each domain subsection is further decomposed into small subdomains (sometimes referred to as patches); in Figure 1 we show the 16 subdomains into which the SMP 1 has been partitioned. The grid arrays associated with each subdomain have storage for fake zones (also known as halo or ghost cells) which exploit redundant computations

[2]Our collaborator Paul Woodward (Director of the LCSE) and his team, including David Porter, B. Kevin Edgar and Steve Anderson, performed the largest fluid dynamics calculation (1024x1024x1024 grid points) on an array of SGI Challenge machines made available to them at SGI's Mountain View manufacturing facility. SGI Chief Executive Officer Ed McCracken later stated that SGI's decision to introduce the Power Challenge array was given a strong boost by the success of this calculation.

[3]It should be pointed out that MPP machines such as the Cray T3E and Fujitsu VPP500 are a subset of the SMP cluster class of machines for which each "SMP" has 1 processor. Uniprocessors are a subset with only 1 processor and 1 "SMP". Hence, we argue that this class of machines is fully general and therefore the most appropriate as a "defensive" programming target for portable and efficient code.

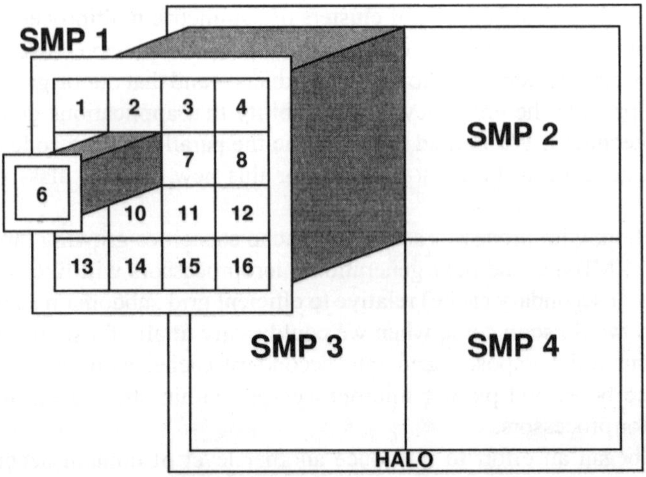

Figure 1: The two-level domain decomposition used in SC-MICOM.

and memory storage to reduce communication and synchronization frequency [7, 27, 28]. Any work thread in the SMP can compute any subdomain, while a single I/O thread handles MPI communication between SMP boxes and serial code within a box.

For the arrays, the first indices are the physical location within the subdomain. The last index is the subdomain number. The physics and dynamics routines never see the subdomain number; array variables are passed in as subroutine arguments which contain all the data needed to compute on that subdomain as shown in Figure 2. Note that no communication happens directly in these routines. Instead, a separate communications function for each physics routine is passed to the thread generator and is used to indicate which array variables need their fake zones updated. This update function is in fact overloaded to perform three separate tasks:

- copy data for updating fake zones in a specified subdomain
- copy data from a subdomain to an outgoing message buffer
- copy data from an incoming message buffer to subdomain fake zones.

The API for the main timestep control routine is characterized by single-threaded code regions for performing I/O and initialization tasks and calls to the *threadit* routine to accomplish parallel work. The barotropic portions of code are somewhat more involved than the other physics routines that are part of the baroclinic phase. Three barotropic timesteps are performed between communication

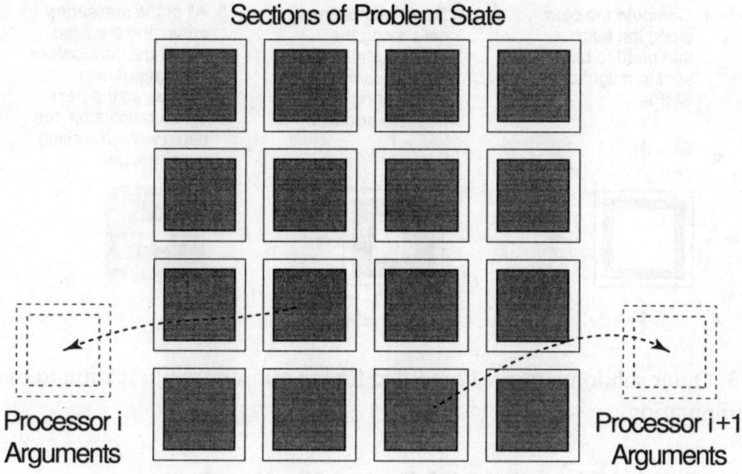

Figure 2: SC-MICOM data layout within the SMP.

events; this number is limited by the number of fake zones being used. The *threadit* routine takes a physics function and physics update function as arguments. Within *threadit*, an atomic fetch-and-add is used in a loop to assign subdomains to work threads. After a work thread finishes a subdomain, it notifies the I/O thread and starts computing the next available subdomain[4]. When all subdomains for that *threadit* call have been computed, the subdomain fake zones are updated.

Note that within an SMP box, subdomains that must be computed before sending messages to other SMPs (i.e., the patches on the perimeter of the subsection assigned to the SMP) are computed first as shown in Figure 3. When all subdomains for a message are finished, a single large message is packed up and sent. Once all messages have been sent, the SMP control code shifts to start receiving messages that neighboring SMPs are sending. These are single, large message transfers containing all the arrays needed for updating neighboring fake zones along the entire subsection edge on the SMP. This approach maximizes the potential for overlapping computation and communications in the parallel code. In one experiment across 2 SGI Power Challenges connected by a 10-megabit Ethernet connection we found this overlapping improved performance by 25% compared to no overlap.

[4]This is often referred to as self-scheduling or dynamic scheduling.

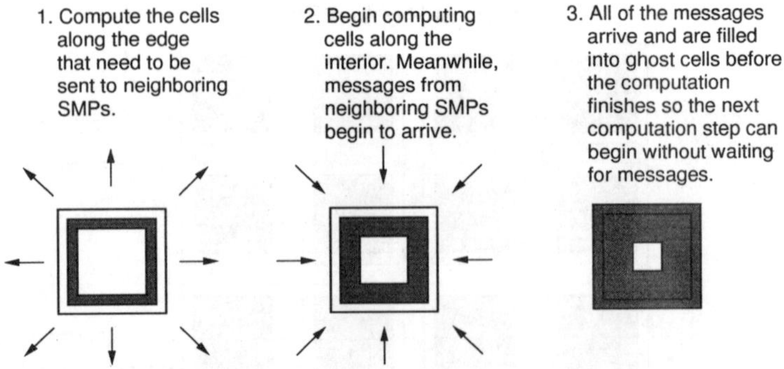

Figure 3: Outer subdomains are computed first to allow communication to overlap with computation.

3.1 SC-MICOM on the SGI Power Challenge Array

We measured SC-MICOM performance on both the SGI Power Challenge and the new SGI Origin2000 machines. All performance measurements are stated in C90-relative Mflops: we measured floating-point operations and computation time on the C90 and then used the relative time on the other architectures to scale the Mflops number appropriately. Both architectures use the R10000 MIPS microprocessor.

SC-MICOM has performed well on current SMP clusters yielding good speedups within and across SMPs even on slow networks, and achieving nearly 9 times (about 87 Mflops sustained) the single node performance on the latest MIPS microprocessor, the 195-MHz MIPS R10000, compared to our earlier parallel MICOM runs on the Alpha 21064 and MIPS R4400 chips. We expect further improvements in processor clock frequency (from 195-MHz to 275-MHz), reductions in networking overhead and latency, and improvements in the MIPS compilers and our own parallel software libraries to increase this performance in the future.

As noted previously, the "inner" decomposition of an SMP subsection into subdomains allows the size and aspect ratio of a patch to be chosen for best performance. For SC-MICOM using a perfect 800x800x16 "bathtub" grid on an R10000-based SGI Power Challenge machine requiring 1.86 GBytes of memory, the best patch configuration turned out to be roughly 130x40 (with 16 layers) which yielded 87 Mflops per processor. In contrast, when a single 500x500x16 SC-MICOM grid was computed on the same machine but without an inner decomposition it achieved only 42 Mflops, less than half the speed of the machine compared to the performance when a well-chosen inner decomposition was applied.

Though further study is necessary, we found that patches that are long in the x-direction—the stride-1 direction—achieved the best performance, presumably due to better cache re-use. Performance increased by about 23% when larger 2-MByte

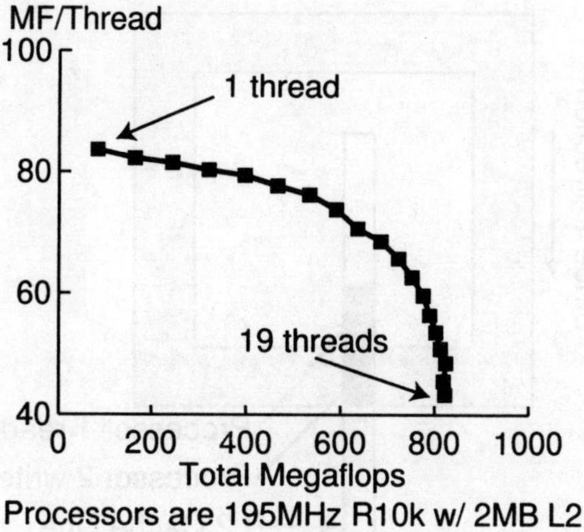

Figure 4: Performance per thread as number of processors (threads) increased.

Figure 5: The parallel efficiency of the internal fake zone updates is poor compared to that of the numerical computations.

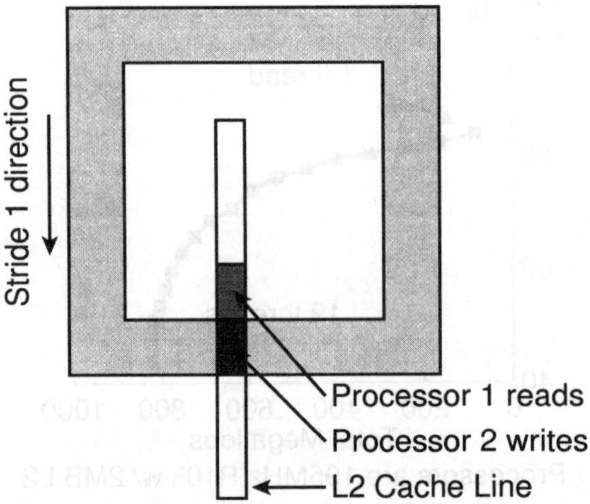

Figure 6: Illustration of false sharing within L2 cache lines in MICOM.

secondary caches were used instead of 1-MByte caches. About 325 Mflops were achieved on 4 processors while 16 processors yielded over 800 Mflops within a single SMP box. Scaling with increasing processors inside the SMP was limited primarily, we believe, by cache coherence, bus bandwidth and synchronization overhead as the number of processors increases. Figure 4 dramatically illustrates the limited scaling past about 12 threads.

Through the use of the R10000 hardware counters and some analysis of the code, we have traced the scalability problem to a problem with false sharing [15] of cache lines. As described earlier, data is copied between subpatches within each SMP to update the fake zones of all the subpatches. Figure 5 shows how the internal fake zone update part of SC-MICOM scales poorly compared to the numerical computation part. Figure 6 illustrates how multiple processors can be reading and writing the same L2 cache line at the same time, thus causing excessive numbers of cache invalidations during this phase of the computation. This problem could be eliminated through the use of an additional buffer to store data destined for fake zones as it is computed, so the processors would only be writing to the cache line illustrated in Figure 6.

Scaling across multiple SMP boxes was measured on a 4-SMP SGI Challenge Array at the University of Minnesota. Each Challenge machine had four R10000 (2-MByte L2 caches) processors active; the 4 machines were interconnected via a fast HiPPI network. On a 750x750x16 bathtub grid decomposed across the 4 SMPs (375x375x16 grid points per box) we achieved 990 Mflops across the 16-processors in these 4 machines, nearly 200 Mflops more than that achieved on a similar grid within a single SMP with 16 processors.

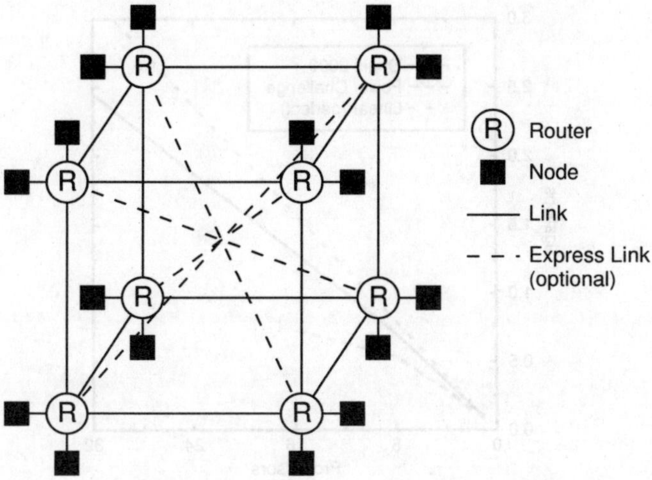

Figure 7: Internal network for a 32-processor SGI Origin2000 system. Each node contains one or two processors and a memory.

Further experiments showed that the per-SMP performance decreased by 10% when going from 1 to 2 SMPs (doubling the problem size) and a further 14% when going from 2 to 4 SMPs (doubling the problem size again). These tests were performed without the latest SGI software for fast transfers over HiPPI known as "HiPPI bypass" which bypasses buffering the data in kernel space and TCP/IP processing, both of which slow transfers and require significant processor resources and bandwidth. These scaling results show both the efficiency and potential scalability of our programming model for SMP clusters.

3.2 SC-MICOM on the SGI Origin2000 Machine

We also performed several tests on the latest SGI machine known as the Origin2000. This architecture differs from the Power Challenge primarily in that it does not use a single bus to interconnect processors but instead uses a hypercube-like network augmented with additional links to increase bisection bandwidth, as shown in Figure 7. In addition, the Origin machines support distributed shared memory via a scalable cache-coherence mechanism implemented with distributed directories. As shown in Figure 8, a 32-processor Origin machine (unlike the 16-processor Power Challenge) scales nearly linearly. We believe this improved scaling is due to the fact that both the memory bandwidth and cache coherency directories of the Origin2000 increase as you increase the number of processors.

The 2.5 gigaflops performance achieved equals our previous MP-MICOM results on 256 processors of the Cray T3D [24]. This was made possible primarily

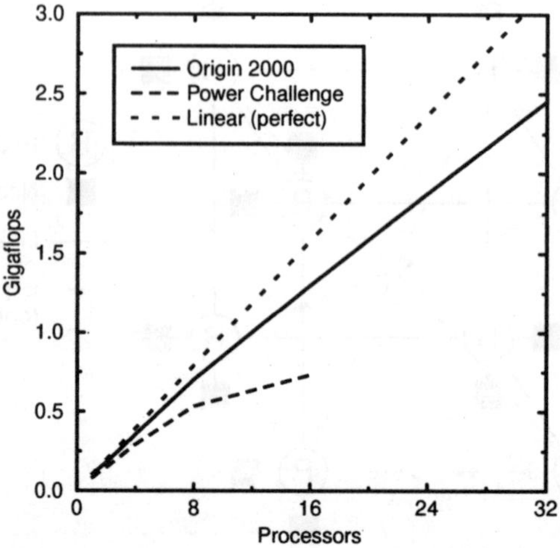

Figure 8: A comparison of the performance and scalability of SC-MICOM on the Power Challenge and Origin2000 systems.

by the increased single processor performance of the MIPS R10000, which is about a factor of 10 faster than the T3D processor when running MICOM. Though we don't expect another factor of 10 increase in the next two years, it is clear that the performance increases in each new generation of microprocessors means that they will be competitive with other supercomputer architectures of the future.

4 Related work

Aggressive optimization of general ocean circulation codes has a long history due to the extraordinary computational demands in fully eddy-resolving regimes. The original Bryan-Cox GFDL models were structured so that computations were performed on 2D slabs (either horizontal or vertical) in main memory while the full 3D state of the models was kept out on disk or on some other secondary store (such as Cray's SSD). Semtner and Chervin used this kind of approach to achieve the first eddy-resolving global calculation [16, 17]. Semtner provides a colorful and fascinating history of ocean modeling from its beginnings in the 1960s up to today [18].

Related work in SMP cluster programming includes a team at CSRD [20, 21] who developed a cluster version of the GFDL Bryan-Cox ocean model for the CEDAR architecture, an early SMP cluster architecture built at the University of Illinois. Unlike recent SMP clusters designs, CEDAR multiprocessors (constructed from Alliant FX/8 machines) had access to a shared global memory across a multistage network through which messages and data could be passed. The CEDAR version of the GFDL model executed 2D vertical slabs within a single Alliant machine, executing one dimension concurrently. The 4 Alliant machines in CEDAR would then execute multiple slabs concurrently; alternative partitionings that were either in the east-west, north-south, or both dimensions were tried with the latter approach (2D decomposition across the 4 Alliant's) yielding the best results. In their approach the CEDAR global memory acts like secondary storage, holding the 3D state of the model. The CEDAR version of the GFDL model uses two levels of parallelism but unlike SC-MICOM its data decomposition is not self-similar (i.e., the same both within and across SMPs). We believe that a self-similar decomposition exploits more locality and parallelism, especially within an SMP, and that this will be important in the most common supercomputing SMP cluster configurations: small number of SMP nodes, each with many processors.

We were fortunate to work closely with Woodward's team as they pioneered self-similar data decompositions for SMP clusters on PPM gas dynamics applications on the SGI Power Challenge array so that we could learn from both their successes and failures [13]. It should be noted that Woodward's early successes using the SMP cluster architecture for large hydrodynamics calculations led the U.S. Department of Energy to use PPM as the benchmark for the ASCI (Advanced Strategic Computing Initiative) Blue program procurement.

Many other groups are developing parallel ocean models. A good survey by Ashworth of the state-of-the-art as of 1992 is found in [22]. Ashworth also developed a simple parallel shallow-sea model to benchmark several parallel machines. Wait and Harding [23] also parallelized a shallow-sea model on a Transputer array using overlapped fake zones to simplify programming (but not to reduce communication frequency) and overlapping between communication and computation. Smith, Dukowicz, and Malone [19] with help from C. Kerr at GFDL developed a parallel version of the Semtner-Chervin ocean model called POP (Parallel Ocean Program) for the CM-2 and CM-5. POP provided a mechanism for executing on both data parallel and message-passing architectures. Smith, Dukowicz, and Malone also worked to improve the barotropic solver in POP which had become a significant bottleneck on the CM-2 and CM-5 MPP architectures. Though data parallel versions of POP and MICOM were developed for the Connection Machine architecture [19, 4], performance problems with the compilers combined with a lack of flexibility and machines supporting the data parallel programming paradigm has resulted in its demise within the computational oceanography community.

Recent work in parallel ocean modeling reported at the 1994 ECMWF conference and elsewhere [9, 10, 24, 25, 26] show that most investigators are using message-passing and two-dimensional decompositions, resulting in parallel columns of seawater being mapped directly to processors or threads. Oberhuber [25] reports impressive single-node performance and very good speedups in the parallel version of his isopycnal model.

5 Conclusions and Future Work

Though sophisticated and therefore complex, SC-MICOM has been structured so that the APIs for the physics and dynamics routines and the model timestep are simple and consistent. These routines can be written with little knowledge of the underlying parallel design of the code. We believe this code design paradigm and programming model can be applied to other parallel models and that it represents a good compromise between code complexity, efficiency and portability.

We believe that SMP clusters and DSM microprocessor-based hardware will provide cost-effective, scalable, user-friendly parallel hardware platforms that will initially compete with but ultimately surpass parallel vector architectures. For scientific codes, single node microprocessor performance is increasing by factors of 4-10 every 3 years. Given appropriate code designs, the SMP advantages in load balancing and efficiency means that they will overtake massively parallel message-based architectures.

Future work on SC-MICOM will include additional performance optimizations for the SGI Origin and other DSM architectures, such as reduction of false sharing. We are attempting to identify performance-critical code features as they relate to the architecture, including the relationship between subdomain size and secondary cache size. As with MP-MICOM, the SC-MICOM code will be transitioned to production mode with the help of the MICOM team at the University of Miami.

References

[1] Matthew O'Keefe, Terence Parr, B. Kevin Edgar, Steve Anderson, Paul Woodward and Henry Dietz, "The Fortran-P Translator: Towards Automatic Translation of Fortran 77 Programs for Massively Parallel Processors," *The Journal of Scientific Programming*, vol. 4, no. 1, pp. 1-22, Spring 1995.

[2] Aaron Sawdey, "Using the Parallel MICOM on the Cray T3D and SGI Challenge Multiprocessor," Technical Report, University of Minnesota, 1994. Available on the WWW at: `http://www-mount.ee.umn.edu/~okeefe/micom`

[3] P. Woodward, "Interactive Scientific Visualization of Fluid Flow," *IEEE Computer*, vol. 26, no. 10, pp. 13-26, October 1993.

[4] Rainer Bleck, Sumner Dean, Matthew O'Keefe and Aaron Sawdey, "A Comparison of Data-Parallel and Message-Passing Versions of the Miami Isopycnic

Coordinate Ocean Model," *Parallel Computing*, vol. 21, no. 10, pp. 1695-1720, 1995.

[5] David Dent, Lars Isaksen, George Mozdzynski, Matthew T. O'Keefe, Guy Robinson and Fritz Wollenweber, "Performance Measurements of the ECMWF Integrated Forecast System," Proceedings of the Sixth ECMWF Workshop on the Use of Parallel Processors in Meteorology, Reading, England, November 1994. Proceedings published by World Scientific Publishers (Singapore) in **Coming of Age**, edited by G-R. Hoffman and N. Kreitz, 1995.

[6] Kelvin K. Droegemeier, Ming Xue, Kenneth Johnson, Matthew T. O'Keefe, Aaron C. Sawdey, Gary W. Sabot, Skef Wholey, Kim Mills, and Neng-Tan Lin, "Weather Prediction: A Scalable Storm-Scale Model," published in High Performance Computing, Addison-Wesley Publishers, ed. by G. Sabot, 1995.

[7] Matthew T. O'Keefe and Aaron C. Sawdey, "Translation Tools for High Performance Computing," Proceedings of the Les Houches Workshop on High Performance Computing in the Geosciences, Les Houches, France, June 1993. Proceedings published by Kluwer Academic Publishers in **High Performance Computing in the Geosciences**, ed. by Francois X. Le Dimet, 1995.

[8] Aaron C. Sawdey and Matthew T. O'Keefe, "A Software-level Cray T3D Emulation Package for SGI Shared-memory Multiprocessor Systems". This software and reference are available on the WWW at:
`http://www-mount.ee.umn.edu/mcerg/T3Dem.html`

[9] Alan Wallcraft, "The Navy Layered Ocean Model," presentation at the First International Workshop on Software Engineering and Code Design in Parallel Meteorological and Oceanographic Applications, Inn at Semi-ah-moo, Blaine, WA, September 16-19, 1996.

[10] A. Wallcraft, D. R. Moore, "A Scalable Implementation of the NRL Layered Ocean Model," Tech. Report NRL CR 7323–96-0005, 1996.

[11] Aaron Sawdey and Matthew O'Keefe, "A Modular and Parallel Isopycnic Ocean Model," presentation at the First International Workshop on Software Engineering and Code Design in Parallel Meteorological and Oceanographic Applications, Inn at Semi-ah-moo, Blaine, WA, September 16-19, 1996.

[12] Robert Numrich, "F–: A Parallel Fortran 90," presentation at the First International Workshop on Software Engineering and Code Design in Parallel Meteorological and Oceanographic Applications, Inn at Semi-ah-moo, Blaine, WA, September 16-19, 1996. Also available as a technical report from Cray Research, Eagan, MN.

[13] P. R. Woodward, "Perspectives on Supercomputing: Three Decades of Change," *IEEE Computer*, invited paper, October 1996 (special 50th anniversary issue), in press.

[14] Robert Numrich, "F–: A Parallel Fortran Language," technical report, Cray Research, Eagan, MN, April 1994. To be published in *The Journal of Scientific Programming*, 1996.

[15] John Hennessy and David Patterson, **Computer Architecture: a Quantitative Approach**, Morgan Kaufmann Publishers, San Francisco, CA, 1996, p. 669.

[16] A. J. Semtner and R. M Chervin, "A Simulation of the Global Ocean Circulation with Resolved Eddies," *J. Geophys. Res.*, vol. 93, pp. 15502-15522.

[17] A. J. Semtner and R. M Chervin, "An Ocean Modeling System for Supercomputer Architectures of the 1990s," in Climate-Ocean Interaction, M.E. Schlesinger, ed., Kluwer Academic Publishers, Dordrecht, The Netherlands, 1990, pp. 87-95.

[18] A. J. Semtner, "Modeling Ocean Circulation," Science, v. 269, #5231, 1995.

[19] R. Smith, J. Dukowicz, and R. Malone, "Massively Parallel Global Ocean Modeling," Tech. Rep. LA-UR-91-2583, Los Alamos National Laboratory, Los Alamos, NM, 1991.

[20] L. DeRose, K. Gallivan, E. Gallopolous, "Experiments with an Ocean Circulation Model on CEDAR," CSRD Report No. 1200, Center for Supercomputing Research and Development, University of Illinois at Urbana-Champaign, February 1992.

[21] L. DeRose, K. Gallivan, and E. Gallopolous, "Status Report: Parallel Ocean Circulation on CEDAR," **Parallel Supercomputing in Atmospheric Science**, G.-R. Hoffman and T. Kauranne eds., World Scientific, Singapore, 1993, pp. 157-172. Paper appeared in Fifth ECMWF Workshop in Use of Multiprocessors in Meteorology, European Centre for Medium Range Weather Forecasts, Nov. 1992.

[22] M. Ashworth, "Parallel Processing in Environmental Modeling," *ibid*, pp. 1-25.

[23] R. Wait and T. J. Harding, "Numerical Software for 3D Hydrodynamic Modeling using Transputer Arrays," *ibid*, pp. 453-464.

[24] Aaron C. Sawdey, Matthew T. O'Keefe, Rainer Bleck, and Robert W. Numrich, "The Design, Implementation, and Performance of a Parallel Ocean Circulation Model," Proceedings of the Sixth ECMWF Workshop on the Use of Parallel Processors in Meteorology, Reading, England, November 1994. Proceedings published by World Scientific Publishers (Singapore) in **Coming of Age**, edited by G-R. Hoffman and N. Kreitz, 1995, pp. 523-550.

[25] J. Oberhuber and K. Ketelsen, "Parallelization of an OGCM on the Cray T3D," *ibid*, pp. 494-504.

[26] C. Gwilliam, "The OCCAM Global Ocean Model," *ibid*, pp. 447-454.

[27] B. Rodriguez, L. Hart, and T. Henderson, "A Library for Portable Parallelization of Operational Weather Forecast Models," *ibid*, pp. 148-161.

[28] T. Henderson, C. Baillie, et al., "Progress Toward Demonstrating the Operational Capability of Massively Parallel Processors at the Forecast Systems Laboratory," *ibid*, pp. 162-176.

THE LOS ALAMOS PARALLEL OCEAN PROGRAM (POP) AND COUPLED MODEL ON MPP AND CLUSTERED SMP ARCHITECTURES

PHILIP W. JONES
Theoretical Division, Los Alamos National Laboratory,
Los Alamos, NM 87545, USA

The Los Alamos Parallel Ocean Program has been developed and used on a variety of parallel computers. Implementation and portability of the model is described and performance results are presented. Recent improvements for efficient performance on cache-based microprocessors and SMP architectures are also described. Efforts to couple POP with a new sea ice model and atmospheric general circulation models on parallel computers are discussed.

1 Introduction

The Parallel Ocean Program (POP) is an ocean general circulation model developed specifically for use on parallel computers. With the the large memory and processing power of massively-parallel computers, POP has been used to simulate ocean circulation at very high resolution. Such eddy-resolving simulations have been very successful at reproducing many aspects of the global ocean circulation [1]. In addition, POP is being coupled to a new sea ice model and existing atmospheric models to examine the coupled climate system with an eddy-resolving ocean model.

This paper will describe the POP model and recent efforts to couple it with sea ice, atmospheric and land surface models. Because detailed descriptions of model physics and numerics have been published elsewhere, only a brief description will be included here to provide background. The following sections will describe the implementation of POP and the coupled model on parallel computers with emphasis on recent architectures and performance on cache-based microprocessors.

2 POP Model

2.1 Physics and Numerics

The POP model was developed by Smith, Dukowicz and Malone [2] based on earlier models by Bryan [3], Cox [4], Semtner [5] and Chervin [6]. The code integrates the primitive equations using a B-grid for the horizontal discretization and depth (z) as the vertical coordinate. The primitive equations are split into baroclinic and barotropic modes with the baroclinic equations advanced in time using a leap-frog scheme.

A major improvement in POP over the original Bryan-Cox formulation is the reformulation of the barotropic equations to solve for surface pressure rather than the streamfunction [2]. The surface-pressure formulation eliminates the need for a line integral around island boundaries and is thus far more efficient on parallel computers. Such a reformulation also allows any number of islands at virtually no additional computational cost and permits much more realistic bottom topography. Another change introduced in POP is the elimination of the rigid-lid approximation in favor of an implicit free-surface method for the barotropic mode [7]. This change permits a much faster convergence for the iterative solution of the barotropic mode. In addition, the free-surface method computes the surface height directly and solutions can be compared easily with altimetry data [8].

The horizontal grid in POP can be an arbitrary logically-rectangular grid. Most recently, the displaced-pole grids of Smith *et al.* [9] (see Fig. 1) have been used. Such grids can be used for simulations that include the polar regions without the need for Fourier filtering or complicated data structures. As a beneficial side-effect, displaced-pole grids with the pole displaced into the North American continent have higher resolution in the North Atlantic where many important physical and dynamic processes occur.

A variety of parameterizations are available within the POP model for sub-grid-scale physical processes. The Gent-McWilliams parameterization [10] for mixing along isopycnal surfaces is included as well as the vertical mixing schemes of Pacanowski and Philander [11] and Mellor and Yamada [12]. The KPP parameterization [13] for vertical mixing is currently being installed and tested. Desired parameterizations are chosen at compile time through preprocessor switches.

Ocean circulation in POP can be forced either by restoring to climatology or by using atmospheric fluxes. The latter is important for coupling to other component models in a fully-coupled climate system. Forcing using a variety of fluxes and methods have been fully examined with the model for standard cases [14]. Sea ice is not simulated directly in the POP model. A new sea ice model has been developed for use with POP (see Section 3).

2.2 Code Design

From the beginning, POP was designed to be portable among a variety of machines and programming models. In particular, POP currently supports the data-parallel and message passing programming models for use on machines

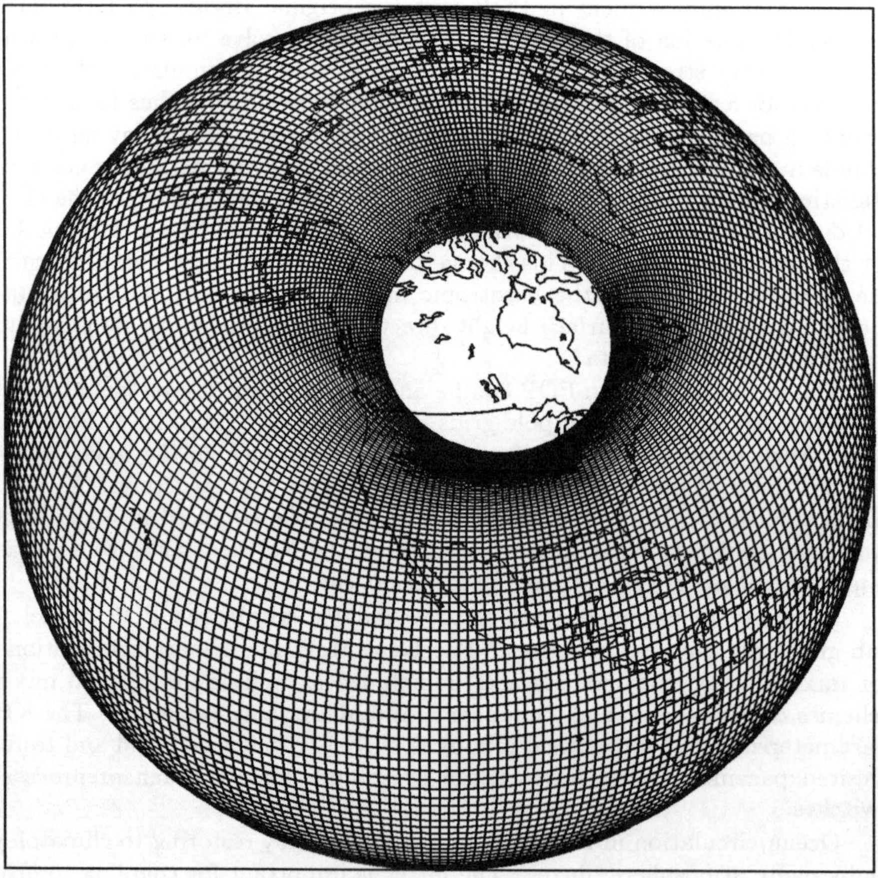

Figure 1: An example of a displaced pole grid used in POP.

ranging from parallel vector processors to massively-parallel processors. Such portability is achieved by isolating routines that depend on the architecture or programming model and implementing these routines as libraries optimized for each architecture. The number of special routines is quite small and includes only routines necessary for communication, I/O and timers. All communications are isolated in a set of stencil routines and global reduction routines. For a data-parallel machine like the Connection Machine, these routines are wrappers around the Fortran 90 intrinsics like CSHIFT, SUM and MAXVAL. On other machines, message-passing and shared-memory versions are available. Aside from these few routines, the remainder of the code is identical for all programming models and architectures. For all programming models and machines with distributed memory, the domain is block-decomposed in both horizontal directions with the vertical dimension always declared local to the processor.

Because a data-parallel programming model is supported, the code makes extensive use of array syntax and array constructs. High Performance Fortran (HPF) directives for data distribution are also included. Array syntax often requires allocation of conformable arrays. In order to reduce memory use, only essential arrays are three-dimensional; there is an outer loop over the vertical coordinate in the baroclinic solver so that any allocated temporary arrays need only be two-dimensional. The baroclinic routines then work over one horizontal slice through the domain at a time. Array syntax carries performance penalties on machines without mature optimizing Fortran 90 compilers. We have found on most machines that WHERE constructs are less efficient than equivalent MERGE functions. An extreme example occurs on Cray parallel vector machines where the compiler is able to translate the MERGE function to the highly-optimized conditional vector merge, while the equivalent WHERE construct is not so translated. Replacing the WHERE constructs with MERGE functions resulted in a factor of two improvement locally and a 15% improvement in the total model CPU time on these machines. Some compilers still do not provide optimized code for Fortran 90 intrinsics like CSHIFT. Because the CSHIFT only appears in the machine-specific stencil routines, they can be easily replaced with a Fortran 77 equivalent, resulting in another 12% improvement in the total CPU time on some machines.

Parallel computer manufacturers in the United States are concentrating on building parallel machines by linking commodity microprocessors together. These microprocessors make use of a memory cache to compensate for reduced memory bandwidth compared to vector-processing machines. Performance of POP on such machines is limited by the memory bandwidth and we have recently begun to make improvements that will reduce the memory traffic and improve cache use. For example, the leap-frog scheme requires three time levels for each prognostic variable. In the original POP code, these three levels were stored in separate arrays and at the end of a time step, several array copies were necessary to place data into the appropriate time level array. This construct created unnecessary data motion and was eliminated by adding a time dimension to all prognostic variable arrays and simply changing the time index to point to the appropriate time slice. On cache-based machines, this resulted in a 15% improvement in the total CPU time.

Another example of coding style which can negatively affect performance on cache-based microprocessors is shown in the following piece of code from an advection routine in POP:

```
This piece computed earlier:

UUE = e_shift(UUW)
VUN = n_shift(VUS)

The advection routine contains:

FC  = 0.5/UAREA
FVN =  VUN*FC
FVS = -VUS*FC
FUE =  UUE*FC
FUW = -UUW*FC

FC = FVN + FVS + FUE + FUW

call fivept(ADVUK,FC,FVN,FVS,FUE,FUW,UUU)
call fivept(ADVVK,FC,FVN,FVS,FUE,FUW,VVV)

The five-point stencil looks roughly like:

subroutine fivept(AOUT, CC, CN, CS, CE, CW, AIN)

AOUT = CC*AIN + &
       CN*CSHIFT(AIN, DIM=2, SHIFT=+1) + &
       CS*CSHIFT(AIN, DIM=2, SHIFT=-1) + &
       CE*CSHIFT(AIN, DIM=1, SHIFT=+1) + &
       CW*CSHIFT(AIN, DIM=1, SHIFT=-1)

end subroutine fivept
```

This form of the code requires the creation of five coefficient arrays, some of which are simply shifted forms or sums of other coefficients. This coding style is fine for vector-based machines where storing of intermediate results and creating coefficient arrays does not carry the same performance penalty as on cache-based machines where memory bandwidth is limited. An optimized loop form of the same code might look like:

```
do j=2,jmt-1
   do i=2,imt-1
      fcscalar = VUS(i,j+1) - VUS(i,j) + &
                 UUW(i+1,j) - UUW(i,j)

      ADVUK(i,j) = 0.5*( fcscalar*UUU(i  ,j  ) + &
                         VUS(i,j+1)*UUU(i  ,j+1) - &
                         VUS(i,j  )*UUU(i  ,j-1) + &
                         UUW(i+1,j)*UUU(i+1,j  ) - &
                         UUW(i  ,j)*UUU(i-1,j  ))/UAREA(i,j)
      ADVVK(i,j) = 0.5*( fcscalar*VVV(i  ,j  ) + &
                         VUS(i,j+1)*VVV(i  ,j+1) - &
                         VUS(i,j  )*VVV(i  ,j-1) + &
                         UUW(i+1,j)*VVV(i+1,j  ) - &
                         UUW(i  ,j)*VVV(i-1,j  ))/UAREA(i,j)

   end do
end do
```

with some additional work to take care of the boundaries. In the latter case, there are only two coefficient arrays required and the ratio of load/store operations to arithmetic operations is much smaller. The second version is much more efficient on cache-based machines and can be as much as five times faster, but does not encapsulate the communications and is more difficult to support in a portable code. A compromise between the two versions is to support more specific forms of stencils. For example, a five-point stencil that required only two coefficient arrays would reduce the memory traffic at the expense of supporting a slightly larger stencil library. This alternative form, combined with compilers which can do procedure in-lining or inter-procedural analysis, would greatly benefit POP performance when memory-bandwidth is limited.

In addition to an increased use of cache-based microprocessors, computer vendors are beginning to offer networked clusters of symmetric multi-processor (SMP) nodes. Several programming models can be used in such a system. One possibility is to use a message-passing model between SMP nodes and multi-processing directives or explicit thread programming for shared-memory processors within a node. The current approach for POP is to use a message-passing model for all processors and rely on the vendor to supply a message passing library that can distinguish between shared and distributed memory. This approach avoids mixing programming models in a single source code, but relies on a vendor to supply an efficient implementation of a message-passing library. A result of this approach is shown in the next section.

2.3 Performance

The POP model runs successfully on a variety of machines. On parallel vector processors, the model is run in data-parallel mode and achieves approximately 50% of peak performance (450 Mflops on a single Cray C90 processor). Using auto-tasking, multi-processor performance exhibits a speed-up of 12.6 on 16 processors of a Cray C90. Auto-tasking is less efficient for small problems where there is not enough work to amortize the cost of auto-tasking.

Most of the production runs for POP simulations have utilized Thinking Machines computers (CM-200 and CM-5). The CM-5 uses vector processors on each node and the nodes are connected by a network that is relatively slow compared to other distributed-memory machines. Performance on the CM-5 is between 10-15 Mflops per node for large problems where the vector units are used efficiently and the ratio of computation to communication is high. As the problem size is reduced, the ratio of computation to communication decreases and performance suffers due to the relatively slow network.

The Cray T3D is in many ways opposite to the CM-5. On the T3D, the communications network is quite fast, but the processors suffer from poor memory bandwidth and a very small cache. The T3D therefore performs much better for small problems that fit better in cache; the fast network can support the increased ratio of communications to computation, particularly when using the highly-efficient shared-memory communications (SHMEM) library. Figure 2 shows the performance for several different horizontal resolutions of the POP model on a Cray T3D. For all of these runs, 20 vertical levels were used. The dashed lines in the figure are results from an older version of POP which used the SHMEM library for message-passing. The solid lines are results from the latest version of POP which includes improvements for cache performance described in the previous section, but used the MPICH MPI library for message-passing. The relatively poor scaling using the MPICH library is immediately evident, particularly for large processor numbers where there is not enough work on processor to mask the MPI overhead. For low processor numbers, the total time is dominated by node CPU time and the improvements for cache performance have resulted in a substantial reduction in CPU time for the new version compared to the older version. The latest code achieves between 8-10 Mflops per node. An implementation of the new version with the SHMEM calls will result in improved scaling.

Figure 3 shows timing results for a series of simulations on the recently-announced SGI/Cray Origin 2000 machine. This is a distributed-shared--memory (DSM) machine with a cache-coherent shared memory for up to 128 processors. In our initial tests with this machine, we have chosen not to use the machine as a shared-memory machine, but instead have used the message-passing version of the POP model and relied on the shared-memory implementation of MPI for communications between processors. The scaling with processor number is quite poor. Also, simulations which have identical subgrids (*e.g.* 128x64 on 4 nodes and 256x128 on 16 nodes) actually take longer to complete at higher processor number. The current MPI implementation on shared memory appears to add a communications cost which increases with processor number. A future version of MPI is planned which will eliminate some of this cost. The scaling with problem size is also relatively poor. For small problems where the model fits into the 4 Mbyte cache, the model achieves 58 Mflops/node (out of a theoretical peak of 400 Mflops/node). For larger problems, the performance drops to 30 Mflops/node. This poor scaling is due to high memory traffic in the POP model. In POP, the ratio of loads and stores to floating point operations is greater than two so the code is clearly memory-bandwidth limited. Further work to reduce this ratio as discussed in the previous section will be necessary.

3 Coupled Model

In order to simulate the Earth's climate system, the POP model must be coupled to atmosphere and land surface models. As mentioned in Section 2.1, a sea ice model must also be coupled to the ocean model. A new sea ice model by Hunke and Dukowicz [15] has been developed for use with POP. This sea ice model uses an elastic-viscous-plastic model for sea ice dynamics which results in a fully-explicit in time ice model ideally-suited for parallel computers. We have chosen to use the CCM3 atmospheric model [16] from the U.S. National Center for Atmospheric Research (NCAR) for the atmosphere and land surface models in our coupled climate system.

For coupling each of the component models together, we have adopted the NCAR Flux Coupler framework [17]. In this framework, each component model runs autonomously while periodically passing information to the flux coupler. The flux coupler is responsible for interpolating the information to

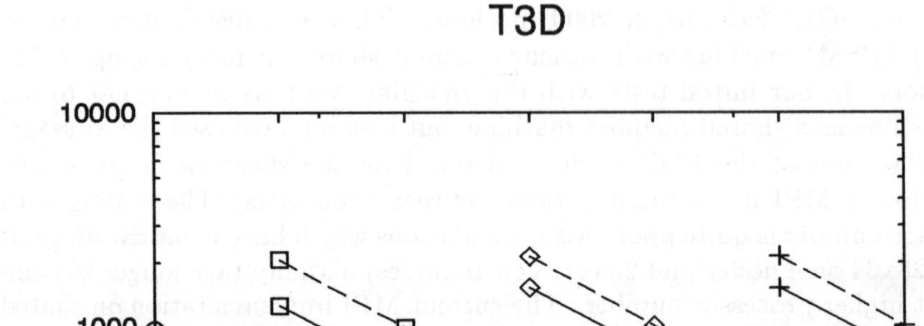

Figure 2: Performance of POP on a Cray T3D. Different horizontal resolutions are denoted by different symbols. CPU time is the measured time for 100 time steps. Dashed lines are results from an older version of POP using SHMEM for message-passing. Solid lines are results from a newer version using the MPICH MPI library for message passing. The dashed line with no symbols shows an ideal scaling for linear speedup.

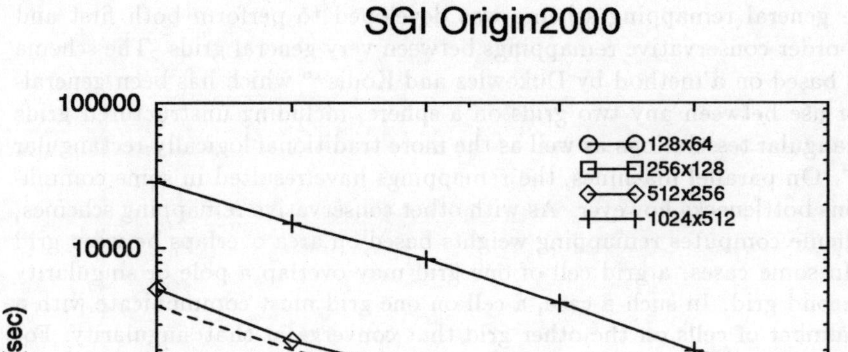

Figure 3: Performance of the message-passing version of POP on a SGI/Cray Origin 2000. CPU time is the measured time for 100 time steps. The dashed line shows an ideal scaling for linear speedup.

the finest grid (typically the ocean), computing surface fluxes, and passing the information back to each component on the appropriate grid. The flux coupler provides a message-passing interface to a communications library for passing information between individual component models.

Because the POP ocean grid uses the displaced pole grid shown in Figure 1, a more general remapping scheme was developed to perform both first and second-order conservative remappings between very general grids. The scheme used is based on a method by Dukowicz and Kodis [18] which has been generalized for use between any two grids on a sphere, including unstructured grids and triangular tesselations as well as the more traditional logically-rectangular grids [19]. On parallel machines, the remappings have resulted in some communications bottlenecks however. As with other conservative remapping schemes, this scheme computes remapping weights based on area overlaps between grid cells. In some cases, a grid cell of one grid may overlap a pole or singularity on a second grid. In such a case, a cell on one grid must communicate with a large number of cells on the other grid that converge at that singularity. For example, the ocean grid shown in Figure 1 has one grid cell overlapping the north pole. If we couple to a T42 atmosphere grid, the ocean cell overlapping the pole must communicate with all 128 grid points of the atmosphere grid that converge at the north pole. Elsewhere, ocean grid cells more typically communicate with four or fewer atmosphere cells. This creates a severe load imbalance if a block decomposition for each component model grid is used. Methods for mitigating this problem include different data decompositions or running the flux coupler on a single processor (or set of processors with shared memory) where distributed memory does not create such a bottleneck.

Another difficulty with the flux coupler framework occurs when all the components are run in the same partition of a given machine. In some cases, a time-sharing daemon may be giving a time slice to a component model even if the component model is simply waiting for a message from the flux coupler. This results in severe inefficiencies in the use of the computing resource as up to 60% of the cycles are idle. Because each model is autonomous, there is no method for letting the operating system know that a model is simply waiting. A more efficient method is to run each model in a different partition of a parallel machine. This allows the number of processors allocated to each component model to be adjusted for the best load balance between models. Such an allocation may also require changes in administrative policies to allow multiple jobs to be submitted by a single user and to allow communication between partitions of a parallel machine.

Despite the above difficulties, the POP model has been successfully coupled with the NCAR CCM model and run for a period of one year. As a result of this initial success, we are currently spinning up the POP model using CCM3 fluxes in preparation for a production run of the fully-coupled climate system.

4 Future Improvements

The POP model has been a very useful tool for studying ocean circulation at eddy-resolving resolution. Future changes are planned to further improve both the model itself and its performance on parallel computers.

Currently, new advection schemes are being tested for use in POP. Also, a generalized vertical coordinate is planned. An Arbitrary Lagrangian-Eulerian (ALE) [20] coordinate is being examined that would allow a Lagrangian vertical coordinate in the deep ocean similar to isopycnal coordinates, while simultaneously allowing Eulerian coordinates near the surface where layers are well-mixed.

Performance of the POP model on cache-based microprocessors is disappointing and continued work to reduce memory traffic is necessary. This may require a mixture of explicit thread-based programming and traditional message passing. We will also continue a gradual addition of advanced Fortran 90 features as they become more generally available.

Acknowledgments

The author would like to thank John Dukowicz, Elizabeth Hunke, Bob Malone, Mat Maltrud and Rick Smith for the opportunity to join them in the further development of POP. This work was supported by the Computer Hardware, Advanced Mathematics and Model Physics (CHAMMP) program of the U.S. Department of Energy. We gratefully acknowledge the Advanced Computing Laboratory at Los Alamos National Laboratory where most of the simulations have been carried out.

References

1. M.E. Maltrud, R.D. Smith, A.J. Semtner Jr. and R.C. Malone, in preparation.
2. R.D. Smith, J.K. Dukowicz and R.C. Malone, *Physica D*, **60**, 38 (1992).
3. K. Bryan, *J. Comput. Phys.* **4**, 347 (1969).
4. M.D. Cox, *GFDL Ocean Group Tech. Rept. No. 1*, GFDL/NOAA, Princeton University, Princeton, NJ (1984).
5. A.J. Semtner Jr. in *Advanced Physical Oceanographical Modelling*, ed. J.J. O'Brien (Reidel, Dordrecht, 1988).

6. R.M. Chervin and A.J. Semtner Jr. in *Proc. of the NATO Advanced Research Workshop on Climate-Ocean Interaction*, ed. M. Schlesinger (Kluwer, Dordrecht, 1988).
7. J.K. Dukowicz and R.D. Smith, *J. Geophys. Res.* **99**, 7991 (1994).
8. L.-L. Fu and R.D. Smith, *Journ. Amer. Met. Soc.* (1996), in press.
9. R.D. Smith, S. Kortas and B. Melz, *J. Comput. Phys.*, submitted.
10. P.R. Gent and J.C. McWilliams, *J. Phys. Oceanogr.* **20**, 150 (1990).
11. R.C. Pacanowski and S.G.H. Philander, *J. Phys. Oceanogr.* **11**, 1443 (1981).
12. G.L. Mellor and T. Yamada, *Rev. Geophys.* **20**, 851 (1982).
13. W.G. Large, J.C. McWilliams and S.C. Doney, *Rev. Geophys.* **32**, 363 (1994).
14. M.E. Maltrud and C.A. Lai, private communication (1996).
15. E.C. Hunke and J.K. Dukowicz, *J. Phys. Oceanogr.*, submitted.
16. T.L. Acker, L.E. Buja, J.M. Rosinski and J.E. Truesdale, *NCAR Technical Note TN-421*, National Center for Atmospheric Research, Boulder, CO (1996).
17. F.O. Bryan, B.G. Kauffman, W.G. Large and P.R. Gent, *NCAR Technical Note TN-424*, National Center for Atmospheric Research, Boulder, CO (1996).
18. J.K. Dukowicz and J.W. Kodis, *SIAM J. Sci. Stat. Comput.*, **8**, 305 (1987).
19. P.W. Jones, in preparation.
20. C.W. Hirt, A.A. Amsden and J.L. Cook, *J. Comput. Phys.* **14**, 227 (1974).

A FEEDBACK BASED LOAD BALANCE ALGORITHM FOR PHYSICS ROUTINES IN NWP

J. M. BULL, R. W. FORD
Centre for Novel Computing, The University of Manchester, U.K.
and
A. DICKINSON
Meteorological Office, Bracknell, U.K.

ABSTRACT

In recent years, Numerical Weather Prediction (NWP) has made increasing use of parallel machines with large numbers of processors. The physics portion of NWP is particularly amenable to large scale parallelism as, in most cases, each grid column can be independently computed. However, grid columns have varying amounts of work associated with them, so, a simple partition of grid columns can leave some processors with much less work than others. As the number of processors grows, this effect becomes increasingly important. Such imbalance can be time invariant (such as computation based on orographic features), vary predictably in time (such as short wave radiation), or vary unpredictably in time (such as convection and precipitation). This paper presents a load balance algorithm which is particularly aimed at the last of these cases. The feedback mechanism uses the time taken and number of grid columns for each processor at the previous time-step, to provide an improved partition for the current time-step. Simulation results are presented both for synthetic workloads, and using data from ECMWF's IFS model.

1 Introduction

The existence of load imbalance in the physics phase of NWP models is well known. For example, in [4], Michalakes gives a detailed analysis of the load in all the physics routines in the NCAR Community Climate model. The drawbacks of using standard loop scheduling techniques such as Guided Self-Scheduling [5] were noted in [3]. The essential idea of such methods is to break the set of grid columns into *chunks*, where there are more chunks than processors. Once a processor completes a chunk, it is assigned a new chunk which has not yet been processed. On vector architectures, small chunks result in small vector lengths, thus losing performance. On message passing architectures, such schemes result in unpredictable communication patterns, possibly with short messages, which is also inefficient. Other load balancing techniques exist, but many of these are designed for situations where the number of (identical) parallel tasks required is unpredictable, rather than the time taken for a known number of tasks.

The load balancing algorithm presented here is an extension of the work described in [3]. The key observation is that the load on grid column tends to vary slowly in time. We can therefore use the load on the current time-step as a prediction of the load on the next, and split the grid columns into exactly p contiguous chunks, where p is the number of processors. Unfortunately, it is not practicable to measure the load on every grid column, and thus we use an approximation to the load instead.

In Section 2 we introduce the load balancing algorithm, and in Section 3 we define the metric we will use for measuring load imbalance. Section 4 gives the results of simulations based on a synthetic workload, and on data derived from an instrumented run of ECMWF's IFS model [1]. Since some models, such as the UK Meteorological Office's Unified Model [2], require processors to synchronise and exchange data in between physics routines, we address the problem of balancing individual routines, as well as the physics as a whole. Section 5 draws some conclusions and points the way to future work in this area.

2 Algorithm

We consider only a one-dimensional partitioning of grid columns. Such a partitioning is frequently used to maximise vector length. Suppose there are N grid columns and p processors. A contiguous partitioning on grid columns can be described simply by giving lower and upper loop bounds for each processor. On time-step t we suppose that processor j, $j = 1, \ldots, p$ computes grid columns l_j^t to h_j^t, where $l_1^t = 1$, $h_p^t = N$ and $l_{j+1}^t = h_j^t + 1$, $j = 1, \ldots, p-1$. If the measured execution time for processor j on time-step t is T_j^t, then its mean load per grid column is

$$\frac{T_j^t}{h_j^t - l_j^t + 1}.$$

This gives us a piecewise constant approximation $\hat{w}_i^t$ to the true workload w_i^t on grid column i:

$$\hat{w}_i^t = \frac{T_j^t}{h_j^t - l_j^t + 1} \quad \text{if} \quad l_j^t \le i \le h_j^t, \ j = 1, \ldots, p.$$

We then choose new bounds for each processor to use for the next time-step, based on an (approximate) equipartition of this piecewise function:

$$\sum_{i=l_j^{t+1}}^{h_j^{t+1}} \hat{w}_i^t \approx \frac{1}{p} \sum_{i=1}^{N} \hat{w}_i^t = \frac{1}{p} \sum_{j=1}^{p} T_j^t, \ j = 1, \ldots, p.$$

To compute the equipartition, we treat i as a continuous variable, and find real values $\hat{l}_j^{t+1}$, such that

$$\int_{\hat{l}_j^{t+1}}^{\hat{l}_{j+1}^{t+1}} \hat{w}_i^t \, \mathrm{d}i = \frac{1}{p} \sum_{j=1}^{p} T_j^t, \; j = 1, \ldots, p-1,$$

and then set $l_j^{t+1} = [\hat{l}_j^{t+1}]$, $j = 2, \ldots, p$. By keeping a running total of the areas under each constant piece of $\hat{w}_i^t$, this can be achieved in $O(p)$ time. Figure 1 illustrates the algorithm for $p = 4$.

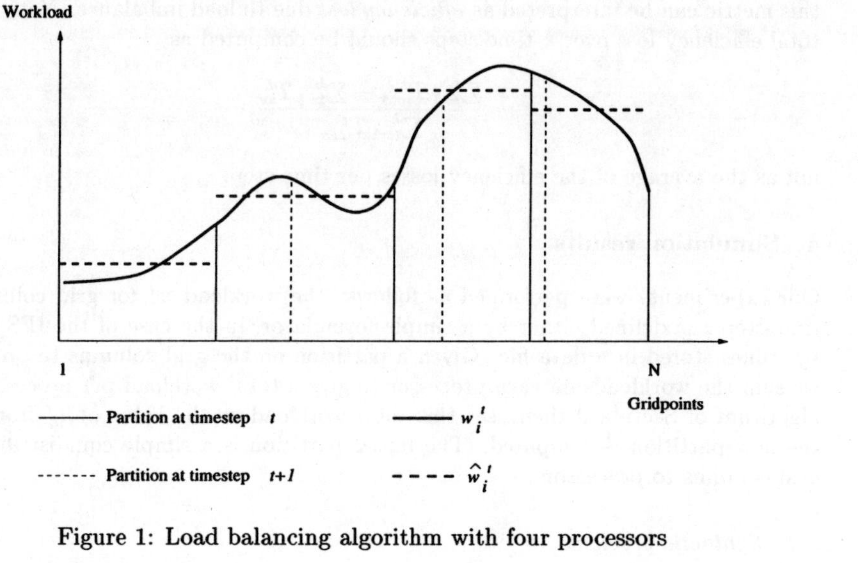

Figure 1: Load balancing algorithm with four processors

3 Load balance metrics

If we define

$$T_{\mathrm{av}}^t = \frac{1}{p} \sum_{j=1}^{p} T_j^t$$

and

$$T_{\max}^t = \max_{1 \le j \le p} T_j^t$$

then there are two obvious choices for a load balance metric:

$$\frac{T_{\max}^t - T_{\mathrm{av}}^t}{T_{\mathrm{av}}^t} \text{ and } \frac{T_{\max}^t - T_{\mathrm{av}}^t}{T_{\max}^t}.$$

The first of these is perhaps the more commonly used, but there are two reasons why the second, which we will use in the results of Section 4, is to be preferred. The second metric is lies in the interval $[0, 1]$, whereas the first can take values greater than 1. Furthermore, since pT_{av}^t is the sequential execution time (T_1), and $T_{\max}^t$ is the execution time on p processors (T_p), the second metric is equivalent to

$$1 - \frac{T_1}{pT_p}.$$

The second term in this expression is just the usual formula for parallel efficiency, so this metric can be interpreted as *efficiency loss* due to load imbalance. Note that the total efficiency loss over k time-steps should be computed as

$$\frac{\sum_{t=1}^{k} T_{\max}^t - \sum_{t=1}^{k} T_{\mathrm{av}}^t}{\sum_{t=1}^{k} T_{\max}^t},$$

not as the average of the efficiency losses per time-step.

4 Simulation results

Our experiments were performed as follows: the workload w_i^t for grid column i on time-step t is defined either by a simple formula or, in the case of the IFS physics, by values stored in a data file. Given a partition on the grid columns to processors, we sum the workloads on each processor to give a total workload per processor. The algorithm of Section 2 then uses this total workload as the value of T_j^t from which the new partition is computed. The initial partition is a simple equidistribution of grid columns to processors.

4.1 Synthetic problems

We present results for a synthetic workloads consisting of a Gaussian distribution whose midpoint oscillates sinusoidally, defined by

$$w_i^t = \exp - \left(\frac{i - (N/2 + A\sin(2\pi t/T))}{b} \right)^2$$

where N is the number of grid columns, A is the amplitude of the oscillation, T is the period of the oscillation and b determines the width of the Gaussian. In all cases we take $N = 10000$.

It is frequently instructive to apply dynamic load balancing algorithms to a static load. Figure 2 show the load balance evolving with successive time-steps for a stationary Gaussian ($A = 0$, $b = 2000$), for different numbers of processors. On two processors, the initial distribution is exactly balanced, by the symmetry of the workload. However, this situation is unstable, and load imbalance increases with successive

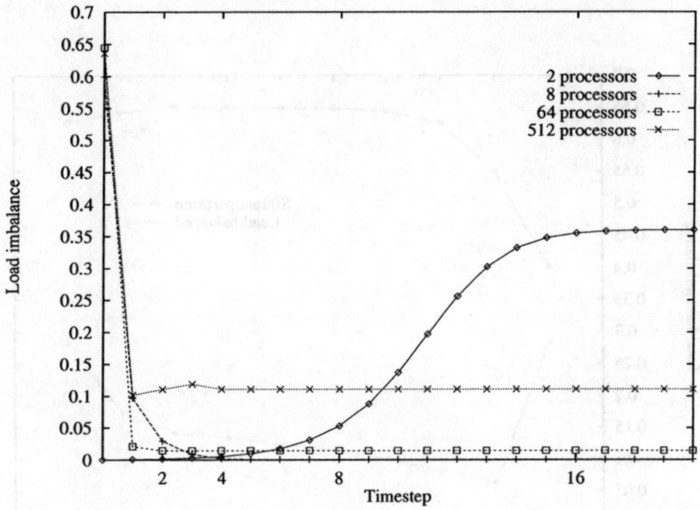

Figure 2: Stationary Gaussian, $b = 2000$

time-steps. This occurs because the algorithm has very little information (only two measurements) about the workload. For larger numbers of processors, the initial load imbalance is removed in the first four or five time-steps. For large numbers of processors, the requirement that each processor has a whole number of grid columns prevents the load imbalance from dropping below a certain level.

This fast response, gives us hope that the algorithm may be able to track dynamically changing loads. To test this, we apply the algorithm to the oscillating Gaussian. Figure 3 shows the total load imbalance against number of processors over one complete oscillation with $b = 2000$, $A = 2500$ and $T = 200$. For comparison, the load imbalance resulting from a simple partition of grid columns is shown. We see that apart from the case of two processors, significant decreases in load imbalance are achieved. As the number of processors increases, the load imbalance achieved by the algorithm increases, as the discretisation effects seen in the static case become more important.

We now consider the effect of changing the period of the oscillation. Figure 4 shows the total load imbalance against the period of oscillation over one complete oscillation with $b = 2000$, $A = 2500$ and $p = 64$. As we might expect, short period oscillations cannot be tracked well by the algorithm, but as the period increases, the load imbalance decreases. Nevertheless, useful decreases in execution time would be observed for periods as short as 50 time-steps.

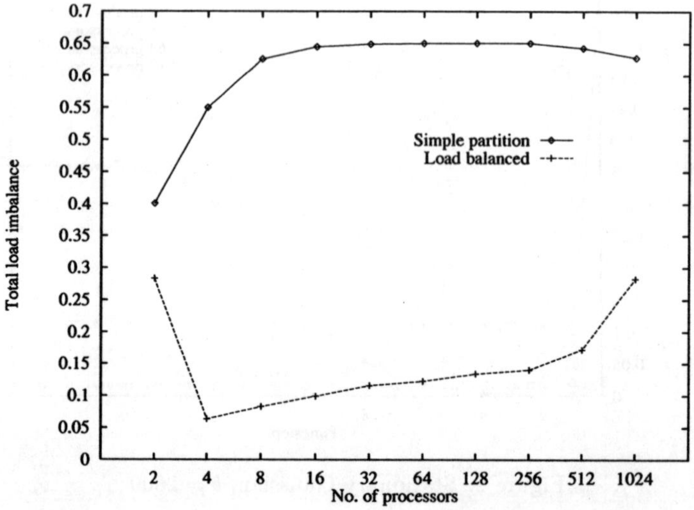

Figure 3: Oscillating Gaussian, $b = 2000$, $A = 2500$, $T = 200$.

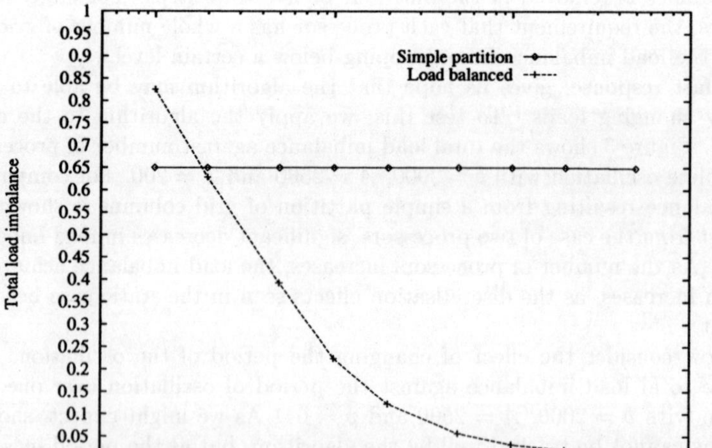

Figure 4: Oscillating Gaussian, $b = 2000$, $A = 2500$, $p = 64$.

Finally we can consider the effect of changing the width of the Gaussian. Figure 5 shows the total load imbalance against the width parameter b over one complete oscillation with $A = 2500$, $T = 200$ and $p = 64$. We observe that for both the simple

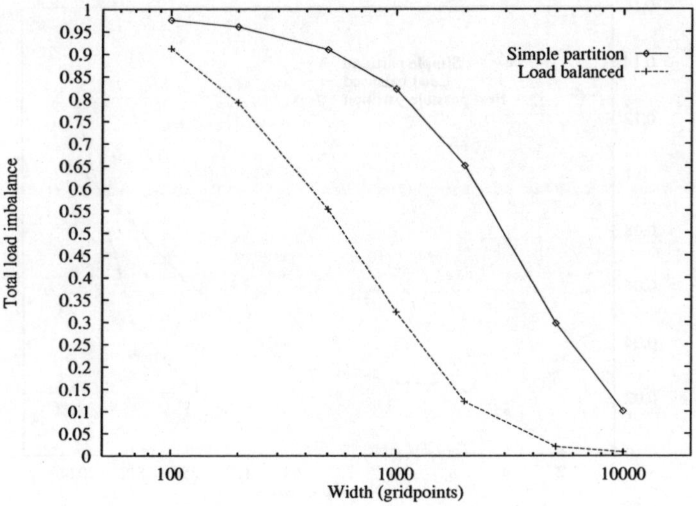

Figure 5: Oscillating Gaussian, $A = 2500$, $T = 200$, $p = 64$.

partition and the load balanced partition, load imbalance decreases as the Gaussian becomes wider. For all widths, however, the load balance algorithm would give useful reductions in parallel execution time.

4.2 IFS physics

The data used for the simulations in this section was obtained from a specially instrumented run of IFS at T63L19 resolution on a Cray C90. At this resolution there are 13280 grid columns, and the data consists of a time for each major physics routine for every grid column for each of 96 time-steps, corresponding to one day of forecast time. To avoid spin-up effects, the first forecast day is ignored and results are recorded for the second day. A complicating feature is that the very expensive radiation routines are only called every 12 time-steps. One possible solution is to use a different partition for the radiation routines than for the remainder of the physics. Indeed, if current plans to use a reduced grid for the radiation routines are implemented, this seems unavoidable. Our approach here, however, is to observe that the load variation in the radiation routines in both space and time is very small. We can therefore simply add a constant value to $\hat{w}_i^t$ if the subsequent time-step is a radiation step, and subtract it if the previous time-step was a radiation step, thus maintaining a single partition for the whole of the physics.

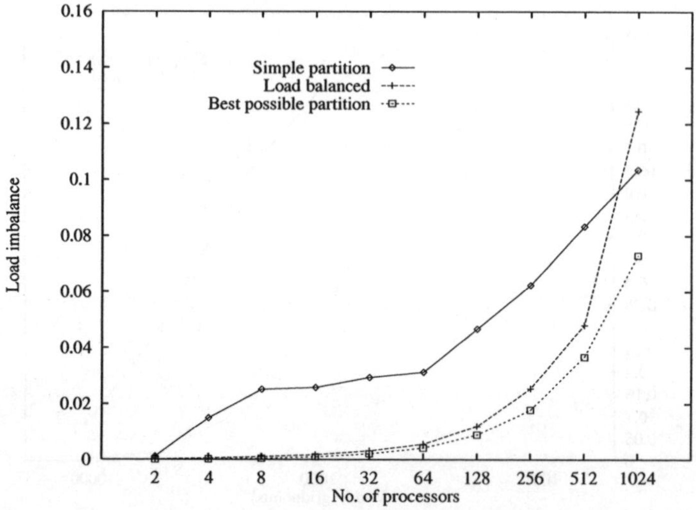

Figure 6: Load imbalance for IFS full physics.

Figure 6 shows the total load imbalance against number of processors over the 96 time-steps. For comparison, the load imbalance resulting from a simple partition of grid columns is shown, as is the best possible partition. This is the partition we would compute if we knew in advance the workload on every grid column for the next time-step.

Our first observation is that a simple partition results in quite small values of load imbalance. Nevertheless, we see that for small and medium numbers of processors, the load balancing algorithm comes close to the best possible solution. As the number of processors becomes large, the imbalance becomes noticeably larger than the best possible, and on 1024 processors, the algorithm cannot improve on the simple partition.

In order to simulate the effects of treating radiation separately, we can simply ignore the radiation routines, and attempt to balance the workload for the remainder of the physics. Figure 7 shows the total load imbalance against number of processors over the 96 time-steps, ignoring radiation. Without radiation, the overall load imbalance for the simple partition is higher. The load balance algorithm achieves values close to best possible, though again there is some degradation as the number of pro-

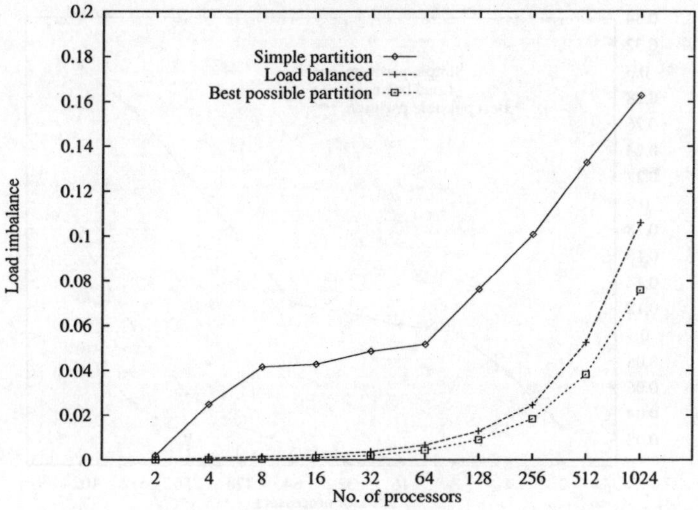

Figure 7: Load imbalance for IFS physics without radiation.

cessors becomes large, and hence the number of grid columns per processor becomes small. On 1024 processors, some processors have fewer than 10 grid columns.

Finally, since we are interested in forecast models where synchronisation between physics routines may be necessary, and in local area models, where load imbalance tends to be more severe than in global models, we consider balancing only the convection routine, which is the most significant source of unpredictable load imbalance in IFS. Figure 8 shows the total load imbalance against number of processors over the 96 time-steps for convection only. Efficiency losses for convection can be quite significant—over 30% on 1024 processors. Our load balancing algorithm gives significant reductions, though imbalances are not as close the optimum values as in they are in Figure 7. This is to be expected, as balancing convection alone is a more challenging problem, but for example on 256 processors an efficiency loss of 22% with a static partition is reduced to under 5.5%, compared to a best possible of just over 2%.

5 Conclusions and Future Work

We have presented a practical load balancing algorithm, based on measured execution time on a previous time-step, which has the potential to reduce load imbalance in the physics phase of NWP. We have presented simulation results based on data from IFS which show that useful reductions in execution time may be possible.

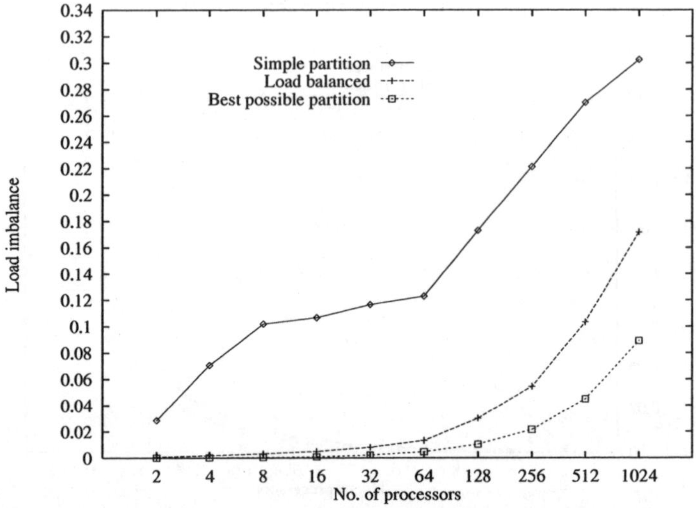

Figure 8: Load imbalance for IFS convection.

Future work will focus on implementing the algorithm in real models, and testing it under the more demanding conditions of local area forecasting. Practical issues, such as implementing the necessary non-uniform data distribution will be addressed. We will also extend the algorithm to two-dimensional partitioning of the iteration space, and explore other theoretical issues, such as using timing information from more than one time-step, modelling the cost of data redistribution, and using higher order functions to approximate the workload.

6 Acknowledgements

The authors would like to thank David Dent and Fritz Wollenweber for making available the data from the instrumented runs of IFS.

References

[1] Barros, S. R. M., D. Dent, L. Isaksen and G. Robinson (1994) *The IFS model—overview and parallel strategies* in Proc. of Sixth Workshop on the Use of Parallel Processors in Meteorology (G.-R. Hoffman and N. Kreitz, eds.), World Scientific, pp. 303–318.

[2] Bell, R. S. and A. Dickinson, (1990) *The Meteorological Office Unified Model for data assimilation, climate modelling and NWP and its implementation on a*

Cray Y-MP, in Proc. of Fourth Workshop on the Use of Parallel Processors in Meteorology, World Scientific.

[3] Ford, R. F., D. Snelling and A. Dickinson, (1994) *Controlling load balance, cache use and vector length in the UM*, in Proc. of Sixth Workshop on the Use of Parallel Processors in Meteorology (G.-R. Hoffman and N. Kreitz, eds.), World Scientific, pp. 195–205.

[4] Michalakes, J. G., (1991) *Analysis of workload and load balancing issues in the NCAR Community Climate Model*, Tech. Memo. ANL/MCS-TM-144, Argonne National Laboratory.

[5] Polychronopoulos, C. P. and D. J. Kuck, (1987) *Guided self-scheduling: a practical scheduling scheme for parallel supercomputers*, IEEE Trans. Comp. vol. C-36, no. 12, pp. 1425–1439.

Transitioning from MPP to SMP: Experiences with a Navier-Stokes solver

Chris N. Hill
Center for Meteorology and Physical Oceanography, Massachusetts Institute of Technology, Cambridge, MA 02139, USA. (e-mail: cnh@plume.mit.edu)

Andrew Shaw
Laboratory for Computer Science, Massachusetts Institute of Technology, Cambridge, MA 02139, USA. (e-mail: shaw@abp.lcs.mit.edu)

High-end, mainstream symmetric multiprocessor (SMP) computers are beginning to offer numerical modelers realistic alternatives to traditional vector or massively parallel processing (MPP) supercomputers. We report on the implementation of the MIT ocean-atmosphere dynamical kernel on a DEC 8400 SMP. The SMP implementation is contrasted with an existing optimized version of the same model on the CM-5 parallel vector supercomputer. We discuss the numerical methods and programming styles appropriate to the two machines. To characterize the differences in the architectures, we compare the performance of key kernels of the model which are typical of those in many computational fluid dynamics (CFD) codes. To render the model readily accessible to individual researchers and students, it is written in High Performance Fortran, which emphasizes both high performance and a high-level programming environment. Our study highlights cache memory architecture and compilation systems as particularly critical technologies in determining the overall utility of a mainstream SMP system for the oceanographic and meteorological community. Notwithstanding these caveats, we demonstrate that the performance of "high-end mainstream" SMP's is rapidly converging on that of "low-end" computer center facilities for real scientific applications.

1 Introduction

During the last twenty years, the application of supercomputers to the numerical study of fluids, in particular the atmosphere and ocean, has been an unquestionable success, paving the way for vast improvements in the fidelity of computational fluid simulations. Throughout this period, low-volume, high-end vector supercomputers have clearly led in absolute performance and vector processing has proved an indispensable tool for tackling the virtually insatiable demands of computational fluid dynamics (CFD). Recently, however, commodity microprocessor-based multiprocessors have reached a level of performance where they can be seriously considered as an affordable alternative to specialized, purpose-built, and hence generally expensive, vector supercomputers.

Today there is a widely articulated viewpoint that mainstream symmetric multi-processor (SMP) architectures, based on the same commodity micropro-

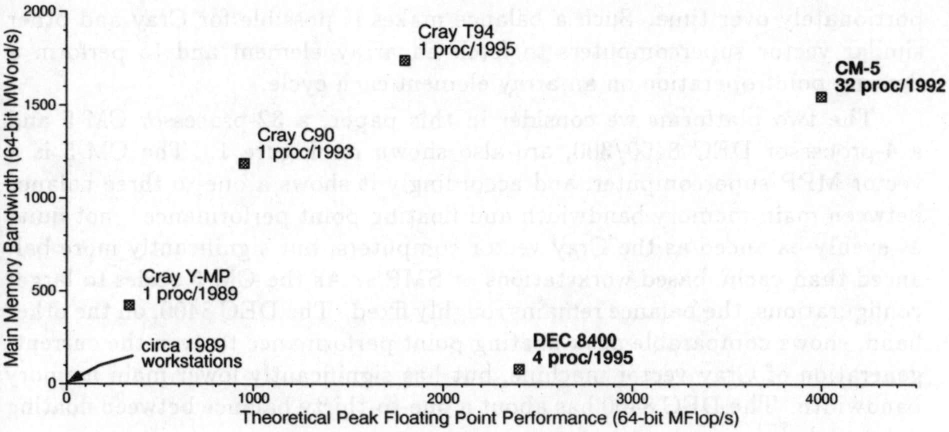

Figure 1: Peak floating point performance versus main memory bandwidth. Peak floating point performance is the maximum floating point operations which can be issued in a cycle multiplied by the cycles per second. Main memory bandwidth is measured using the STREAM benchmark.

cessor technology used in desktop workstations, will become the building blocks for a new generation of high-end, scalable supercomputer platforms. Computer architects now enthuse over the possibility of multiprocessor supercomputers based on commodity microprocessors delivering significant performance to numerical applications at a fraction of the cost of traditional vector supercomputers. Skeptics question the soundness of this conviction, arguing that vector processing is a natural paradigm for core supercomputer fields such as computational fluid dynamics, and that many legacy applications do not perform well on current generation commodity platforms[1,2]. The skeptic's argument is compelling. Current generation commodity microprocessor platforms all rely on hierarchical cache-based memories to give them high performance – however, this memory organization performs poorly with the vector-centric numerical procedures that lie at the heart of many supercomputer applications, including most meteorological and oceanographic applications.

Figure 1 illustrates one important architectural difference between vector, massively parallel processor (MPP), and mainstream workstation/SMP platforms. The Cray vector computers (Y-MP, C90, and T94) are a yardstick for the performance of vector supercomputers. They show a close to one-to-one balance between main memory bandwidth (as measured by the STREAM benchmark[2]) and peak floating point performance, with both increasing proportionately over time. Such a balance makes it possible for Cray and other similar vector supercomputers to fetch an array element and to perform a floating point operation on an array element each cycle.

The two platforms we consider in this paper, a 32-processor CM-5 and a 4-processor DEC 8400/300, are also shown on Figure 1. The CM-5 is a vector MPP supercomputer, and accordingly it shows a one-to-three balance between main memory bandwidth and floating point performance – not quite as evenly balanced as the Cray vector computers, but significantly more balanced than cache-based workstations or SMP's. As the CM-5 scales to larger configurations, the balance remains roughly fixed. The DEC 8400, on the other hand, shows comparable peak floating point performance to even the current-generation of Cray vector machine, but has significantly lower main memory bandwidth. The DEC 8400 has about a one-to-thirty balance between floating point and main memory performance.

Generally, oceanographic and atmospheric models, either grid-point or spectral, are relatively straightforward to map efficiently to architectures like the Crays, with their even balance between main memory bandwidth and floating point performance. Typically, the models use low-order finite difference schemes, which can be expressed naturally using vector-processor-friendly constructs with a roughly equal ratio of floating point operations to memory references. As Figure 1 shows, such constructs are not well-suited to the DEC 8400 or other cache-based microprocessor platforms, because the main memory bandwidth soon becomes the limiting factor.

Nevertheless, there are numerical codes for which the performance of a 4-processor DEC 8400 is comparable with a top-of-the-line vector platform. The LINPACK[3,4] benchmark rates a 4-processor DEC 8400/300 at 1351 MFlop/s and a 1-processor Cray T94 at 1603 MFlop/s, for the hand-optimized LINPACK 1000 test. A workstation of the same era as the Cray Y-MP, on the other hand, had both low floating point and low main memory performance, and would have been unsuitable for any form of large-scale computational fluid dynamics. The rapidly increasing performance of mainstream microprocessor computers, along with their relatively low costs, makes it tempting to investigate the role they might play in large-scale meteorological and oceanographic modeling, and to anticipate how that role might evolve in the future.

In this paper, we report on the design, implementation and evaluation of the hydro-dynamical kernel of the MIT ocean-atmosphere model on a mass-market SMP platform, the DEC 8400 5/300. We compare this implementation with an existing implementation on a CM-5 MPP currently used for daily production runs of the MIT model. A vision of "personal supercomputers" as effective research and teaching tools underlies this study, and so we limit our experiments to high-level programming environments that would be readily accessible to researchers and students who are not necessarily expert scientific programmers or numerical modelers. Our study employs a hierarchy of problems, culminating in the MIT model hydro-dynamical kernel, to probe the strengths and weaknesses of the platforms, and to identify promising implementation strategies.

2 The MIT Model

The MIT ocean-atmosphere model is built around a versatile incompressible Navier-Stokes solver developed at MIT for the numerical study and analysis of rotating fluids. A hydro-dynamical kernel, designed to be sensitive to the transition from non-hydrostatic to hydrostatic dynamics, lies at the heart of the model, providing computational efficiency across a broad spectrum of simulation regimes. A review of the model's utility for research in a range of problems related to ocean and climate dynamics can be found in Marshall et.al.[5]. Implementations of the model targeted to the MPP CM-5 and to CRAY vector supercomputers are being actively used by researchers at MIT[6,7,8,9,10,11,12].

The discrete algorithm uses a finite volume[13], time-stepping formulation which is rich in data parallelism and can handle the intricate, order one variations in geometry of ocean basins. The CM-5 code achieves sustained floating point throughput of over 10 GFlop/s on a 512 node CM-5. Daily production runs of the model sustain between 500 MFlop/s and 1 GFlop/s on a 32 node CM-5, and between 100 MFlop/s and 200 MFlop/s on a CRAY Y-MP.

2.1 The numerical scheme

The numerical procedure used in the model is described in detail in Marshall et.al.[14] The scheme is a variant on the theme set out by Harlow and Welch[15], in which a pressure correction to the velocity field is used to guarantee non-divergence.

The numerical kernel is a discrete form of the Boussinesq incompressible Navier Stokes fluid equations: it can be written in semi-discrete form to second order in time Δt thus:

$$\frac{\underline{v}^{n+1} - \underline{v}^{n}}{\Delta t} = \underline{G}_v^{n+\frac{1}{2}} - \nabla p^{n+\frac{1}{2}} \tag{1}$$

$$\nabla . \underline{v}^{n+1} = 0 \tag{2}$$

Equations 1 and 2 describe how the fluid evolves ($\underline{v} = (\underline{v}_h, w)$ is the velocity in the horizontal and the vertical) in response to forcing due to $\underline{G}$ (representing inertial, Coriolis, metric, gravitational, and forcing/dissipation terms) and to the pressure gradient ∇p. The same kernel algorithm is used by us to study both the atmosphere and the ocean - the anelastic approximation as described in Brugge et.al.[16] and in Miller[17] is used to render the same model suitable for the study of both fluids.

The pressure field required in Equations 1 and 2 is found by separating p into hydrostatic, surface and non-hydrostatic parts. Vertical variations in p can be computed hydrostatically yielding p_{hy} using only information in each column of the fluid. Horizontal variations in p_s are diagnosed by solving an elliptic equation which ensures non-divergent depth integrated flow:

$$\nabla_h . H \nabla_h p_s = \nabla_h . \overline{\underline{G}_{v_h}^{n+\frac{1}{2}}}^H - \nabla_h . \overline{\nabla_h p_{hy}{}'}^H \tag{3}$$

($\overline{\quad}^H$ indicates the vertical integral over the local depth, H, of the domain and the subscript $_h$ denotes horizontal). In the non-hydrostatic limit a three-dimensional elliptic equation is inverted for the non-hydrostatic pressure. In this study we consider the hydrostatic case involving the two-dimensional problem, Equation 3. More details of the numerical procedure can be found in Marshall et.al.[5,14].

2.2 Program structure

Figure 2 shows the high-level program structure of the MIT model. Almost all of the execution time is spent in the time-stepping loop, which is divided into two parts – **Forward_Step** and **Inversion_Step**. The **Forward_Step** comprises a series of nearest-neighbor computations involving two time levels of the model state variables and their time derivatives. This stage computes explicit time derivative terms for the velocity and tracer equations, steps them forward in time and updates the density field through an equation of state. There is considerable latitude for different orderings of the computations within **Forward_Step**, and, as discussed in Shaw et.al.[18], this stage contains an abundance of both data parallelism and functional parallelism. The **Inversion_Step** requires global communication to solve a set of simultaneous equations to ensure that the flow remains non-divergent.

3 The computer platforms

We compare the implementation of the MIT model on two very different platforms: the CM-5 vector MPP, and the DEC 8400 cache-based SMP. The CM-5

```
BEGIN
  Initialize
    Define topography, initialize flow field, tracer distributions
  FOR each time step DO
    Forward_Step
      Calculate velocity time derivatives
      Calculate tracer field time derivatives
      Step forward state variables
      Update density field
    Inversion_Step
      Solve elliptic problem to ensure non-divergent flow
  END FOR
END
```

Figure 2: High-level structure of the MIT model. The model iterates repeatedly over a time-stepping loop, comprised of two main blocks, Forward_Step and Inversion_Step.

is well-suited for scientific simulation applications, and it, along with the Cray Y-MP, is the mainstay of our day-to-day production runs. We will be testing a 32-processor configuration of the CM-5.

The DEC 8400 is a very high-end mainstream SMP system, designed for use in a broad variety of both technical and non-technical situations. Digital has already sold several thousand 8400 systems, while in contrast, Thinking Machines shipped less than 100 CM-5 systems. We will be testing a 300 MHz, 4-processor configuration of the DEC 8400.

3.1 Machine architectures

Both systems have theoretical peak floating point performance well in excess of 100 MFlop/s, as illustrated in Figure 1. However, the two systems have radically different architectures, which we discuss here in more detail.

The CM-5 MPP[19] was one of the first parallel computers which could routinely out-perform traditional vector supercomputers for real scientific problems. The machine is built from relatively modest vector processors connected through two communications networks, as shown in Figure 3. Each vector unit runs at 16 MHz and can issue a multiply-add instruction each cycle, giving it a peak performance of 32 MFlop/s. Each CM-5 node contains 4 vector units, yielding a peak performance of 128 MFlop/s per node. In real applications, however, sustained performance of 10-30 Mflop/s per node is more typical.

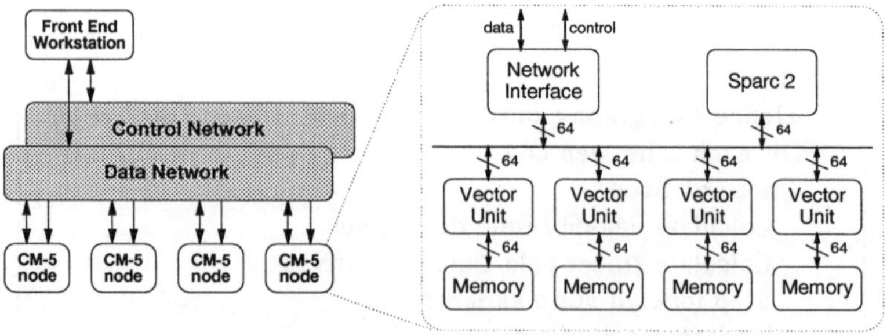

Figure 3: High-level architecture of the CM-5

The key to exploiting the CM-5's vector units is to express computation in terms of data-parallel vector operations, which is a natural paradigm for regular finite difference/volume models. The vector units act as both memory controllers and vector floating point units – each vector unit can read or write one 64-bit word of memory every cycle, as well as performing one floating point operation.

Main memory on the CM-5 is distributed across the vector units and so any sort of inter-processor communication on the CM-5 is relatively expensive, motivating the programmer to structure codes to minimize inter-processor communication. Often, this entails a tradeoff of extra memory use for less communication.

The architecture of the DEC 8400 [20] is typical of the current generation of microprocessor-based SMP's. As shown in Figure 4, the DEC 8400 consists of nodes connected by a shared system bus. Sharing of data between the processors can only occur through reads and writes to memory. Each node of the 8400 consists of an Alpha 21164 microprocessor [21] with three levels of cache – the first two levels of cache are on-chip, and the third level of cache is off-chip, but on the same board. The caches are small, fast-access memories which mirror sections of the computer's larger, slower access main memory. Computation using data from cache is much faster than using data from main memory. As the caches are situated farther from the processor core, they are successively larger and slower.

Other SMP's and even successive generations of DEC SMP's may have different sized caches, or even a different number of levels of caches. However, all microprocessor-based SMP's have one or more levels of cache and a shared view of main memory.

The 8400 and other mainstream SMP's rely on caches to get good perfor-

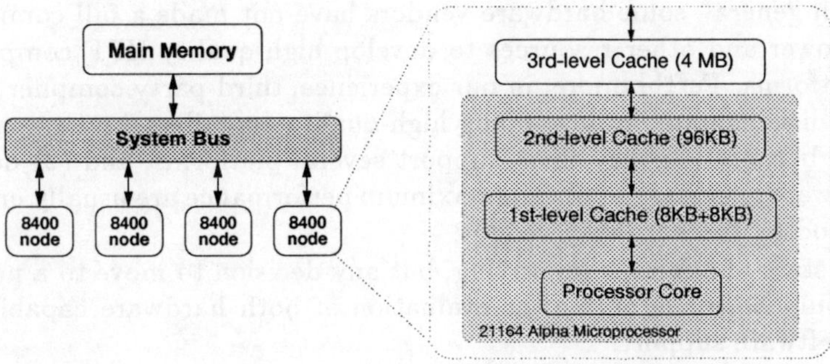

Figure 4: High-level architectures of the DEC 8400 SMP

mance. If a memory reference from the processor is returned by the 1st-level cache, the processor can continue working on the next cycle. References which are returned by the 2nd or 3rd level cache cause the processor to wait longer, and references to the main memory can cause the processor to wait for tens of cycles. Each Alpha microprocessor runs at 300 MHz, and if all of the operands are available, the Alpha can perform one floating point operation (which may be a multiply-add) per cycle, thus having a peak performance of 600 MFlop/s.

The shared memory architecture of the 8400 allows for lower communication costs than on the CM-5. However, if the problem does not fit into the 8400's caches, performance may be very bad, so it is important to minimize memory use. It should be noted, however, that main memory itself can be very large.

3.2 Programming environments

Our goal is to make a model that is accessible to the general scientist, as well as to specialist scientific programmers and numerical modelers. A high-level, intuitive scientific programming environment is therefore of interest to us.

Motivated by this goal, we implemented the model on the CM-5 in CM Fortran (CMF) [22], which is a high-level, data-parallel dialect of Fortran. For the DEC 8400, we implemented the model in High Performance Fortran (HPF) using Digital's native High Performance Fortran (HPF) compiler.[23] HPF is very similar to CMF because the design of the HPF language standard was strongly influenced by CMF.

HPF compilers are available for most SMP platforms, provided either by the hardware vendor or third-party compiler vendors. However, the quality of HPF compilers can vary widely – although HPF compiling technology is fairly mature in general, some hardware vendors have not made a full commitment in manpower and other resources to develop high-quality HPF compilers for their platforms. Furthermore, in our experience, third-party compiler vendors are at a disadvantage in developing high-quality compilers for any particular platform because (1) they must support several platforms, and (2) details of the hardware necessary to obtain maximum performance are usually only well-understood by the hardware vendor.

The state of affairs is improving, but any decision to move to a new platform should be made only after evaluation of both hardware capabilities as well as software support.

4 Performance profiling

To obtain high performance for the MIT model on a new computer platform, we take an incremental approach, beginning with performance profiling of elementary kernels, and moving towards profiling of the full application. At each stage, we gain a better quantitative understanding of the architecture, programming environment, and algorithmic match to the machine. The successive tests give us a step-by-step guide to the design of the full application.

4.1 **DAXPY** *test*

The **DAXPY** test measures the performance of a system for vector-style operations. **DAXPY** is a primitive of the BLAS1 [24] linear algebra library which performs the operation $\underline{y} = a\underline{x} + \underline{y}$, where $\underline{y}$ and $\underline{x}$ are 64-bit floating point arrays.

Figure 5 shows the floating point performance as a function of the length of the input arrays for the 32 node CM-5, and for different processor counts of the DEC 8400. Examining this graph, we clearly see the impact of the cache size on performance. For the 8400 system, performance tails off sharply when cache size is exceeded, which is indicated at the "cache shoulders". As the number of processors increases, the total amount of cache memory also increases, as seen by the cache shoulder moving towards the right as more processors are used in the 8400.

For the CM-5, the performance increases with the size of the problem, because the CM-5 can more fully use its vector units on larger sized problems. In comparing the absolute performance of the CM-5 to the 8400, we should try to determine what region of interest is delineated by the curves. Once the size of the problem is too large to fit into the caches of the 8400, the performance is uncompetitive with the CM-5, or like machines.

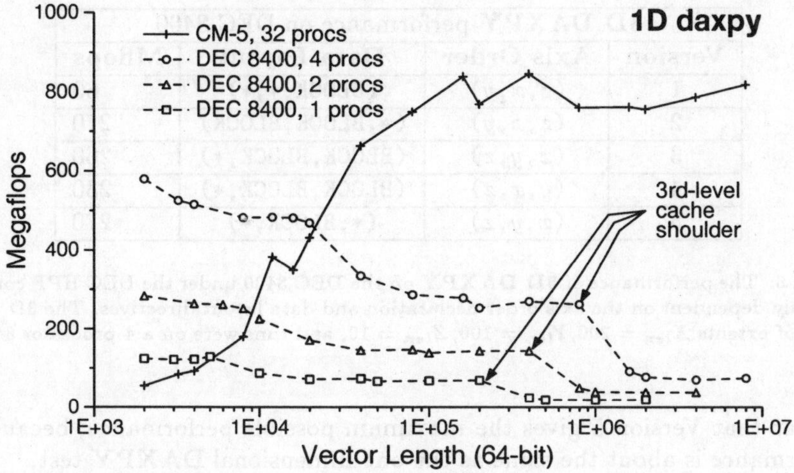

Figure 5: Performance of DAXPY BLAS1 primitive for different lengths of vector on one, two and four processors of a DEC 8400, and on a 32 node CM-5 computer.

In some ways, **DAXPY** is the worst-case situation for the SMP, because no data is reused, and therefore, the caches are poorly exploited. We can attempt to alleviate the effects of computations like **DAXPY**, by reusing data that is in the cache. However, traditional vector supercomputers favored applications which were structured as compositions of DAXPY-like operations, so programming and compiling for microprocessor-based SMP's requires a re-thinking of the mapping of an algorithm to the computer.

4.2 **3D DAXPY** *test*

3D DAXPY is a three-dimensional version of **DAXPY**. Although the **3D DAXPY** computation seems extremely straightforward, there are actually some subtleties about programming even such a simple piece of code. Figure 6 shows some of the results obtained when we ordered the axes of the arrays differently, and when we used different HPF data layout directives for the same code using a fixed problem size and a fixed number of processors. We ran the code with arrays of size 200 × 100 × 10, which is representative of runs we do daily on the CM-5.

There is almost a factor of 5 difference between the performance of Version 1 and Version 5. Version 2 actually uses the same axis order and layout as the CM-5 version of the MIT model, but it is not optimal on the 8400. It

3D DAXPY performance on DEC 8400			
Version	**Axis Order**	**Data Layout**	**Mflops**
1	(z,x,y)	(BLOCK,*,*)	60
2	(z,x,y)	(*,BLOCK,BLOCK)	210
3	(x,y,z)	(BLOCK,BLOCK,*)	260
4	(y,x,z)	(BLOCK,BLOCK,*)	230
5	(x,y,z)	(*,BLOCK,*)	270

Figure 6: The performance of **3D DAXPY** on the DEC 8400 under the DEC HPF compiler is highly dependent on the axis order declaration and data layout directives. The 3D arrays are of extents $X_{len} = 200, Y_{len} = 100, Z_{len} = 10$, and runs were on a 4 processor 8400.

is clear that Version 5 gives the maximum possible performance, because its performance is about the same as the one-dimensional **DAXPY** test.

4.3 **Forward_Step** *test*

The **Forward_Step** part of the code consists of updates to a three dimensional grid of double-precision floating point numbers representing the state of the fluid. In our study, in keeping with our goal of a highly intuitive and readable code, the grid is mapped directly to three-dimensional arrays. Figure 7 shows the performance of a representative fragment from **Forward_Step,** which calculates the G terms in Equation 1. Problem sizes were chosen to be wide and thin, which is representative of geometries we use for our research. Times were gathered by extracting the section of code from the model, and using the optimal axis ordering and data layout as determined by the **3D DAXPY** test.

The results in Figure 7 represent the performance after modest optimizations were made to the DEC 8400 code relative to the CM-5 code. The CM-5 version actually uses about 40 arrays, some of which are used to minimize communication. With the DEC 8400 shared-memory architecture it is more important to reduce the overall memory usage, so some of these buffer arrays are removed. Memory requirements were further reduced by storing time invariant terms as 32-bit rather than 64-bit values.

Note that the performance of the DEC 8400 does not drop off as much as for **DAXPY** with this more representative code extract, but that the drop-off occurs at an earlier point. The improvement in asymptotic performance reflects the higher ratio of floating point operations to memory references of the **Forward_Step** test (which has a ratio of $\approx 26:8$) relative to the **DAXPY** tests (where the ratio is $1:1$). This higher ratio translates into increased op-

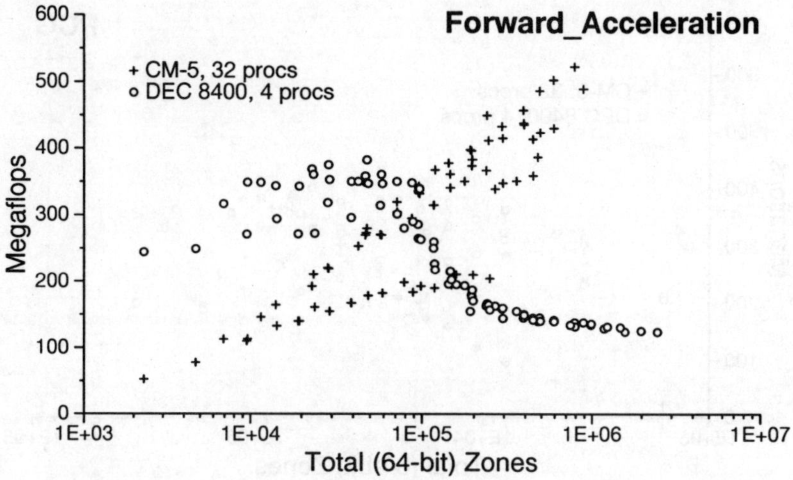

Figure 7: Performance of the **Forward_Step** test as a function of problem size. The shoulder for the DEC 8400 has moved to about 10^5 elements, because many more arrays are used for **Forward_Step** than for **DAXPY**.

portunities for reusing data loaded from main memory into cache. The degree to which this is exploited is highly dependent upon optimizations performed by the compiler and the way in which the code is written by the user.

The earlier drop-off point relative to **DAXPY** is the result of using more arrays – whereas **DAXPY** uses 2 arrays, this code segment uses the equivalent of 24 64-bit arrays, and accordingly, the drop-off point is about a factor of 12 earlier than for **DAXPY**. The full MIT model requires about the same number of arrays (to hold six state variables, their time derivatives at two time levels and geometric terms), so we expect that the rest of the **Forward_Step** code should behave as this code fragment does. To put this problem size in perspective, it is about equivalent to a global simulation with a four degree horizontal grid and twenty vertical layers.

4.4 **Inversion_Step** *test*

The other important stage in the model algorithm is the **Inversion_Step** which solves an elliptic problem to ensure non-divergent flow. In discrete form this inversion can be written:

$$\underline{\underline{A}}\,\underline{p} = \underline{f} \tag{4}$$

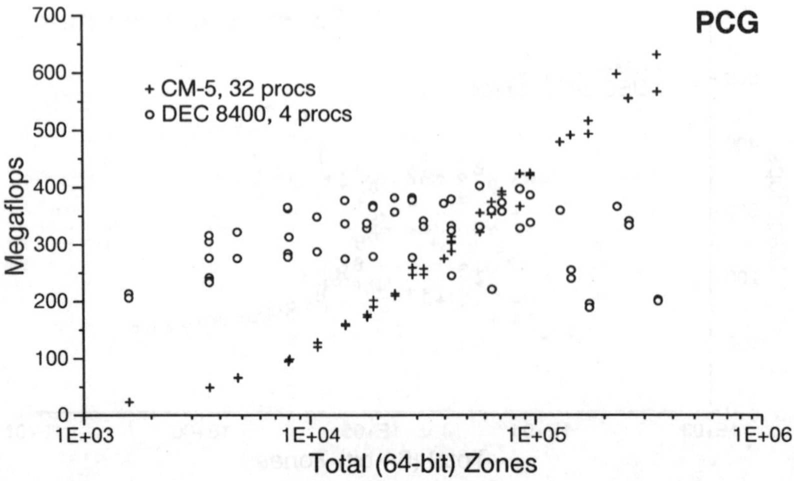

Figure 8: The preconditioned conjugate gradient algorithm used to solve for the pressure is competitive to the CM-5 over the region of interest (10^4 to 10^5).

in which, for hydrostatic simulations, $\underline{\underline{A}}$ is the horizontal Laplacian operator, a matrix with five diagonals, and $\underline{p}$ and $\underline{f}$ represent the surface pressure and the right-hand side term of Equation 3 respectively.

In the CM-5 implementation of the model, Equation 4 is solved using an iterative pre-conditioned conjugate gradient (PCG) scheme. PCG is well suited to data parallel implementations and the same algorithm was implemented in HPF on the DEC 8400.

Figure 8 shows the performance of the PCG algorithm on the CM-5 and DEC 8400. Note that because we are solving a two-dimensional problem instead of a three-dimensional problem (required in the non-hydrostatic limit formulation of the model), the problem sizes we consider are much smaller, and thus the performance of the DEC 8400 does not show the characteristic cache shoulder of the previous tests.

In addition, we perform further optimizations to minimize memory use. In the CM-5 implementation of PCG, to reduce communication we duplicate data structures that hold the diagonals of $\underline{\underline{A}}$ shifted by one point in x and y. On the DEC SMP system, this duplication is not necessary, and we exploit the symmetry of $\underline{\underline{A}}$ to eliminate another array by recalculating its value at run-time. Finally, we also store $\underline{\underline{A}}$ in 32-bit precision rather than 64-bit. In total, we were able to reduce the memory required to represent $\underline{\underline{A}}$ by a factor

Skyline (Direct) vs. PCG (Iterative) solvers on DEC 8400					
Ocean	**Skyline** (1 proc)			**PCG** (4 proc)	
Size	**Factor**	**Solve**	**Megaflops**	**Solve**	**Megaflops**
100 × 100	1.1 s.	0.1 s.	9.1	0.1 s.	350
200 × 100	2.4 s.	0.2 s.	8.7	2.7 s.	380
200 × 200	23.0 s.	0.9 s.	8.8	1.6 s.	467
400 × 200	49.7 s.	1.9 s.	8.4	4.6 s.	480
400 × 400	238.0 s.	10.0 s.	6.4	25.8 s.	230
800 × 400	633.3 s.	20.9 s.	6.1	83.5 s.	210

Figure 9: The direct solution to $Ax = b$ is divided into a factorization of A and the solve for x. Factorization only occurs once during the run of a program, and its cost is amortized out on long runs. The Skyline direct solver is faster in wall clock time than the PCG iterative solve, even though it has a lower Megaflops rating.

of 5.

The PCG iterative solution can require anywhere between 10 and 200 iterations to converge – the exact number will depend on the physics of the particular simulation. For a domain with a surface of size $NX \times NY$, if we assume that convergence requires *nits* iterations, the number of floating point operations executed is:

$$nits \times 32 \times NX \times NY$$

where 32 is the number of floating point operations performed per grid point per iteration in our PCG implementation.

An alternative approach to an iterative PCG solver is a direct solver. Because $\underline{\underline{A}}$ is symmetric and sparse, the LU decomposition can be stored within the profile of the $\underline{\underline{A}}$ matrix. For any $\underline{\underline{A}}$ representing a closed physical domain, this requires about $NX \times NX \times NY$ elements of storage. Once the LU factorization has been accomplished, we can find $\underline{p}$ using forward-backward substitution. Solving for $\underline{p}$ using this *direct* method requires the following number of floating point operations:

$$3 \times NX \times NX \times NY$$

where 3 floating point operations are performed per word of the LU factorization for the forward-backward substitution.

Because of the way the problem is posed, we can choose to appropriately switch NX and NY to arrive at a problem with a smaller LU decomposition. We should note, however, the lateral aspect ratio of typical geometries is not usually greater than 2 : 1, so that the direct method has an algorithmic complexity $O(n^3)$, whereas the iterative methods generally require about $O(n^2)$.

Despite the higher order complexity of the direct method, the high constant factors in the iterative method make the direct method competitive for a certain range of problem sizes. Suppose that a 100×100 sized problem required 100 PCG iterations to converge – the PCG algorithm would require $100 \times 32 \times 100 \times 100 = 32,000,000$ floating point operations, whereas the direct method would only require $3 \times 100 \times 100 \times 100 = 3,000,000$ floating point operations. However, even though the direct method may execute fewer floating point operations, each floating point operation will be much slower than the direct method because the direct method must stream through the entire LU factorization, performing only 3 operations per element, leading to a fairly bad ratio of floating point operations to memory operations.

To test the effectiveness of a direct approach versus the iterative approach, we solved the same 2D problem using a sparse direct solver included with the DEC DXML scientific library [25]. The solver uses the skyline algorithm, which assumes matrices have elements which are primarily close to the main diagonal. This particular algorithm does not assume any other structure (other than symmetry) within the $\underline{\underline{A}}$ matrix, and currently only runs on 1 processor of the 8400 system – a more specialized or parallel direct solver may be faster than this very general-purpose, single-processor algorithm.

Nevertheless, the results in Figure 9 show that a direct solver could be a competitive solution method for certain problems. For example, Hill and Marshall[26] make a non-hydrostatic model competitive with a hydrostatic model, even in the presence of complex geometry, by separating out, and solving to high-accuracy, the two-dimensional surface pressure problem we examine here. Because the direct solver returns a solution to Equation 4 which is accurate to machine precision, the solver would be ideal for accelerating a non-hydrostatic model.

4.5 Full application performance

We evaluate the full model's DEC 8400 implementation by comparing its performance with the CM-5 implementation for a specific physical simulation. The simulation we chose was an experiment modeling the collapse and break up, through baroclinic instabilities, of an anomalously dense column of fluid in a rotating frame of reference.[9] The chosen test has a computational grid of size $100 \times 100 \times 26$, which is large enough to exceed the cache size of a four processor DEC 8400, but not so large that there are no opportunities for optimization.

Figure 10 shows the relative performance of the CM-5 and DEC 8400 for

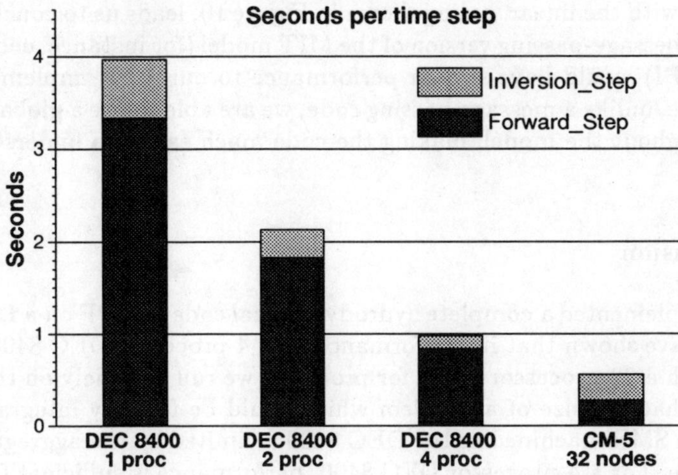

Figure 10: Performance of the full model on a 32 node CM-5 and a DEC 8400 5/300 server.

this simulation. The DEC 8400 four processor system is about one half as fast as the CM-5. From our previous tests, we should be able to predict the performance of the DEC 8400 given the problem size. For the **Forward_Step**, referring to Figure 7, the problem size is 2.6×10^5, which shows that the DEC 8400 should be about 2.5 times slower than the CM-5. For the **Inversion_Step**, referring to Figure 8, the problem size is 10^4, so the DEC 8400 should be about 2.5 times faster than the CM-5.

For the problem that we simulated in Figure 10, the time spent in one time step on the CM-5 is about equally divided between the **Forward_Step** and the **Inversion_Step**. This ratio will depend on the physics of the simulation, in particular, the amount of work required by the **Inversion_Step** varies greatly from one problem to another. Notice that our performance prediction was fairly accurate: the time spent in the **Inversion_Step** for the DEC 8400 is indeed about 2.5 times less, and the time spent in the **Forward_Step** is roughly 2.5 times greater than the CM-5 time.

Although the particular integration considered here had a 50/50 balance between forward and inversion steps on the CM-5, other simulations have a different split. For problems that spend a large fraction of time in the inverter, the 4-processor DEC 8400 should out-perform the CM-5.

We have also compared single processor performance of our HPF test codes expressed in Fortran 77, using the DEC compilers. In these checks, the HPF test code performance was always within 10% of its Fortran 77 equivalent.

This, along with the linear scaling shown in Figure 10, leads us to conclude that an explicit message-passing version of the MIT model (for instance, coded using PVM or MPI) would have similar performance to our HPF implementation. Furthermore, unlike a message-passing code, we are able to use a global address space throughout the model, making the code much easier to understand and modify.

5 Conclusion

We have implemented a complete hydrodynamical code in HPF on a DEC 8400 SMP. We have shown that its performance on a 4-processor DEC 8400 is competitive with a 32-processor CM-5 for problems we run routinely on the CM-5. We found that the size of a problem which could be feasibly integrated on a mainstream SMP machine like the DEC 8400 is limited by the aggregate cache memory size. For a 4-processor DEC 8400, performance is sufficient for coarse resolution ($\approx 4°$ horizontal resolution) climatological time scale integrations, or for higher resolution (down to $\approx 1°$) decadal simulation.

To obtain good performance on the SMP, we analyzed and optimized a series of isolated kernels that are the building-blocks of the full model. These test codes highlight factors which affect performance of the full application. For example, we see how performance scales with problem size, machine size, and cache size. Each test also reveals strengths and weaknesses of the compiler, and allows us to explore alternative implementations which may better match the system. These alternatives range from simple axis order re-ordering to completely different numerical procedures such as the substitution of a direct linear solver for an iterative one.

Achieving good performance on microprocessor-based SMP's depends upon many factors, including the number of processors in the system, the size of the caches for each processor, the speed of the individual processors, and the maturity of the compilation system. The DEC 8400 represents a successful combination of these attributes for our application. Other codes, such as spectral models or elaborate physics packages, present different challenges that would require further analysis. Similarly, different SMP platforms vary in the degree to which they provide hardware and software to support the approach we have taken here. Cache memories vary in size from 512KB to 16MB per processor for the current generation of mainstream SMP platforms and, although the performance of our HPF code, with the Digital HPF compiler, is close to the equivalent message-passing code, other HPF compilers can, and often do, produce code which is significantly slower than the equivalent message-passing code.

We are optimistic that microprocessor-based SMP platforms will, for real oceanographic and meteorological applications, bridge the gap between frequently over-burdened, centralized supercomputer center facilities and underpowered desktop workstations. However, the widespread adoption of SMP's is contingent on their affordability – currently, the 4-processor DEC 8400 system is significantly more expensive than four individual workstations. Nevertheless, it seems likely that prices will fall as high-performance SMP technology gains popularity in the marketplace, driven primarily by demands of the business community, as well as the scientific and engineering communities.

Finally, we note that our adherence to a high-level, global programming model has made it significantly easier to automate the complex data flow analysis required to differentiate our forward model code. As a result we are able to apply the TAMC adjoint model generation tool[27] to our HPF code, and generate adjoint and tangent linear models automatically. Furthermore, through appropriate replication of the data distribution directives from the forward model, the automatically derived codes transparently inherit the parallelism of our forward model. Using this technique we are incorporating our model into a systematic optimization framework with relative ease.

Acknowledgments

This research is a joint effort of the MIT Center for Meteorology and Physical Oceanography and the MIT Laboratory for Computer Science. The research of Chris Hill is partially funded by ONR contract N00014-95-1-0967. The research of Andrew Shaw is partially funded by ARPA under ONR contract N00014-92-J-1310.

John Marshall and Arvind read drafts of this paper. We would also like to thank Digital Equipment Corporation and the Pittsburgh Supercomputing Center for giving us access to the DEC 8400 SMP. Experiments on the CM-5 were done under the MIT SCOUT project, ARPA contract #MDA972-92-J-1032.

References

1. S. McKee, Wm. A. Wulf, and T. C. Landon. Bounds on Memory Bandwidth in Streamed Computations. In *Lecture Notes in Computer Science 966*, pages 83–99. Springer-Verlag, Berlin, 1995.
2. John D. McCalpin. Memory Bandwidth and Machine Balance in Current High Performance Computers. *IEEE Computer Society Technical Committee on Computer Architecture (TCCA) Newsletter*, December 1995.

http://www.computer.org/tab/tcca/news/dec95/dec95_mccalpin.ps.

3. J. Dongarra. The LINPACK Benchmark: An Explanation. *Supercomputing*, pages 10–14, Spring 1988.
4. J. Dongarra. Performance of Various Computers Using Standard Linear Equations Software in a Fortran Environment. Technical Report TR MCSRD 23, Argonne National Laboratory, March 1988.
5. J. Marshall, C. Hill, L. Perelman, and A. Adcroft. Hydrostatic, quasi-hydrostatic and non-hydrostatic ocean modeling. *J. Geophys. Res.*, in press 1997.
6. B. Klinger, J. Marshall, and U. Send. Representation and parameterisation of deep convective plumes by mixing. *J. Geophys. Res.*, 101 (C8):18175–18182, August 1996.
7. K.G. Speer and J. Marshall. The growth of convective plume at seafloor hot springs. *J. Marine Res.*, 53:1025–1057, 1995.
8. M. Visbeck, J. Marshall, and H. Jones. On the dynamics of convective chimneys in the ocean. *J. Phys. Oceanogr.*, 26:1721–1734, September 1996.
9. H. Jones and J. Marshall. Restratification after deep convection. *J. Phys. Oceanogr.*, in press 1997.
10. T. Haine and J. Marshall. Gravitational, symmetric and baroclinic instability of the oceanic mixed layer. *J. Phys. Oceanogr.*, in press 1997.
11. D. Menemenlis, T. Webb, C. Wunsch, U. Send, and C. Hill. Basin-Scale Ocean Circulation from Combined Altimetric, Tomographic and Model Data. *Nature*, in press 1997.
12. D. Menemenlis and C. Wunsch. Linearization of an Oceanic Circulation Model for Data Assimilation and Climate Studies. *Journal of Atmospheric and Oceanic Technology*, in press 1997.
13. A. Adcroft, C. Hill, and J. Marshall. On the Treatment of Topography in Numerical Ocean Models. *Mon. Wea. Rev.*, in press 1997.
14. J. Marshall, A. Adcroft, C. Hill, and L. Perelman. A finite-volume, incompressible Navier-Stokes model for studies of the ocean on parallel computers. *J. Geophys. Res.*, in press 1997.
15. F. H. Harlow and J. E. Welch. Numerical calculation of time-dependent viscous incompressible flow. *Phys. Fluids*, 8:2182, 1965.
16. R. Brugge, H.L. Jones, and J. Marshall. Non-hydrostatic ocean modelling for studies of open-ocean deep convection. In *Proceedings of the Workshop on Deep Convection and Deep Water Formation in the Ocean*, pages 325–340. Elsevier Science Publishers, Holland, 1991.
17. M.J. Miller. On the use of pressure as vertical co-ordinate in modelling. *Quart. J. R. Meteorol. Soc.*, 100:155–162, 1974.

18. A. Shaw, K-C Cho, C. Hill, R. P. Johnson, J. Marshall, and Arvind. A comparison of implicitly parallel multithreaded and data parallel implementations of an ocean model based on the Navier-Stokes equations. Technical Report CSG Memo 364, MIT LCS, 1995.
19. TMC. *The CM-5 Technical Summary.* Thinking Machines Corporation, Cambridge, MA, 1991.
20. D.M. Fenwick et.al. The AlphaServer 8000 Series: High-end Server Platform Development. *Digital Technical Journal*, 7(1), 1995.
21. J.H. Edmondson et.al. Internal Organization of the Alpha 21164, a 300-MHz 64-bit Quad-issue CMOS RISC Microprocessor. *Digital Technical Journal*, 7(1), 1995.
22. Thinking Machine Corporation. *CM Fortran Language Reference Manual, V2.1*, January 1994.
23. J. Harris et.al. Compiling High Performance Fortran for Distributed-memory Systems. *DEC Technical Journal*, 7(3), 1995. ftp://ftp.digital.com/pub/Digital/info/DTJ/v7n3_03_FORTRAN.ps.
24. C. Lawson, R. Hanson, D. Kincaid, and Krogh F. Basic Linear Algebra Subprograms for Fortran Usage. *ACM Transactions on Mathematical Software*, 5(3):308–323, 1979.
25. D.P. Manley C. Kamath, R. Ho. DXML: A High-performance Scientific Subroutine Library. *Digital Technical Journal*, 6(3), 1994. ftp://ftp.digital.com/pub/Digital/info/DTJ/v3n6-dxml.txt.
26. C. Hill and J. Marshall. Application of a parallel Navier-Stokes model to ocean circulation. In *Parallel Computational Fluid Dynamics: Implementations and Results Using Parallel Computers*, pages 545–552. Elsevier Science Publishers, Holland, 1995.
27. R. Giering and T. Kaminski. Recipes for Adjoint Code Construction. Technical Report 212, Max-Planck-Institute for Meteorology, 1996.

A Modular 3-D Dynamical Core Testbed

Richard D. Loft

Scientific Computing Division
National Center for Atmospheric Research
P.O. Box 3000
Boulder, CO 80307

Abstract

A *modular 3-d dynamical core testbed* is being developed at NCAR to evaluate the suitability of promising new shallow water model numerical techniques for 3 dimensional climate modeling. In addition to the spherical harmonic transform method, two grid based techniques are also being implemented: Baumgardner's method based on spherical triangular grids derived from the icosahedron, and the spectral element method of Taylor, Tribbia and Iskandarani applied to projections of the cube onto the surface of the sphere. A dry baroclinic sigma model is currently being developed. The governing equations are written in the velocity form. The initial goal of this effort will be to force the model with Held-Suarez idealized physics and intercompare the climatological state produced by each dynamical method. The model is also being used to study issues in modular model design using F-90 language constructs including modules, allocatable arrays, interface blocks, pointers, and derived types. These features improve the model's flexibility, efficiency, portability and scalability. The baroclinic model is easily modified to switch between new grid schemes, Eulerian or semi-Lagrange timestepping, and explicit or semi-implicit time integration techniques, through replacement of a set of well defined functional and data modules.

1 Introduction

The motivation for investigating new algorithms for dynamics and transport stems from two modeling considerations. First, many of the new gridding strategies avoid the well known "pole problem" associated with latitude-longitude grids in which the narrow cells (in the longitude direction) approaching the poles impose severe time step restrictions. A triangular grid, for example, can provide an almost uniform subdivision of the spherical surface.

A second motivation arises from liabilities associated with the widely used spherical harmonic transform or spectral method, most notably its computational cost at high resolutions, and non-local nature of the transforms, which is an obstacle to efficient implementation of the spectral method on distributed memory computers. Each spectral coefficient depends on values from every grid point, and each grid point value depends on all the spectral coefficients. This necessitates considerable non-local memory traffic. Although local methods must also use a non-local technique to solve the semi-implicit elliptic equation, derivatives can be computed locally and efficient and scalable elliptic solvers do exist for some grid point methods. There are also purely modeling concerns related to the spectral method. One is the occurrence of ringing when sharp gradients are present. In the context of atmospheric modeling, strong gradients in topography, i.e., mountain ranges, for example, commonly lead to spurious zones of upward and downward motion. This behavior is observed in most spectral models west of the Andes[1]. On the other hand, local methods should avoid such difficulties.

The introduction of standard test cases by Williamson, Drake, et. al. has provided a framework for the intercomparison of numerical integration schemes for the shallow water equations on the sphere for use in meteorological forecast and climate simulation[2]. These seven standard test cases range from simple advection of a cosine bell to realistic atmospheric initial conditions. These test cases have been widely used to intercompare numerical methods, and have proven useful in sorting out the relative merits and limitations of each[3]. An interesting observation of Taylor, et. al. is that as the "realism" of the shallow water test cases increases, the difference in the numerical accuracy of the different solution techniques decreases. For example the normalized l_2 error for the T42 spectral model running case 2 (Global Steady State Nonlinear Zonal Geostrophic Flow) is approximately five orders of magitude smaller than the twisted icosahedral grid[4], while for the case 7a (500 mb Height and Wind Field Initial Condition) the l_2 errors are about the same order of magnitude. This observation suggests that it may be inappropriate to evaluate the suitability of new numerical schemes to atmospheric modeling based solely on their accuracy properties.

Held and Suarez have proposed a three dimensional dry baroclinic model with idealized physics forcing which can be used as a test case for evaluating, at least to first approximation, the suitability of dynamical schemes to climate modeling[5]. Boer and Denis have also proposed an idealized forcing[6]. Regardless of the their individual merits, the concept of simple three dimensional climatological test cases provides a useful bridge between $O(10^3)$ line 2-d shallow water equations codes and $O(10^5)$ line atmospheric model codes.

At the same time, this dynamical core approach does have limitations. We do not believe that this modeling environment will shed much light on the issue of the overall computational efficiency of methods, or prove that one method is "better" than another. In other words, performing well in the baroclinic sigma model, as in the shallow water equations, is a necessary, but not sufficient test of a particular dynamical scheme's suitability to atmospheric modeling.

In this paper we report on software under development at NCAR in collaboration with John Baumgardner, Mark Tayor and Dave Williamson to facilitate the introduction of promising 2-D methods into 3-D atmospheric models built for high performance computers. This *dynamical core testbed* is being implemented so that it can be easily modified to incorporate different 2-d dynamical algorithms and idealized physics packages. This testbed makes use of modern F-90 features such as allocatable arrays, pointers, and user defined types to achieve modularity and portability, with minimal sacrifice of performance. Great care is being taken from the outset to support extensibility and portability of the model.

The flexibility of the testbed software design has limitations. It is unlikely that the code can be extended to include every conceivable modeling variant in a meaniful way. Nor can it be expected to run optimally on all platforms. Rather the purpose of the design exercise is to show that careful design and use of modern language constructs can make programs *more* reliable and *more* extensible.

The paper is organized in the following way. The testbed's baroclinic equations of motion are discussed in Section 2. Section 3 is devoted to the testbed's design. Features that support different numerical modeling choices are discussed in Section 3.1. Features to accomodate parallel computer architectures are discussed Section 3.2. Section 3.3 discusses portability across different computer architectures. Section 4 discusses details of two grid methods being incorporated into the model, a spherical triangle method and a spectral element method on the sphere. Finally, Section 5 discusses plans to extend the testbed concept towards an implementation of the complete Community Climate Model physical parameterization package. This development will allow comparison of numerical methods in a complete model environment as well as in isolated dynamical cores with idealized forcings.

2 The Baroclinic Equations in Sigma Coordinates

The model testbed uses the velocity form of the 3 dimensional dry baroclinic equations of motion in sigma coordinates, because it is more natural to express Semi-Lagrangian methods in terms of ($\vec{v}$) than in the vorticity-divergence ($\zeta - \delta$) form.

The momentum equation of the model is given by

$$\frac{d\vec{v}_k}{dt} = -\left[f\hat{k} \times \vec{v}_k + \vec{\nabla}\left(\Phi_s + RH_{kl}T_l\right) + RT_k \vec{\nabla} lnp_s\right] + \vec{F}_k \tag{1}$$

The thermodynamics equation is

$$\frac{dT_k}{dt} = \kappa \left[T\left(\frac{\omega}{p}\right)\right]_k + Q_k \tag{2}$$

The continuity equation in the σ coordinate system is

$$\frac{d(lnp_s)_k}{dt} = -\delta_k - \left(\frac{\partial\dot{\sigma}}{\partial\sigma}\right)_k \tag{3}$$

Note that Eq. (3) applies at each level k. There is a simple relationship which can be derived for $\left(\frac{\omega}{p}\right)$ which is also needed; see for example equation (1.d.6) in Williamson, et. al.[7].

$$\frac{\omega}{p} = \vec{v} \cdot \vec{\nabla} lnp_s - \frac{1}{\sigma}\int_0^\sigma \left(\delta + \vec{v} \cdot \vec{\nabla} lnp_s\right) d\sigma \tag{4}$$

In the equations above, repeated indices imply a summation, e.g.,

$$H_{kl}T_l = \sum_{l=k}^{K} H_{kl}T_l.$$

Also, $\vec{\nabla}$ is the two-dimensional horizontal gradient operator, $\hat{k}$ is the radial unit vector outward, and K is the number of levels.

The physical constants of the model are:

- Ω the angular velocity of the earth,
- R the gas constant,
- C_p the specific heat of dry air, and
- κ the ratio of R/C_p.

The prognostic variables of the system are:

- $\vec{v}$ the horizontal velocity field, and
- δ the divergence of $\vec{v}$,
- T the temperature field, and
- lnp_s the natural log of the surface pressure.

There is one diagnostic quantity:

- $\dot{\sigma}$ the vertical velocity in sigma coordinates.

The physical forcing terms are:

- $\vec{F}$ the velocity forcing from physics,and
- Q the net heating rate from physics.

Finally the model is completed by the definition of the time invariant quantities

- f the Coriolis parameter ($\equiv 2\Omega\sin\phi$), where ϕ is the latitude,
- Φ_s the surface geopotential, and
- H_{kl} the hydrostatic matrix.

The form of the idealized physical forcings are generically

$$\vec{F} = -K_v(\sigma)\vec{v}, \tag{5}$$

and

$$Q = K_t(\sigma,\phi)(T_{eq} - T). \tag{6}$$

The detailed form of the coefficients K_v and K_t is described in the paper by Held and Suarez[5].

The discretization of the equations in the vertical follows CCM. The vertical direction is divided into K layers, $k = 1, K$ and $K+1$ interfaces, $k = \frac{1}{2}, K+\frac{1}{2}$. The arrangement of state variables on these levels is illustrated on page 36 of Williamson, et. al.[7]. The difference operator Δ is defined

$$\Delta \equiv (\ )_{k+\frac{1}{2}} - (\ )_{k-\frac{1}{2}}$$

Based on this operator there are some simple relationships to note:

$$\sum_{k=1}^{K} \Delta\sigma_k = 1,$$

$$\sum_{l=1}^{k} \Delta\sigma_l = \sigma_{k+\frac{1}{2}}$$

For example, discretizing Eq. (4) yields

$$\left(\frac{\omega}{p}\right)_k = \vec{v}_k \cdot \vec{\nabla} lnp_s - \frac{1}{\sigma_k} \sum_{l=1}^{k} \gamma_{lk} \left[\delta_l + \vec{v}_l \vec{\nabla} lnp_s\right] \Delta\sigma_l \tag{7}$$

where

$$\gamma_{lk} = \begin{cases} 1 & l < k \\ \frac{1}{2} & 1 = k \end{cases} \tag{8}$$

To see why this final factor of $\frac{1}{2}$ is necessary, one must remember that the integration interval in Eq. (4) above is $(0, \sigma_k)$. It turns out that for energy consistency, the hydrostatic matrix H_{lk} is given in terms of these deffinitions by

$$H_{kl} = \frac{\gamma_{kl}\Delta\sigma_l}{\sigma_l} \tag{9}$$

This relationship is discussed in detail in Williamson and Olson[8].

The explicit time discretization for a timestep Δt, is generically given, at timestep n, by

$$\vec{v}^{n+1} = \vec{v}^{n-1} + 2\Delta t \frac{\partial \vec{v}^n}{\partial t}, \tag{10}$$

$$T^{n+1} = T^{n-1} + 2\Delta t \frac{\partial T^n}{\partial t}, \tag{11}$$

$$lnp_s^{n+1} = lnp_s^{n-1} + 2\Delta t \frac{\partial lnp_s^n}{\partial t}. \tag{12}$$

Where $\frac{\partial \vec{v}^n}{\partial t}$, $\frac{\partial T^n}{\partial t}$ and $\frac{\partial lnp_s{}^n}{\partial t}$ are the right hand side terms of the governing equations given in Eqs. (1-3).

The semi-implicit scheme modifies this situation by introducing linearized terms at timesteps n, $n-1$, and $n+1$, in which T is replaced by a reference temperature, T^r. For example, the thermodynamics equation, Eq. (11), becomes

$$\left[T - \kappa T^r (lnp_s) - \kappa T^r \Delta t \left(\frac{\dot{\sigma}}{\sigma}\right)\right]^{n+1} = T^{n-1} + 2\Delta t \left[\frac{\partial T^n}{\partial t} + \mathcal{L}_1^n + \mathcal{L}_2^{n-1}\right], \tag{13}$$

where $\mathcal{L}_1$ and $\mathcal{L}_2$ are collections of linearized terms at timesteps n and $n-1$ respectively. The resulting system of coupled equations can be solved to produce a single equation in one variable of the form

$$(\nabla^2 - \frac{1}{(\lambda_k \Delta t)^2}) X_k = Y_k. \tag{14}$$

Here Y_k is formed from linear combinations of the right hand sides of the governing equations, λ_k is the kth vertical normal mode of the baroclinic system, and the solution X_k is used to generate all the fields and derivatives at timestep $n+1$ at level k.

3 The Dynamical Core Testbed Software Design

It is often cited as a goal of software design that code be efficient, modular, portable and scalable. It is not often mentioned that there is no "free lunch" in this endeavor. There are *always* some tradeoffs. The magnitude of the tradeoffs between modularity, scalability, efficiency and portability must be measured or estimated as the code is developed, so that judicious compromises can be made at the outset. For example, performance optimizations such as code inlining, loop unrolling and loop fusion clearly damage code modularity and portability but may greatly enhance performance. A simple 4 way unrolling by hand of a matrix vector multiply can more as double performance on the DEC 21164 processor. Such hand optimizations are gained at the expense of model flexibility and readibility. On the other hand, modern compilers increasingly provide facilities for automating these tasks, rendering these optimizations by the model developer increasingly unnecessary. In this project, the view is taken that most short term performance enhancements of this kind have long term bad effects on model flexibility, and should be avoided.

Previous work with legacy codes has shown that it is difficult to parallelize code beyond the highest level of parallelism originally conceived in the code design[9]. Therefore, it is vital that all of the parallelism inherent in the problem is can be made manifest. Designing a code to be scalable from the outset should allow the code to keep pace with the industry wide trends towards increasing levels of processor parallelism in computer architectures.

3.1 *Towards a Flexible Modular Design*

The testbed design objectives include

1. The capability to intercompare new grid point methods directly with the spectral method in the same model. The spherical harmonic transform, or spectral method, is widely used in general circulation models, including CCM3, and provides an important modeling reference for debugging purposes.

2. The capability to perform either Eulerian or semi-Lagrange time integration. Not all grid methods, for example the spectral element method described in section 4.2, are set up to use a semi-Lagrangian time integration scheme. An Eulerian formulation must be provided.

3. The capability to perform explicit or semi-implicit time integration. Again, efficient elliptic solvers do not exist for all methods, for example the spectral element method considered here. An explicit formulation must be provided.

4. Nothing in the design should *preclude* efficient use of traditional vector or cache-based microprocessors.

Supporting efficient and readable Eulerian or semi-Lagrangian and implicit or explicit time integrations schemes for shared or distributed memory computers whith either vector or cache processors potentially involves sixteen different versions of the code. Thus for m different dynamical methods, $16m$ possible different versions of the model could exist, with $16m$ times as much source code, and many more opportunities for bugs. Clearly, supporting all of these different capabilities by maintaining separate codes is prohibitively expensive.

The dynamical core testbed is being constructed with the belief that modularity of design leads to significant code reuse, which in turn increases modeling flexibility and reliability. For this design process to be successfull, the viability of the design must be checked against the requirements of different real numerical methods from the outset. A useful starting point is to develop a timestepping prescription which is generic enough to encompass a diverse set of modeling options.

The timestep begins by assuming that the state variables, physics forcings and the derivative terms δ, $\vec{\nabla} v$, $\vec{\nabla} T$, and $\vec{\nabla} lnp_s$, are all available in grid point space. This allows the spectral and grid point versions of the dynamical core to share the code calculating the right hand sides of Eqs. (1-3) in one subroutine called **buildrhs**. The use of semi-Lagrange time integration will introduce the need for horizontal **interpolation** to departure points for some of the terms in **buildrhs**. Using semi-implicit time integration will introduce extra terms in the equations in **buildrhs**[8].

Next the prognostics in Eqs. (10-12), and the derivatives of the prognostics for use in the next timestep must be calculated. In the case of semi-implicit time integration, this update requires the solution of K elliptic equations in Eq. (14). Eq. (14) can be solved with an elliptic solver routine taylored to the particular grid method. This solver is referred to here generically as **esolver**.

The spectral and grid point methods differ dramatically in how the timestep $n+1$ quantities are calculated from the output of **buildrhs** . Because the velocity is not single valued at the poles, one must transform the velocity $\vec{v}$ to a new variable $\vec{V} \equiv \cos(\phi)\vec{v}$ for the spectral method. In the spectral method, horizontal derivatives, diffusion and elliptic solve are all performed in spectral space where they take an especially simple form. Because of these differences, all of these calculations are hidden under a routine called **update** in routine **advance**. In grid point models, applying horizontal derivatives and diffusion and performing the elliptic solves are complicated functions which require the separate routines **hordif**, **derivative**, and **esolve** under **update**.

Following CCM[10], the model's timestepping scheme is completed with Robert time filtering of the prognostics[11]. A module called **filter**, is provided under **advance**. The analysis of the similarities and differences of spectral and grid point methods has thus resulted in a more complex generic calling tree of *functional modules*:

```
program main
call init_model()
call stepon(
     call buildrhs(
          call interpolation()    ! SLT
          )
     call advance(
          call update(
               call hordif()      ! grid point
               call esolver()
               call derivative()
               )
          call filter()
          )
     )
call end_model()
end program main
```

The testbed shares code in a hierarchy of **MODULEs**. The hierarchy links are constructed with the Fortran 90 **USE** construct. The scope of variables and functions defined and used in the modules is controlled using the **PUBLIC** and **PRIVATE** directives within modules. For example, the Robert time filtering routine **filter** is a module shared by all versions. Grid point methods can share the **update** routine source, but not the derivative functions it uses. For example, **buildrhs** is the same for spectral and all grid methods, but differs markedly between semi-Lagrangian/Eulerian and semi-implicit/explicit versions.

At the lowest level new grid methods interface to the model through a small number of basic routine interfaces:

vector operations

kcross($\vec{a}$,$\vec{b}$) computes $\vec{b} = \hat{k} \times \vec{a}$.

dotproduct($\vec{a}$,$\vec{b}$,s) computes $s = \vec{a} \cdot \vec{b}$.

derivatives

grad2s(**s**,$\vec{g}$) computes $\vec{g} = \vec{\nabla} s$.

grad2v($\vec{v}$,$\vec{g}$(2)) computes $\vec{g}_i = \vec{\nabla} v_i, i = 1, 2$.

vorticity($\vec{v}$,ζ) computes $\zeta = \hat{k} \cdot (\vec{\nabla} \times \vec{v})$.

div2($\vec{v}$,δ) computes $\delta = \nabla \cdot \vec{v}$.

semi-Lagrange interpolation

interp2s(s,$\vec{d}$,s^D) computes departure point value of s, s^D, from the departure point offset $\vec{d}$.

interp2v($\vec{v}$,$\vec{d}$,$\vec{v}^D$) computes departure point value of $\vec{v}$, $\vec{v}^D$, from the departure point offset $\vec{d}$.

elliptic solver

esolver(X,m,Y) solves $(\nabla^2 - m^2)X = Y$.

3.2 *Data Structures for Parallel Computing*

The functional modules share information about the model's fields and physical constants through a series of associated *data modules.* Fields in the data modules are F-90 **ALLOCATABLE**, to permit resolution to be set at run time through the initial or restart data file. This feature eliminates the need for recompiling the model to change the resolution, and the possibility of the model and its startup data files having an inconsistent definitions of resolution. In performance experiments comparing the use of allocatable fields to statically declared fields, no measurable effects on performance were detected.

The model expresses a 3-D physical array with N_z vertical layers as the product of the vertical axis with one horizontal axis of extent N_h: i.e. the temperature field is $T(N_h, N_z)$.

Note that the horizontal dimension contains no explicit horizontal graph connectivity information. This connectivity information is held private within modules related to computing horizontal derivatives, interpolations and solving the 2 dimensional elliptic equation in the semi-implicit time integration. Using this design, the the model can support grid methods with a completely unstructured horizontal grid. Additionally, no assumptions are made about whether or not this array of points fully covers the sphere or are even physically contiguous. This is essential so that the model can transparently handle an SPMD parallel programming style on distributed memory computers. It also facilitates load balancing of model physics on parallel machines. Finally, the domain of extent N_h is decomposable into $NTASK$ *task subdomains.* A task subdomain is an array subsection of the N_h axis. It is a piece of work which can be performed by an individual thread on a multiprocessor. Task subdomains do not have to be of equal size. The memory bandwidth overhead for this array subsectioning arrays of shape (N_h, N_z) is paid only once outside an autotasking loop, and can be controlled by use of the **INTENT(IN)** and **INTENT(OUT)** arguments in the F-90 **INTERFACE** block.

This domain decomposition technique allows the testbed to execute on a cluster of multiprocessors. A multiprocessor cluster consists of N nodes with P processors per node, and is the natural generalization of both the MPP and the PVP architecture:

1. a 1024 processor SGI/Cray T3E MPP is an SMP cluster with a $P = 1$ and $N = 1024$,
2. an SGI/Cray J916 is a trivial cluster with $P = 16$ and $N = 1$,
3. the desk-top workstation is the $P = 1$, $N = 1$ trivial case.

Despite the development of the HPF language in recent years, message passing, in the form of PVM and MPI libraries, remain the predominant parallel programming paradigm for distributed memory MPP's. In the shared memory world, parallelization of multiprocessors is generally accomplished using shared memory X3H5 standard autotasking directives. It appears that the natural programming style for the multiprocessor cluster will be a *hybrid* of the two defacto standards: MPI message passing between nodes and autotasking directives in parallel regions on each multiprocessor. The dynamic core testbed is being designed assuming that software written in this hybrid style will be the most portable.

There are some restrictions currently on this programming model. For example, thread parallel communications between nodes and parallel I/O may not possible on many cluster systems for some time. Therefore, the testbed design restricts internode communications from single threaded regions of the SPMD code.

It is desirable to have a single routine interface that handles the communication of multiple data types between nodes. The parallel communation routine interfaces in the model can be overloaded with multiple data types, through the use of F-90 **INTERFACE** blocks. This increases the readability and modularity of the communication code.

3.3 *Portability: Vector vs Cache Processors*

Unfortunately, the high performance computer market place is sufficiently muddled between vector and cache-based processors that it is currently impossible to ignore one or the other entirely. Therefore the baroclinic sigma model design is being tested both on cache microprocessors (e.g. workstations and Sun Enterprise and DEC 4100 servers), and traditional vector shared memory computers (SGI/Cray J9's). On the DEC 4100 server it has proven to be possible to obtain, because of cache size effects, performance improvements of about 1.5-2.0, in selected kernels, by transposing the axis order to (N_z, N_h). On vector computers, these optimizations would prove disasterous to performance because of the short vector length in the z direction. It would also require grid point operators, developed for the 2-d shallow equations, to be modified to include N_z layers. Finally, to build a complete climate model, legacy physics of CCM3 will have to be added to the testbed. In CCM3, the physics is all designed around arrays of shape (N_h, N_z). For these reasons, this optimization of placing the vertical axis first is not implemented.

The model design also ignores that fact that caches in different chip architectures are constructed with distinct depths, widths, set associativities and sizes. It is hoped that this sort of detailed performance tuning for specific chip architectures will increasingly be the job of future optimizing compilers.

4 Horizontal Methods

In this section we describe the two grid point methods to be evaluated in the testbed. The two have important similarities and differences in their computational characteristics.

4.1 *Spherical Triangles on the Icosahedral Grid*

The motivation for investigating dynamical methods on spherical triangular grids derived from the icosahedron has long been their almost uniform subdivision of the spherical surface, as shown in Figure 1. The most uniform triangular grids are obtained by projecting the regular icosahedron onto the sphere and then subdividing each of the resulting twenty spherical triangles into smaller spherical subtriangles. Such a discretization yields a data structure corresponding to ten logically square "diamonds" that cover the sphere, a data structure that maps readily onto both vector and parallel machines.

The history of spherical triangular grids in the context of global atmospheric modeling dates back to at least the 1960's. Early formulations displayed noise problems[12,13,14]. Cullen later employed a finite element method which treated the noise problem but displayed a prominent wave number five error pattern[15]. Because of the persistence of this error, and the rapid emergence of the spectral transform method at that time, the icosahedral approach was abandoned.

In the early 1980's Baumgardner[16] applied the icosahedral mesh to model the dynamics of the earth's mantle using the finite element method using a new type of piecewise linear spherical element. He also developed a highly efficient multigrid algorithm for solving the large system of elliptic equations that arise by exploiting the natural hierarchy of nested grids that result from successive refinements of the icosahedron projected onto the sphere[16,17].

Recently, Baumgardner has investigated the suitability of triangular grid methods for ocean and atmosphere applications using the shallow water equations as a convenient testbed[18]. He has introduced several innovations that lead to improved accuracy relative to the formulations of the early 1970's. These include placing each grid point at the equator of its own spherical coordinate system and generating least-squares optimum differential operators based on these coordinates. He has evaluated several transport schemes and has obtained exceptional accuracy with a semi-Lagrangian method[18]. Based on tests using the shallow water benchmark suite of Williamson et. al.[2] this semi-Lagrangian scheme using the icosahedral mesh appears to be competitive with spectral transform method. Baumgardner has applied the

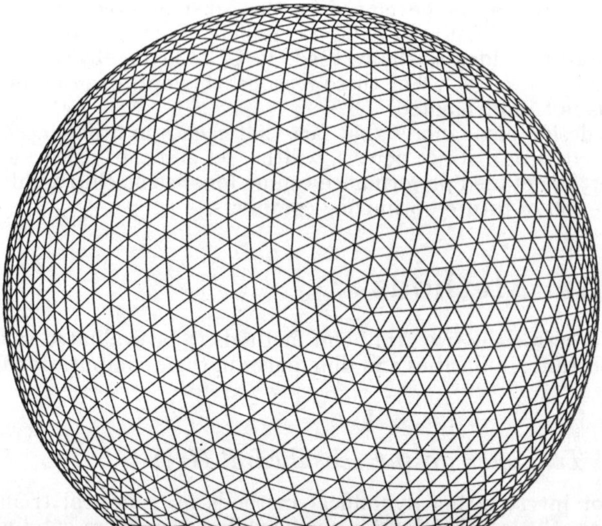

Figure 1: The nearly uniform tiling of the sphere derived from successive refinement of grids derived from the icosahedron is apparent from this image.

fast multigrid solver to obtain a semi-implicit version of the semi-Lagrangian shallow water scheme.

The stencil communication required by Baumgardner's method for parallel derivative calculations is inexpensive and simple. In the semi-Lagrangian time integration scheme, however, Baumgardner has chosen to limit the timestep so that departure points come from nearest neighbor cells. This avoids the more complex generalized departure point cell computations that would result from taking a large timestep. A parallel version of the mantle dynamics code utilizing PVM currently demonstrates less than 10% total parallel overhead on a variety of parallel platforms, from workstation clusters to MPP's such as the Cray T3D[19]. 3-D mantle calculations with 10 million grid points and 50 km spatial resolution are routinely performed on 64 nodes of a Cray T3D[20]

4.2 *Spectral Element Method*

In the spectral element method, an integral form of the governing equations is derived by multiplying both sides of the equations by test functions and integrating over the entire computational domain. This integral is then divided into a sum of integrals over rectangular sub-regions called elements[21]. Within each of these elements, variables are approximated by polynomial expansions of order N and element integrals approximated using Gauss-Lobatto quadrature.

The spectral element method has been used extensively in oceanography in complicated ocean basin geometries[22,23]. Taylor, Tribbia and Iskandarani have studied the shallow water equations on the sphere using a spectral element method similar to that developed by Iskandarani[23]. They have initially tiled the sphere with six square domains by projecting a cube onto the sphere. This tiling with $N = 8$ is shown in Figure 2. Each face of the projected cube can then be further subdivided into smaller

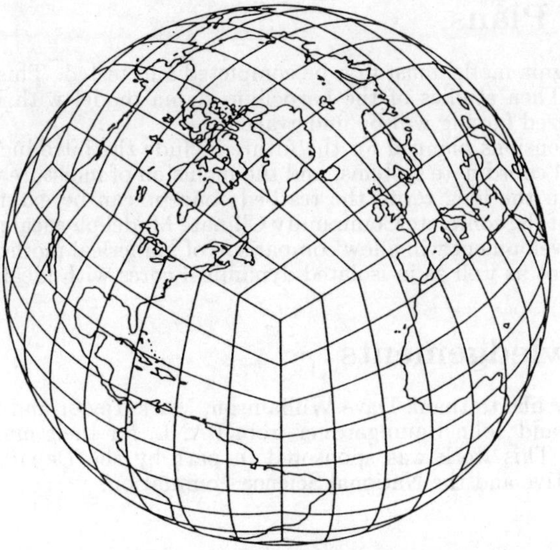

Figure 2: The $M = 6$, $N = 8$ spectral element grid formed from the projection of the cube onto the sphere. Again the "pole problem" is eliminated.

domains. If there are M domains, an operation, such as a gradient, on elements of order N has a computational cost per timestep of $O(MN^2)$. Studies of the tradeoffs between the accuracy and the numerical cost for solving the shallow water equations using the spectral element method, as a function N, have determined that $N \approx 8$ is optimal[3].

The advantages of using the spectral element method on the sphere are clear. The spectral element method converges exponentially under refinement, an attractive feature of the spherical harmonic transform method. The projected cube tiling also avoids the pole problem.

Derivative calculations within the element elegantly reduce to Legendre transforms with $O(N^2)$ computational cost. These are essentially matrix vector multiplies which perform well on high performance computing platforms. Communication between elements required for derivative calculations occurs via the element boundaries, which are shared in common. Elements merely exchange edge values and update them with a weighted average in a way completely analogous to stencil communications in a finite difference code.

One drawback associated with the spectral element method is the lack of efficient spectral element elliptic solvers. This prevents the acceleration of the simulation through the use of semi-implicit time integration. The development of an efficient elliptic solver is vital to further maturation of the technique and remains an active area of research.

5 Future Plans

The baroclinic sigma model must first be completed and tested. This effort is already well underway. Then studies of the baroclinic sigma model with Held-Suarez and Boer Denis idealized forcing will be undertaken.

Model extensions planned for the future include the introduction of hybrid η, and other vertical coordinate systems, and the inclusion of moisture thermodynamics and advection. In the long term, the testbed concept can be extended towards an implementation of the complete Community Climate Model physical parameterization package. This development will allow comparison of numerical methods in a complete model environment as well as in isolated dynamical cores with idealized forcings.

6 Acknowledgements

I would especially like to thank Dave Williamson, Mark Taylor and Steve Hammond all from NCAR, and John Baumgardner from LANL, for their many contributions to this project. This work was sponsored in part by the Department of Energy CHAMMP initiative and the National Science Foundation.

References

[1] H. Ritchie and H. and M. Tanguay *Monthly. Weather Review*, **124**, (1996) 167.

[2] D. L. Williamson, J. B. Drake, J. J. Hack, R. Jakob and P. N. Swarztrauber, *J. Comp. Phys.*, **102**, (1992) 211.

[3] M. Taylor, J. Tribbia and M. Iskandarani, to appear in *J. Comp. Phys.*, 1997.

[4] R. Heikes and D. A. Randall, *Monthly. Weather Review*, **123**, (1995) 1862.

[5] I. Held and M. J. Suarez *Bul. Amer. Meteor. Soc.* (1994) 1825.

[6] Boer and Denis, submitted to *Climate Dynamics*.

[7] Williamson, D. L., J. T. Kiehl, V. Ramanathan, R. E. Dickinson, and J. J. Hack, Description of NCAR Community Climate Model (CCM1), NCAR Tech. Note, NCAR/TN-285+STR, 112 pp., National Center for Atmospheric Research, Boulder, Colo., 1987, Gov't ordering no. NTIS PB87-203782/AS.

[8] D. L. Williamson and J. G. Olson, *Mon. Wea. Rev.*, **122**, (1989) 1594.

[9] S. W. Hammond, R. D. Loft, J. M. Dennis, and R. K. Sato, in *Proceedings of Seventh SIAM Conference on Parallel Processing for Scientific Computations*, ed. D. H. Bailey [et. al.], (1995) p. 125.

[10] J. J. Hack, B. A. Boville, B. P. Briegleb, J. T. Kiehl, P. J. Rasch, and D. L. Williamson, *Description of the NCAR Community Climate Model (CCM2)*. Technical Report NCAR Technical Note/TN-382+STR, Climate and Global Dynamics Division, National Center for Atmospheric Research, PO Box 3000, Boulder, CO 80307, 1993.

[11] A. J. Robert, *J. Meteor. Soc. Japan*, **44**, (1966) 237.

[12] R. Sadourny, A. Arakawa, and Y. Mintz, *Monthly Weather Review*, **96**, (1968) 351.

[13] D. Williamson, *Tellus*, **20**, (1968) 642.

[14] D. L. Williamson, *J. Comp. Phys.*, **7**, (1971) 301.

[15] M. Cullen, *Quart. J. Roy. Met. Soc.*, **100**, (1974) 555.

[16] Baumgardner, J. R., *A three-dimensional finite element model for mantle convection*, Ph.D. thesis, UCLA, 1983.

[17] J. R. Baumgardner and P. O. Frederickson, *SIAM J. Numer. Anal.*, **22**, (1985) 1107.

[18] J. Baumgardner, *An Implicit Semi-Lagrange Method for the Shallow Water Equations*, LA-UR-94-2328 (1994).

[19] Bunge, H.-P. and J. R. Baumgardner, *Computers in Physics*, **9** (1995) 207.

[20] H. P. Bunge, M. A. Richards, and J. R. Baumgardner, *Science*, in press, 1995.

[21] Y. Maday and A. T. Patera, in *State of the Art Surveys in Computational Mechanics*, ed. A. K. Noor (ASME,New York,1988).

[22] H. Ma, *J. Comp. Phys.*, **109**, (1993) 133.

[23] M. Iskandarani, D. L. Haidvogel and J. P. Boyd, *Int. J. Num. Meth. Fl.*, **20**, (1995) 393.

CLIMATE AND WEATHER MODELLING USING DIFFERENT PARALLEL COMPUTERS

S.K.DASH
Centre for Atmospheric Sciences
Indian Institute of Technology, Delhi
New Delhi 110016, India

ABSTRACT

A spectral general circulation model had been parallelised and implemented at different horizontal and vertical resolutions on a number of processors of an indigenous parallel computer based on low-cost transputers. The parallel spectral model at horizontal resolution T42 with 9 sigma levels in the vertical was used on 32 transputers to conduct a set of sensitivity experiments on the seasonal time scale to study the interannual variability of Indian summer monsoon circulation and rainfall. Each experiment consisting of 120 days model integration took about 62 hours. In order to conduct more numerical experiments on longer time scales using higher resolution parallel model, state-of-the-art processors are being used. An optimum parallel climate model is also being developed.

1. Introduction

Parallel processing has definitely made a mark in climate and weather modelling. Nearly a decade back it was not possible to conduct numerical experiments with General Circulation Models (GCMs) without the help of a supercomputer. Because of the advent of parallel architectures and the inherent parallelism in GCMs, at present one can plan to study an atmospheric phenomenon through numerical experimentation without depending on the prohibitively expensive supercomputers. Ofcourse, there are problems in parallelising a large code written by some one else. For a researcher who, otherwise, is not in a position to carry out his study because of lack of adequate supercomputing facility, these difficulties are worth undertaking. In an environment of heterogeneous computers which may include mainframe computers, workstations and parallel computers, the issue is somewhat different because of wide choice of computing. Nevertheless, parallel processing is a technique rather than a computing system only. One can apply this technique to an application software to take advantage of many a computing platforms and complete the job faster.

The present paper deals with the issue of using low-cost parallel processors for numerical experimentation in a situation where a supercomputer is not available to the researchers. The T21 spectral model of ECMWF was parallelised in order to examine the feasibility of running the GCM on 16 transputers of C-DOT (Centre for Development of Telematics) High Performance

Parallel Computer (CHIPPS) built in India[1] and on 17 processors of Meiko T800. The parallel model was integrated for many days and the results were verified with those obtained from the corresponding sequential model code integrated on Cray XMP. Comparison of timings indicated that 16 transputers were about 8 times slower than that of one processor of Cray XMP in integrating the T21 spectral model. The parallel GCM at horizontal resolutions T21 and T42 was successfully implemented on 16 and 32 transputers[2] of Meiko respectively and the theoretical maximum speedup and parallel efficiency were computed. The parallel model at resolution T63L09 was also implemented on 48 processors of CHIPPS and it was shown that using vector global sum technique for adding the contributions of different latitude pairs to spectral coefficients, the time for sequential computations can be reduced and thereby the theoretical maximum speedup and parallel efficiency of 48 processors[3] can be enhanced to about 41% and 85% respectively. The primary aim of these exercises was to develop an environment for integrating the GCM for months together without depending on supercomputers.

The present text deals with a number of sensitivity experiments concerning the interannual variability of Indian summer monsoon conducted on 32 transputers using the T42L09 parallel spectral model, the efforts made to use state-of-the-art processors in different PARAM parallel computers and the development of an optimised parallel climate model on 192 transputers of CHIPPS.

2. Sensitivity Experiments with T42L09 Parallel Model

The parallel spectral GCM at resolution T42L09 with 42 waves in the triangular truncation scheme in the horizontal and 9 sigma levels in the vertical has been used on 32 nodes of Meiko T800 to conduct a number of sensitivity experiments concerning the interannual variations in monsoon in 1987 and 1988. Physical processes in the parallel model consist of dry and moist adiabatic adjustment, large scale precipitation, planetary boundary layer fluxes of heat, moisture and momentum, GFDL radiation scheme and Kuo convection scheme. The parallel model has been used to conduct sensitivity experiments for examining the role of orographic representation, sea surface temperature and deep convection scheme in the monsoon simulation.

In total 10 sets of sensitivity experiments of 120 days each were conducted on 32 nodes. The prescribed initial conditions were the analysed data of 1 June 1987 and 1988 prepared by ECMWF. For each of these initial conditions two different sets of integrations were performed; one with climatological Sea Surface Temperature (SST) and another with observed SST. For these runs, COADS global climatological SSTs and the observed SSTs for 1987 and 1988 from ECMWF analysis were used. In one group of 4 experiments mean orography was prescribed in the model and in the other 4 experiments envelope orography of one standard deviation had been used. In 1988 two more experiments had been conducted by enhancing the orography to twice the standard deviation. Each run covered the four monsoon months from June to September. In each run, the seasonal average wind field and geopotential at different levels, sea level pressure and rainfall have been computed in order to analyse the results.

The time for parallel computation, overheads such as sequential computation and data communication across the nodes and the total elapsed time are shown in Table 1. Since a single 120 days run takes about 62 hours, it was convenient to use restart facility and complete the job in many runs of short duration. The theoretical maximum speedup and parallel efficiency of 32 processors have been computed based on the concept of Amdahl[4]. The following simple formulae have been used for the purpose.

$$Speedup = \frac{p\,t_e}{t}, \qquad Efficiency = \frac{t_e}{t} \tag{1}$$

where, $t = t_e + t_o$

Here the total elapsed time $t = t_e + t_o$, t_e is the time of execution on a single node and t_o is the total time for overheads on the p-node system.

As shown in Table 1, the speedup and efficiency are less in case of T42L09 model on 32 processors obeying Amdahl's law, but the values are enhanced after using the vector global sum technique in case of T63L09 and T80L18 parallel models. In the latter cases there is infact overlap of computation and communication.

3. Climate Modelling using PARAM systems

In order to conduct more sensitivity experiments with T42L09 parallel spectral model on the time scale of years or with enhanced spatial resolution of the model, it is essential to reduce the elapsed time for the model integration. PARAM system of parallel computers of the Centre for Development of Advanced Computing (C-DAC) consists of a range of parallel machines such as PARAM OPOUS, 9000SS, 9000AA and 8600 based on Ultra, Supersparc20, Alpha 21064 and i860 respectively. The parallel spectral model at different resolutions is being implemented on all these parallel systems in order to compare the relative performance of the different types of processors while integrating GCM on longer time scales. The theoretical maximum speedup and parallel efficiency are being computed for all the systems.

The total time for model run is divided into parallel, sequential and communication times and the formula based on Amdahl's approach is being used to compute the theoretical maximum speedup and hence the parallel efficiency. The following relation[5] is used.

Performance of a number of nodes in running parallel GCMs of different spatial resolutions on different time scales

	120 days run of T42L09 parallel model on 32 nodes of Meiko T800	10 days run of T63L09 parallel model on 48 nodes of CHIPPS	1 day run of T80L18 parallel model on 128 nodes of CHIPPS
Time for parallel execution (t_e) in hours	17.5	4.5	2.3
Time for overheads (t_o) in hours	45.0	1.0	1.1
Total elapsed time (t) in hours	62.5	5.5	3.4
Speedup	8.96	39.27	86.53
Parallel Efficiency (%)	28.0	81.8	67.6

$$Speedup = \frac{1}{f + \frac{1-f}{p} + \frac{t_c}{t_s}} \qquad (2)$$

Here f is the fraction of the sequential code that can not be parallelised to the total sequential code, t_c is the communication overhead in the p-processor system and t_s is the total execution time for the sequential code on a single processor. It is known that increasing the number of processors beyond a certain limit need not be beneficial in the context of decreasing the elapsed time of model integration. Thus an examination of a possible relationship between model resolution and the number of processors to get the maximum throughput is warranted.

4. Development of an Optimised Parallel Climate Model

The spectral model of the National Centre for Medium Range Weather Forecasting (NCMRWF), Delhi at resolution T80L18 has been parallelised and implemented on 128 processors of CHIPPS. This parallel model takes about 3 hours and 25 elapsed minutes (Table 1) for one day integration. Considering the fact that a climate model is run for a long time, it is almost impossible to use this set up for climate studies. Most of the climate models are having resolution at about T42 in the horizontal. Hence the T80L18 parallel model can be scaled down to T42L18 for long time integrations. Thus the recent version of the operational model of NCMRWF is being parallelised on CHIPPS-192 situated at the climate modelling division of the Indian Institute of Tropical Meteorology (IITM), Pune. The approach to parallelisation is two dimensional which includes dissection along latitude lines and in the vertical. The 64 Gaussian latitudes in the T42L18 model are assigned to 64 processors with computations of one latitude assigned to one processor. The 18 vertical levels are equally split amongst the processors so that one group of 64 processors compute for a group of 6 levels simultaneously. Thus all the 192 transputers are being used for running the model.

It is essentially a joint venture between the IIT Delhi, IITM Pune, C-DOT Bangalore and NCMRWF Delhi funded by the Department of Science and Technology, Government of India. Optimisation of the code consists of modifying\rewriting the code when necessary so that the parallel computation increases and the overhead decreases. Emphasis is given on the overlap of communication and computation in the code of GCM. No optimisation constructs exist in the CHIPPS environment. Hence each portion of the model code is examined carefully with the aim of reducing the elapsed time of model integration.

5. Conclusions

In the present study integrations of general circulation models on seasonal time scale are made on a number of transputers in the absence of a supercomputer. A number of sensitivity experiments have been conducted on such a computing environment in order to study the interannual variability of monsoon. Efforts are going on to use state-of-the-art processors to complete the model runs in less time.

Weather prediction and climate study by numerical models are done in various stages which comprise of data analysis, initialisation, model integration and post-processing of model results using graphics. The whole job is computer intensive and also time consuming. With the wide range of computers such as mainframes, workstations and parallel computers available to the users at present, it may be convenient and useful to depend on heterogeneous systems for such study. The researcher now-a-days has a wide choice. The problem is the large size of the model code and its complete dependence on the computing environment. No researcher wants to devote time for converting the code from one system to the other. Hence it may be worth rewriting the code of current GCMs into various independent different modules consisting of the whole job. Not only the physical parameterisation schemes but also the dynamical portions of the model needs to be split along latitude lines, longitudes and vertical levels and written into different subroutines, so that the model can be used on any computing platform by plugging in the necessary subroutines.

Acknowledgements

The ongoing projects and the results discussed in this text are the outcome ofjoint collaborative research between different Indian organisations such as IIT Delhi, C-DOT, IITM, C-DAC, NCMRWF and funded by the Department of Science and Technology, Government of India and the All India Council of Technical Education. The author is thankful to all the concerned scientists and officials of these organisations for their continuous encouragement and support. The author is specially thankful to Dr. S. Selvakumar for some of the results reported here.

References

1. S.K. Dash, B.Jha, S. Selvakumar, M. Periasamy, M.V. Pitke and V.K. Misra, in *Parallel Supercomputing in Atmospheric Science,* ed. G.R. Hoffmann and T. Kauranne (World Scientific,1993) p.60.

2. S.K. Dash, S. Selvakumar and B.Jha, *Atmos. Env.,* 29 (1995) 2001.

3. S.K. Dash, M. Periasamy, S. Selvakumar and B.Jha, in *High Performance Computing in the Geosciences,* ed. F.-X.LeDimet, (Kluwer Academic Publishers,1995) p.257.

4. G. Amdahl, *AFIPS Conf. Proc.,* SJCC, Vol 3 (1967) P.483.

5. S.K. Dash and B.Jha, *Comput. Syst. Sci. and Eng.* 11(1996) p.93.

Optimization and Parallelization of the FSU Spectral Weather Forecasting Model on the IBM-SP

ZAPHIRIS CHRISTIDIS

IBM T. J. Watson Research Center
Yorktown Heights, New York 10598, USA

JIN ZHIYAN, WEI YU

National Meteorological Center
46 Baishiqiao Rd., Beijing 100081, P.R. China

ZHANG ZHAN

Department of Meteorology, Florida State University
Tallahassee, Fl 32306-304, USA

Abstract

We are optimizing the Florida State University (FSU) global spectral model to execute efficiently on the IBM POWER2 architecture by improving various computational schemes, through alternate programming techniques. In the process, we are using various tools and algorithms which accelerate and monitor computations. Finally we are parallelizing the code, by employing domain decomposition and message passing techniques on the IBM SP2 Scalable Parallel System.

1 Introduction

Weather forecasting via numerical techniques, imposes high demands on computer processing time and other computer resources. The use of high resolution numerical techniques for the discretization of the modeling domain in space, the incorporation of more complex representations of physical processes or the exploitation of longer time scale phenomena can easily consume the resources of a large computer system. The advent of parallel processing opens a new era in numerical weather prediction. The basis of parallel computing lies in the concept of partitioning the computational domain among parallel processors. Domain decomposition techniques are usually dictated by the weather model's numerical algorithms and the nature of physical processes defined in the model.

In this paper we describe the methodology and techniques used to optimize and parallelize the FSU Spectral Tropical Weather model on the IBM SP distributed memory parallel system. In section 2 we describe the spectral transform method. In section 3 we present a brief overview of IBM SP system and the FSU global spectral model, while in section 4 we discuss the steps taken to optimize the FSU model on a single SP node. Section 4 describes the techniques used to port the FSU spectral model on the SP, while results are presented in section 6. Finally we conclude with suggestions of potential areas of future research, targeted towards further improving the performance of the FSU weather model on the IBM SP.

2 The Spectral Transform Method

A fundamental approach in modeling atmospheric motions is to solve numerically a system of time dependent differential equations that is based on mathematical statements of conservation laws for air mass, momentum and energy in the three-dimensional space. These equations are known as the "Primitive Equations of Motion"[10]. Over the past years a number of numerical methods have been developed for the solution of the primitive equations of motion. The object of many of these methods has been to minimize numerical errors in the approximations for spatial derivatives. A widely known method with high accuracy is the *spectral* method [8], which consists of expanding the unknown dependent variables of an equation into a series of global, orthogonal and complete set of functions (basis functions), and demanding that the equation be exactly satisfied at a set of points in the space domain (grid or collocation points). The spherical harmonics [14] is the appropriate basis set for the solution of the primitive equations on a sphere. Thus, any meteorological variable, c, can be expanded as[11]:

$$c(\lambda, \mu) = \sum_{m=-M}^{M} \sum_{n=|m|}^{N(m)} c_n^m P_n^m(\mu) e^{im\lambda}, \tag{1}$$

where λ is the longitude, $\mu = \sin\theta$ is the sine of latitude θ, $P_n^m(\mu)$ is a Legendre associated function of degree n, with m and $n - m$ denoting the longitudinal and latitudinal wavenumbers respectively ($n \geq m$). The spectral coefficients c_n^m of a variable $c(\lambda, \mu)$ can be obtained from the equation

$$c_n^m = \int_{-1}^{1} \left[\frac{1}{2\pi} \int_0^{2\pi} c(\lambda, \mu) e^{-im\lambda} d\lambda \right] P_n^m(\mu) d\mu. \tag{2}$$

In Eq. (1), M is defined as the highest Fourier mode, and $N(m)$ as the highest degree of the associated Legendre function. To evaluate the spectral coefficients c_n^m numerically, a Fast Fourier Transform (FFT) is used to compute the integral $d\lambda$ for any given μ over a set of grid points λ_i ($i = 1, \ldots, I$). The Legendre Transform (LT) is used to calculate the integral $d\mu$ using a Gaussian quadrature rule. Denoting the Gaussian or collocation points by μ_j ($j = 1, \ldots, J$) and the Gaussian weights by w_j we obtain,

$$c_n^m = \sum_{j=1}^{J} c^m(\mu_j) P_n^m(\mu_j) w_j. \tag{3}$$

Eq. (3) is also known as the Direct Legendre Transform (DLT). For each m we can define the Inverse Legendre Transform as,

$$c^m(\mu) = \sum_{n=|m|}^{N(m)} c_n^m P_n^m(\mu). \tag{4}$$

When gradients of meteorological variables need to be evaluated, it is quite common to perform direct Legendre and Fourier transforms, evaluate the gradients analytically in spectral space, and then with inverse Legendre and Fourier transforms, return to the grid-point space, having calculated the gradients with high order of accuracy[2].

The horizontal grid space is rectangular with I grid lines evenly spaced along longitude circles, and J grid lines along latitude circles, placed at the Gaussian quadrature points μ_j. Most of the spectral models employ the so-called *Triangular Spectral Truncation* ($N(m) = M$)[13]. Un-aliased spectral transforms require that $I \geq 3M + 1$ while $J = I/2$. The value of M can be used to denote the horizontal grid resolution. Thus for $M = 106$, the term "T106L14" uniformly identifies a spectral model with a triangular truncation of 106 spectral modes, 14 vertical layers, $I = 320$ grid points in longitude, and $J = 160$ grid points in latitude.

3 The FSU Global Spectral Model

The FSU model[12], is a global spectral weather model suitable for tropical weather prediction studies. It's development originated in 1976 under the supervision of prof. T.N. Krishnamurti at Florida State University[3]. The model continuously undergoes changes and modifications towards the improvement of various physical processes and parameterization schemes suitable to weather phenomena in the tropics.

The FSU model uses the basic primitive equations of motion as specified on a spherical coordinate system, where the independent variables are the latitude θ, longitude λ, height σ and time t. The primitive equations prescribe as prognostic dependent variables the vorticity ζ, divergence D, temperature T, dew point depression q_s, and the logarithm of the surface pressure, $\log p_s$. The rest of the dependent variables like the vertical velocity ω ($\dot{\sigma}$), and geopotential height Φ, are prescribed via a sub-set of diagnostic equations[9].

The numerical schemes used in the horizontal and vertical directions are triangular alias-free spectral truncations (T63, T106, etc), and finite difference representation respectively. A semi-implicit time difference scheme with Asselin time filtering is used to advance the dependent prognostic variables in time. The FSU model employs a linear fourth order horizontal diffusion scheme, envelope orography, and kinematic top and bottom boundary conditions ($\dot{\sigma} = 0$). It also uses a nonlinear normal mode physical initialization scheme, while it incorporates a host of physical processes[12]. These processes, collectively termed as "physics" involve computations to simulate a wide range of atmospheric phenomena relating to clouds and precipitation, dry and moist convection, long and short wave radiative transfer, as well as the development and evolution of the planetary boundary layer. All these processes involve independent computations on each vertical column in the model domain. The processes termed as "dynamics" involve computations mainly in the horizontal directions. These computations treat the non-linear and linear terms in the primitive equations used for the solution of the dependent model variables.

Upon solving the primitive equations on the earth's surface, there is a continuous transformation of various meteorological variables between the grid-point and the spectral domains, at every time step. These transformations are achieved via the use of direct and inverse Fast Fourier (FFT) and Legendre Transforms (LT).

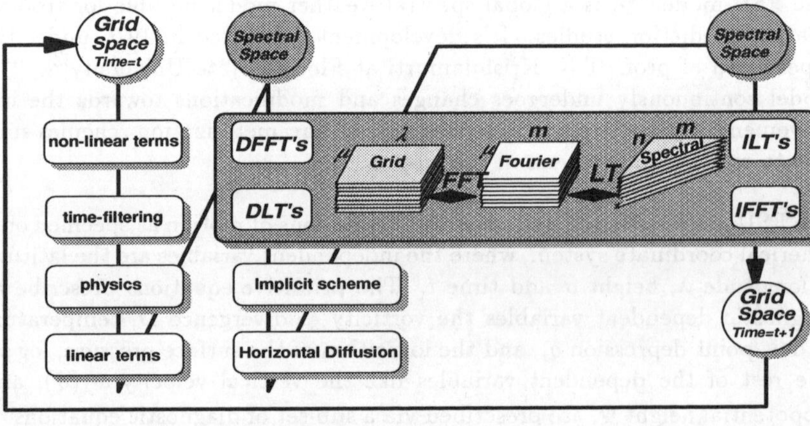

Figure 1: Block diagram of the most significant tasks as executed in the FSU model for each time step.

Figure 1 shows a block diagram involving the main computational processes repeated every time step. The computation of the non-linear and linear terms, as well as the physics and time filtering are performed at the grid point domain. In turn, Direct Fast Fourier Transforms (DFFT's) are applied on each latitude circle and for every vertical layer of the modeling domain, followed by the application of Direct Legendre Transforms (DLT's) on each longitude line and for every vertical layer, to transform the model variables from the grid to the spectral space. The implicit time scheme and the horizontal diffusion are applied in the spectral space, followed by the performance of Inverse Legendre Transforms (ILT's) and Inverse Fast Fourier Transforms (IFT's) to transform the model variables back to the grid-point space.

The FSU model was originally coded in FORTRAN 66, and it was mainly operated on Cray class vector computers. A version of the model containing roughly 15,000 undocumented lines of code, was ported on the RS6000 POWER2 platform[21] and ran in 64 bit precision. Memory requirements for the T106 spectral resolution with 14 vertical layers was close to 16MW. The original code performance was roughly 10 MFLOPS on a single 66.7 MHz POWER2 RS6000.

4 FSU Code Optimization and Tuning

4.1 The IBM SP Parallel System

The FSU model was tuned and parallelized on the IBM SP2 distributed memory parallel system. Tuning of the serial version of the code was performed on a single SP node. The code was tested on a 32 node SP system similar to the one illustrated in Fig. 2. Processor nodes operate at 66.7 MHz and they communicate via a multistage interconnect network known as the High Performance Switch[17], (HPS). Communication is achieved when nodes send or receive message packets through the HPS. The HPS provides a low-latency high-bandwidth communications fabric that can sustain point-to-point bandwidth of over 34 megabytes per second (MB/sec), with latencies close to 40 microseconds for a variety of applications. Each processor node of the SP runs its own copy of AIX (IBM's implementation of Unix). A parallel program, therefore consists of a number of Unix tasks exchanging messages via the HPS. The messages can be encoded using the industry standard Message Passing Interface (MPI), which is a programming interface for FORTRAN and C, and provides with a large number of subroutines supporting task-to-task, as well as collective communication message passing primitives. Execution of the parallel program is achieved via the Parallel Operating Environment (POE)[20] .

Each node of the SP system was configured with 128MB of memory and 2GB of disk space. The SP hardware floating point performance monitor[19] was enabled on each node, to monitor overall code performance during execution.

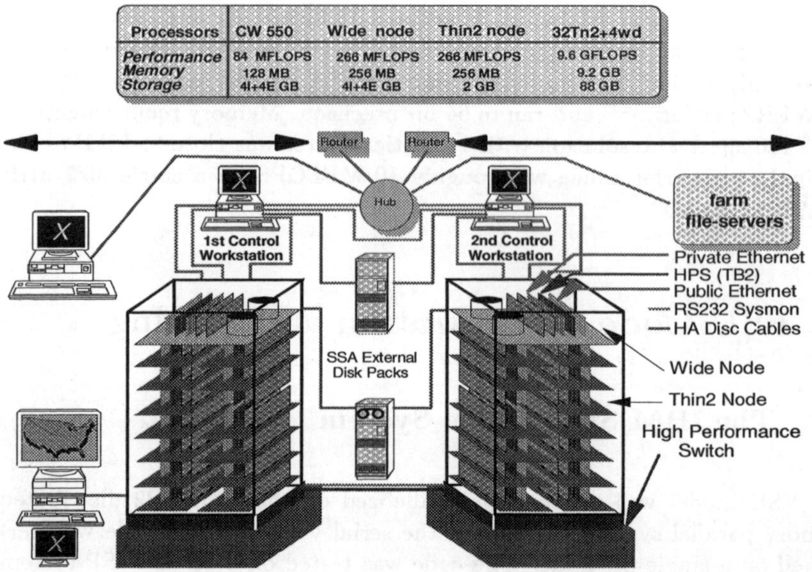

Figure 2: SP2 System Configuration.

4.2 Speedup and Efficiency

If γt_a is the time to execute γ floating point operations on a single node of a parallel system, and βt_c is the time to communicate β bytes between two nodes, then the total elapsed times to execute the FSU model on one, and P processors are:

$$T_1 = G t_a, \qquad T_P = \frac{G}{P} t_a + B t_c \tag{5}$$

where $G = \sum_t \gamma_t$ is the total number of floating point operations performed, while $B = \sum_t \beta_t$ is the total number of bytes communicated. The code Parallel Speedup S_P and Efficiency E_P are defined as:

$$S_P = \frac{T_1}{T_P}, \qquad E_P = \frac{T_1}{P T_P} \times 100\% = \frac{S_P}{P} \times 100\% \tag{6}$$

If the code is completely parallelizable, the computation would be perfectly divisible among processors. Thus,

$$T_P = \frac{T_1}{P}, \qquad S_P = P, \qquad E_P = 100\%$$

Speedup rates deviate from the ideal as P increases mainly due to necessary inter-processor communication, and uneven work-load among processors. When the code is tuned to take into advantage certain CPU architectural features, the overall code execution time is expected to decrease. By letting the superscript τ indicate parameters after code tuning, and by defining ϵ as the reduction of elapsed time due to code tuning ($\epsilon > 1$), we can re-write Eqs. (5) as:

$$T_1^\tau = \frac{G}{\epsilon} t_a, \qquad T_P^\tau = \frac{G}{\epsilon P} t_a + B t_c. \tag{7}$$

The overall effect of code parallelization and tuning can be realized by combining Eqs. (5) and (7) into

$$T_1 = T_P^\tau \frac{\epsilon P}{1+\epsilon\kappa}, \quad \text{where} \quad \kappa = \frac{PB}{G}\frac{t_c}{t_a}.$$

The term κ above, represents the *non-scalable* inter-processor communication effects. By combining Eqs. (5), (6) and (7) we obtain,

$$S_P = \frac{P}{1+\kappa}, \qquad S_P^\tau = \frac{P}{1+\epsilon\kappa} \qquad \text{or} \qquad \frac{E_P}{E_P^\tau} = \frac{1+\epsilon\kappa}{1+\kappa}.$$

Finally by comparing,

$$S_P > S_P^\tau, \qquad \text{or} \qquad E_P > E_P^\tau, \qquad \text{while} \qquad T_P^\tau < T_P. \tag{8}$$

Thus, a tuned version of a parallel code will exhibit worse scalability characteristics than the un-tuned version when the total elapsed time decreases. Our target in optimizing and parallelizing the FSU model was to decrease the overall execution (elapsed) time. Since the code was mainly used as a research tool, not much emphasis was placed towards its efficient execution on scalar and vector processors during development. FORTRAN code tuning yielded significant gains in performance as it can be seen next.

4.3 Single Node Tuning

In tuning standard FORTRAN code for the cache based POWER2 RISC processor architecture[22], we took into account that basic arithmetic operations (addition, multiplication, etc) take a single CPU cycle to execute, while a division operation can take anywhere between 16-19 CPU cycles. The POWER2 processor supports the *multiply-and-add* compound instruction within a single CPU cycle. A total of 4 floating point independent instructions can be executed by the dual floating point unit of the CPU within a single cycle, thus delivering a peak performance of 266 MFLOPS. A raise to a power operation takes roughly 160 cycles to complete, while the computation of a single logarithm or an exponential function averages around 55 cycles[18]

We focused in tuning the serial T106L14 version of the code on a single SP node. Results from the tuned version were directly compared with results from a sample output as obtained from a Cray machine. Initially, aggressive compiler optimization (-O3) resulted in numerically unstable computations. When the code was compiled and executed with no optimization at all, it produced a stable 24 hour simulation with the correct results. The code was executed in 64 bits precision, and it completed a single 7.5 minute time step (with radiation included) in 450 seconds of wall-clock time. A 24 hour forecast was completed in 20 hours of CPU on a single SP node (radiation computations were performed every 3 hours).

It was decided to re-construct the code, by keeping the same programming structure and building blocks, but with eliminating redundancies and "dead-code" segments. In addition, most of the FORTRAN66 constructs were converted to FORTRAN77 with FORTRAN90 extensions. The code was initially profiled, and runtime statistics were obtained in order to focus the optimization efforts to the parts and routines consuming most of the CPU power. One of the primary steps taken, was to replace unnecessary disk I/O with memory to memory copying. The code was executed in double precision (64 bits), in order to achieve maximum floating point performance on the dual 64 bit floating point units of the POWER2 CPU. Various do-loop structures and iterations were modified with the objective to achieve the minimum necessary floating point operation counts. In turn, lengths of crucial do-loops were shortened, in oder to eliminate register "spilling" and to promote better and efficient register re-use. In some occasions the do-loops were even unrolled at a depth of 2, to assist the compiler in generating proper code for the dual floating point units of the POWER2 CPU. Better cache utilization was achieved by inverting several do-loop indices within the code, so data in three dimensional arrays could be accessed with minimal memory strides.

Memory copying routines, as well as the FFT package from the Engineering and Scientific Subroutine Library (ESSL)[16] were used in order to further boost the code performance. A faster but less accurate library of the mathematical intrinsic functions, called Mathematical Acceleration SubSystem[18] (MASS) was used to further reduce the cycle count, in the the standard logarithmic, exponential and raise to a real power intrinsic functions, without any modification to the source code. In addition, vector versions of the exponential and logarithmic functions were used whenever appropriate. In this case, the source code was modified to accommodate calls to the vector MASS libraries. Vector computations of exponentials and logarithms were taking as low as 6-8 cycles per call, resulting in a speedup factor of 8 over the original math library functions. Expressions of the form `y=x**z` were quite common in the code, and were replaced by the equivalent expressions of `y=exp(z*log(x))`, reducing the overall cycle count by 20-40%. Since divisions are relatively expensive to perform as opposed to multiplications, the reciprocals of selected meteorological parameters were computed and stored in memory. These meteorological parameters were mostly static, appearing in denominators of frequently computed expressions. Consequently, divisions were replaced with multiplications for the most part within the code, and whenever a significant amount of cycles could be saved.

Since the FSU model incorporates precipitation processes within the physics, the computation of saturation vapor pressures and it's derivatives for given temperatures T, are often required to be computed. Expressions of the form, $E_s(T) = a \cdot \exp[f(T)]$ where $f(T) = b \cdot g(T)/h(T)$ were evaluated in several routines within the code. The computation of $f(T)$, could take close to 25 cycles, while the computation of the exponential can accumulate up to a total of 50 cycles. Thus, the strategy was to create *lookup tables* by pre-computing E_s for a wide range of temperatures (150-450K), carrying two decimals figures for accuracy and store them in memory as specified by the array `e(it)` (`it=15000,...,45000`). In turn, for a given temperature T, an integer pointer `it` to the array `e()` is produced ($\texttt{it}=100 * T$), in order to fetch from the memory the requested value of $E_s(T)$ =`e(it)`, thus requiring a minimum of 5 CPU cycles. Similarly, in the radiative transfer computation part, transmission functions of the form `fcn(x)=1-a*x**b` were computed several times within a single time step. Each time, this computation required at least 70 CPU cycles. By pre-computing `fcn(x)` for a large range of `x=0.,...500.` and storing it to an array `fc(ix)`, (`ix=0,...,50000`), we were able to retrieve `fcn(x)` from memory as `fc(x*100)`, at the cost of 2 CPU cycles. Emissivities were also computed through linear interpolation from a range of 16 possible radiation regimes, as specified by lookup tables.

Since several approximations were introduced via the use of the lookup tables and the faster intrinsic functions, the accuracy of the computations was monitored by checking the RMS errors of various horizontally integrated model quantities such as energy, momentum, vorticity and divergence. The RMS errors remained sufficiently small during model test simulations. Figure 3

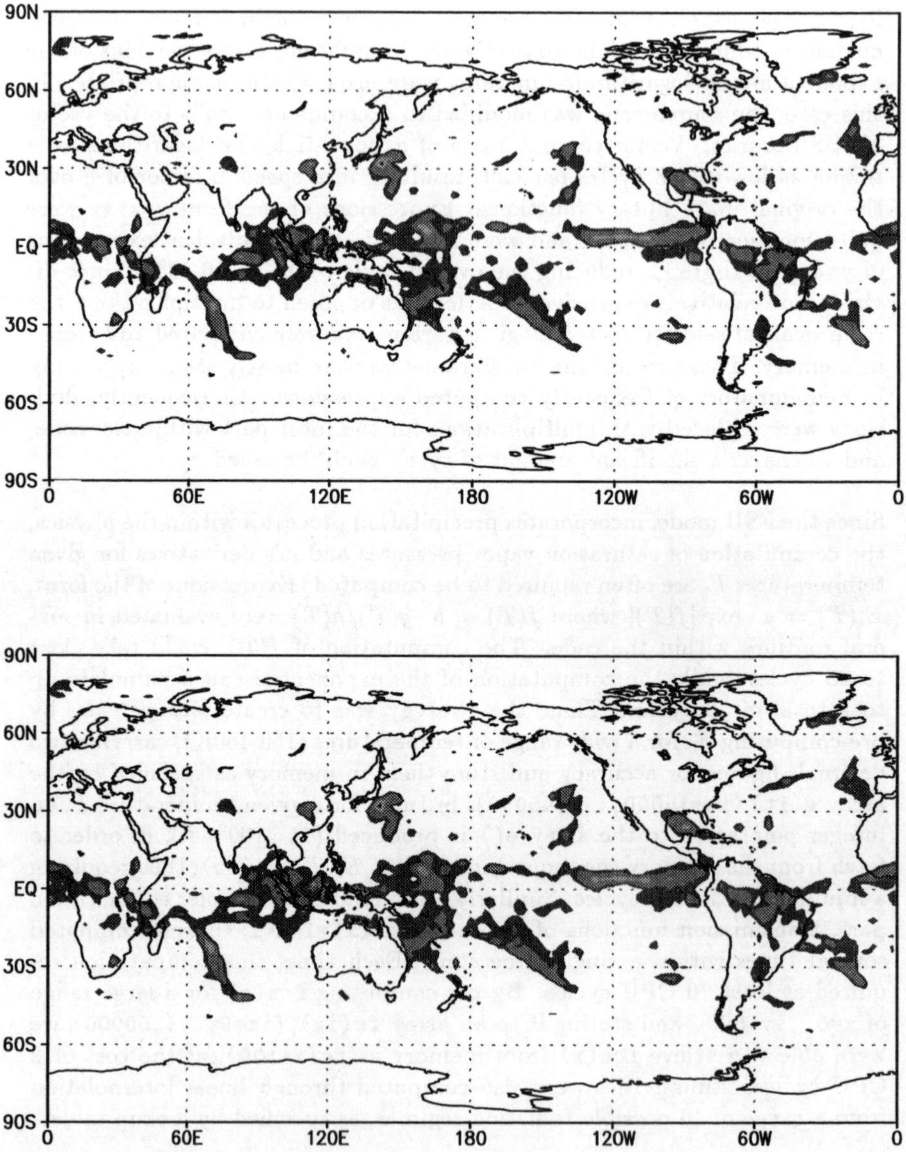

Figure 3: Contours of 24 hour precipitation forecasts with the original version of the FSU model (top), versus the tuned version of the FSU model (bottom, lookup tables included).

displays filled contours of 24 hour precipitation forecasts for the original version of the model (top), versus the tuned version (bottom). As it can be seen, there is practically no difference at all between the forecasted precipitation quantities from the two different serial versions of the code.

Figure 4 displays the reduction in wall-clock time for a single time step, of the serial code. The vertical axis shows time in seconds, while the horizontal axis, indicates the different optimization and tuning steps taken. The ``no-op''

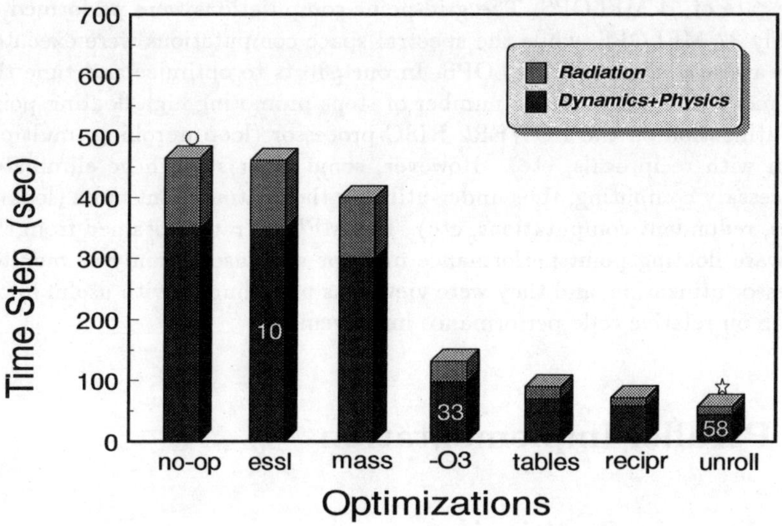

Figure 4: Total elapsed time (sec) of a single time-step (radiation included) for different tuning stages of the FSU T106 model

label on the x-axis indicates that there was no compiler optimization at all, while ``essl'' indicates that the original FFT and memory copying routines were replaced by the equivalent routines from the ESSL library. As expected, there is no significant reduction in elapsed time when using ESSL libraries, due to the small percentage of time taken to execute these routines within the code. The code performance was measured around 10 MFLOPS at that stage. There was a 13% reduction of elapsed time when the MASS intrinsic function libraries were used. The code was further restructured, resulting in the serial model to fully reproduce the original test run results, even under the ''-O3'' compiler option. Using the AIX XL FORTRAN compiler flags **-O3 -qstrict -qrealsize=8 -qarch=pwr2**, the code performed at an average rate of 33 MFLOPS, while taking 135 sec, to execute a single time step. When lookup tables were used (''tables''), there was a further reduction in the total elapsed time by 30% . Upon multiplying with reciprocal quantities, rather than divide (''recipr''), a further reduction of 20% was realized in the overall execution of a single time step. Finally, when unrolling the loops in critical subroutines (''unroll''), the total time was further reduced by 13%, bringing it down to 65 seconds. At this level, a 24 hour forecast was completed in 2.5 hours, while the code was monitored to perform at an average rate of 54 MFLOPS. The grid-point computations were performed at roughly 37 MFLOPS, while the spectral space computations were executed at an average rate of 86 MFLOPS. In our efforts to optimize and tune the FSU model, we have taken a number of steps promoting high floating point unit utilization on the POWER2 RISC processor (loop unrolling, multiplication with reciprocals, etc). However, some other steps have eliminated unnecessary computing, thus under-utilizing the floating point units (lookup tables, redundant computations, etc). The MFLOP rates obtained from the hardware floating point performance monitor were used mainly to monitor processor utilization, and they were viewed as providing us with useful information on relative code performance improvements.

5 Parallel Implementation

5.1 Domain Partitioning

A variety of parallel decomposition schemes and algorithms have been applied in the past on various spectral atmospheric models such as the IFS[4] and the CCM2[5]. The FSU spectral model operates on similar to the IFS and CCM2 three-dimensional data structures, representing grid point, Fourier, and spectral space variables. These data structures are defined in three coordinates. The first two coordinates are latitude and longitude in physical space, Fourier wavenumber m and latitude in the Fourier space, and wavenumber m and degree of of the Legendre associated functions n. The third dimension is the vertical coordinate. On a parallel system these three dimensions can be decomposed over available processors, so parts of the weather model can execute concurrently on every processor. The processors need to communicate with each other at appropriate intervals to exchange data, in order to ensure identical numerical results with the ones obtained from a single processor system. Since there is tight coupling in the vertical direction on various physical processes like radiation, cloud/precipitation and and turbulence, any decomposition over the vertical coordinate would cause high communication overhead. In this work, we focused only in decomposing the horizontal coordinates, as they simplify code development, and provide adequate parallelism on a large number of processors [15].

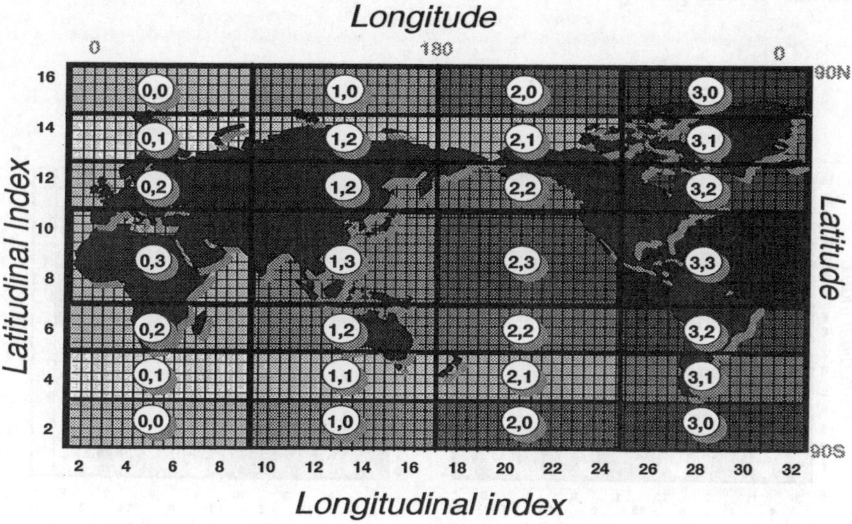

Figure 5: Distribution of the FSU model physical domain over a 4×4 logical grid of processors

The physical or grid-point space was partitioned in two-dimensions, latitude and longitude. The Fourier space was also partitioned in two-dimensions, latitude and wavenumber, while an one-dimensional partition scheme was used in in the spectral space[6]. The horizontal latitude and longitude grid $I \times J$, was viewed as "folded" about the equator, and mapped onto a logical two-dimensional grid of processors $P = P_x \times P_y$. The parameter P_y denotes the number of processors along latitude, while P_x denotes the number of processors along longitude. In specific, the latitudinal dimension was partitioned into $2P_y$ intervals, with each containing $J/2P_y$ Gaussian points along the latitude axis. Each processor row, P_y, was assigned two of these intervals, one from the northern hemisphere, and it's symmetric in latitude from the southern hemisphere. This partition scheme exploited the symmetric properties of the Legendre associated functions, while it had better load-balancing characteristics, as it combined on the same processor northern and southern hemisphere physics computations. Northern hemisphere continental physics are computationally intensive, as opposed to southern hemisphere marine dominated physical processes. The longitudinal dimension was simply partitioned into P_x equal intervals. This decomposition scheme is illustrated in Figure 5, for a $P = 16$ processor system ($P_x = P_y = 4$) and for a total number of 512 grid points ($I = 32, J = 16$). The Fourier and spectral

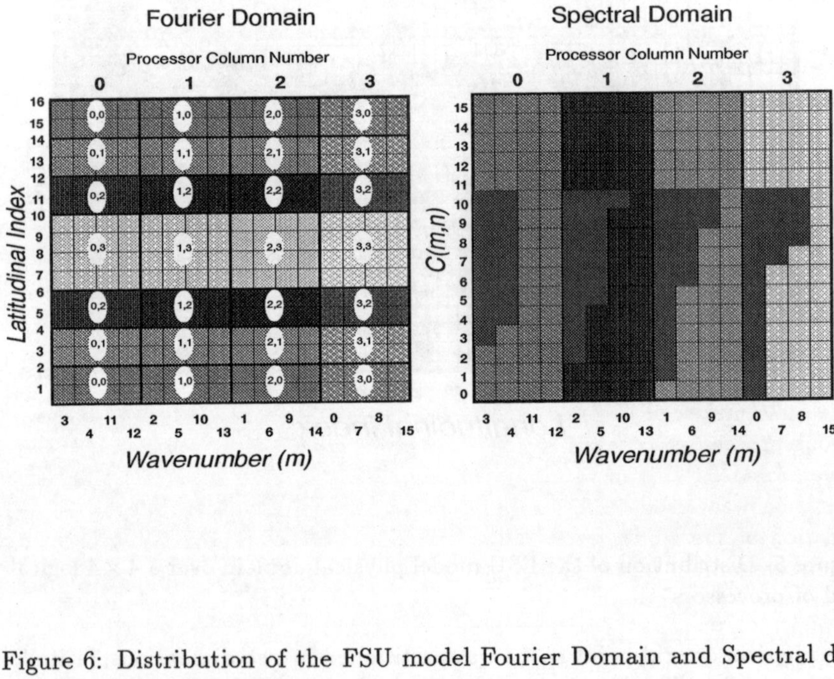

Figure 6: Distribution of the FSU model Fourier Domain and Spectral domain over a 4×4 logical grid of processors

space partitionings are illustrated in Figure 6. In the Fourier partitioning (left), the latitude dimension was partitioned as in the grid-point domain, while the wavenumbers m were partitioned into P_x sub-sets, with each set being assigned to a different processor column. The wavenumbers were reordered in the wavenumber dimension in order to balance the distribution of the spectral coefficients in the spectral space. The spectral coefficients were partitioned in the wavenumber m dimension only. The coefficients associated with a given wavenumber were replicated across the P_x processors to which that wavenumber was assigned (Fig. 6).

The number of the spectral coefficients associated with a given wavenumber m decreases, as the wavenumber increases in a triangular truncation. Without the reordering of the coefficients in the Fourier domain decomposition, noticeable load imbalance would occur in the spectral domain, as processor columns P_x associated with large wavenumbers would have much less spectral coefficients than processor columns associated with small wavenumbers[1].

5.2 Inter-processor Communication

As mentioned before, since most of the physical processes occur in vertical columns in the atmosphere, the two-dimensional data decompositions chosen for the FSU model allowed the physical computations to occur without any communication among processors. However, in order to evaluate gradients of meteorological variables with respect to longitude and latitude, it was necessary to have simultaneous information about all the variables defined on longitude and latitude grid lines. The application of FFTs on longitude lines is a global process, requiring data exchanges among the participating processors P_x. Thus, a matrix transpose operation was used in order to transform the grid point data passed to the FFT from a latitude-longitude decomposition, to a latitude-vertical decomposition. Each processor had locally an equal number of latitude lines, while the vertical layers were partitioned among processors instead. In turn, the processors applied concurrently one-dimensional FFT transforms on local latitude lines. A second transpose operation was used to re-map the Fourier coefficients produced by the application of the FFTs from the latitude-vertical decomposition to the latitude wavenumber m decomposition. This procedure is known as the *Double Transpose FFT* method[6], requiring for each node to send and receive a total number of $2P_y$ messages. The double transpose procedure is illustrated in Figure 7.

Communication was also required for the application of the Legendre transforms in order to go from the Fourier to the Spectral space. For the FSU model, a parallel vector sum was used to accumulate processor contributions of Eq. (3). When applying direct Legendre transforms (DLT), each c_n^m depended only on data associated with the same wavenumber m, which were resident on each processor column, P_x. Each processor concurrently computed its contribution to c_n^m, using data resident in each processor, which depended only on latitude. Finally, the contributions on each processor col-

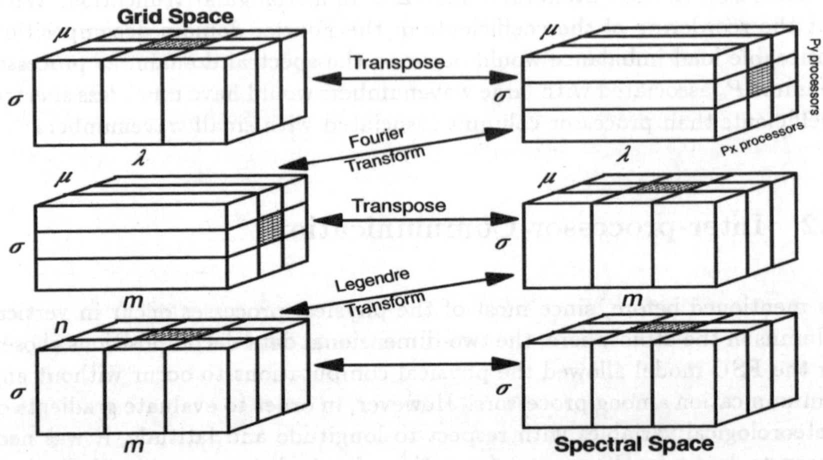

Figure 7: Double Fourier Transposition Strategy

umn were summed, and the results were broadcasted over all P_x processors, hence replicating the spectral coefficients in each processor column. This was achieved via a collective communication primitive, also known as a "combine operation". The Inverse Legendre transform (ILT) phase was performed without any inter-processor communication. The computation of the $c^m(\mu_j)$ required only spectral coefficients associated with the wavenumber m (Eq. 4), all of each were local to every processor, as they were previously replicated in processor columns P_x[7].

The code was parallelized using the MPI message passing library, while the double-transpose FFT was performed on two communication stages for every time step, independed from any latitude or longitude computations.

6 Results

We have measured the performance of two different versions of the paralell code on the 32 node SP system (Fig. 2). The first version was the code including the ESSL and MASS library routines, compiled with the options `-O3 -qstrict -qarch=pwr2 -qrealsize=8`. The second version of code included the lookup tables, the multiplication with reciprocals and the "hand tuned" modifications as described in section 4. The code was allowed to run for a total of 12 time steps (including one radiation time step), thus completing 90 minutes of forecast. The T106 spectral truncation was used, with 14 layers in the vertical direction. Timings for I/O and data intitialization were excluded from our final wall-clock and CPU time measurements. Elapsed times were obtained from P =2 (2×1), 4 (2×2), 8 (4×2), 16 (4×4), 24 (6×4) and 32 (8×4) processor runs ($P = P_x \times P_y$). Speedups and efficiencies were produced by using Eqs. (6) and they are shown in Figure 8,

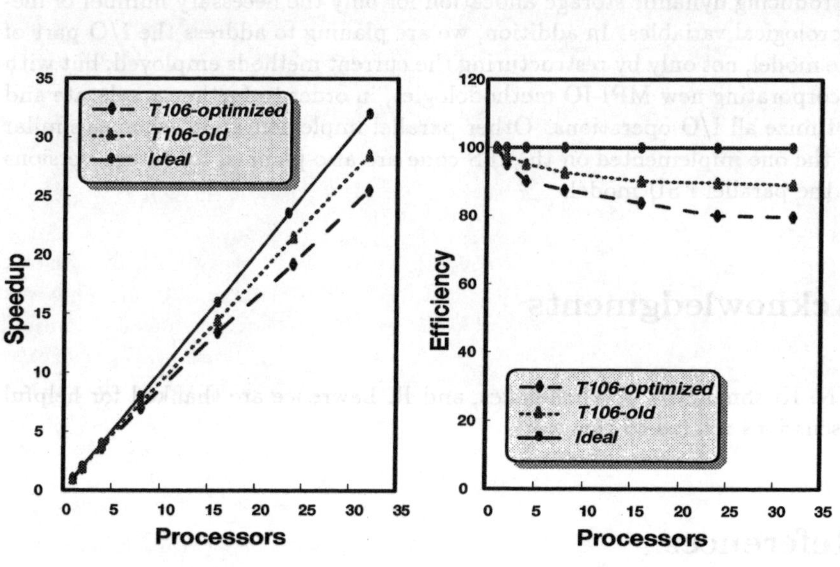

Figure 8: Parallel Speedups and Efficiencies of the T106 FSU model over 32 SP nodes.

as a function of the processors used. It can be seen that the FSU code performance exhibits good scalability characteristics as expected[7]. It is worth noticing that the fully tuned code version exhibits worse scalability characteristics (dashed lines) as compared to the partially tuned version (dotted lines), which is in full agreement with the statement as expressed by Eq. (8).

7 Conclusions

We have parallelized the FSU global spectral model on the IBM SP distributed memory system. The code was completely rewritten and optimized for maximum performance on the the IBM POWER2 processor nodes. Work is still in progress towards additional tuning of the physics computations. We are targeting a further reduction to the total elapsed time on a single CPU by an additional factor of 15%. We are also planning to reduce the overall memory utilization and requirements by further optimizing the code and by introducing dynamic storage allocation for only the necessary number of meteorological variables. In addition, we are planing to address the I/O part of the model, not only by restructuring the current methods employed, but with incorporating new MPI-IO methodologies, in order to further accelerate and optimize all I/O operations. Other parallel implementation schemes similar to the one implemented on the IFS code are also planned for future versions of the parallel FSU model..

Acknowledgments

T.N. Krishnamurti, J. Michalakes, and R. Lawrence are thanked for helpful discussions related to this work.

References

[1] S. Barros T. Kauranne, 1993. On the parallelization of global spectral Eulerian shallow-water models. *Parallel Supercomputing in Atmospheric Science,* Proceedings of the Fifth ECMWF Workshop on the Use of Parallel Processors in Meteorology. G.-R. Hoffman and T. Kauranne, eds., World Scientific Publishing. Co. Pte. Ltd., Singapore, pp. 36-43.

[2] Z.D. Christidis 1989: Parallel Implementation of Advection Calculations in Pseudospectral Atmospheric Hydrodynamical Models, Proceedings of the 3rd Workshop on the Use of Parallel Processors in Meteorology, Edited by G.-R. Hoffmann and D.K. Maretis, Springer Verlag, Berlin Heidelberg New York 1990, ISBN 3-540-51909-2.

[3] Daley, R.C., C. Girard, J. Henderson, I Simmons, 1976: Short term forecasting with a multi-level spectral primitive equation model. Part 1: Model formulation. *Atmosphere*, **14**, 98-116.

[4] D. Dent, 1990. The ECMWF model on the Cray Y-MP8, 1990. *The Dawn of Massively Parallel Processing in Meteorology.* G.-R. Hoffman and D.K. Maretis, eds., Springer-Verlag, Berlin.

[5] J.B. Drake, R. E. Flanery, I. T. Foster, J. J. Hack, J. G. Michalakes, R. L. Stevens, D. W. Walker, D. L. Williamson, and P. H. Worley, 1993. The message-passing version of the parallel community climate model *Parallel Supercomputing in Atmospheric Science,* World Scientific Publishing, Singapore, pp. 500-513.

[6] J. Drake, I. Foster, J. Michalakes, B. Toonen, P. Worley, 1995. Design and Performance of a Scalable Parallel Community Climate Model, *Parallel Computing*, 21(10):1571–1591.

[7] I. T. Foster, W. Gropp, R. Stevens, 1992. The parallel scalability of the spectral transform method. *Mon. Wea. Rev.*, **120**, pp 835-850.

[8] D. G. Fox, and S. A. Orszag, 1973: Pseudospectral approximation to two-dimensional turbulence. *J. Comput. Phys.*, **11**, 612-619.

[9] G. J. Haltiner, and R. T. Williams, 1979: *Numerical Prediction and Dynamic Meteorology.*, Wiley, 477 pp.

[10] Holton, J.R., 1979. *An Introduction to Dynamic Meteorology.*, Academic Press, 391 pp.

[11] D. Gottlieb, and S. A. Orszag, 1977: *Numerical Analysis of Spectral Methods: Theory and Applications.* SIAM, Philadelphia, PA, 200 pp.

[12] T.N. Krishnamurti, D. Oosterhof, N. Dignon, 1989. Hurricane Prediction with a High Resolution Global Model, *Mon. Wea. Rev.*, **117**, 631-669.

[13] T.N. Krishnamurti, H.S. Bedi, V. Hardiker, 1996: Introduction to Global Spectral Modeling, Oxford University Press. (In press).

[14] S. A. Orszag, 1972: Comparison of pseudospectral and spectral approximation.*Stud. Appl . Math.*, **51**, 253-259.

[15] P. H. Worley and I. Foster, 1994. Parallel spectral transform shallow water model: A runtime-tunable parallel benchmark code. *Proc. Scalable High Performance Computing Conf.*, IEEE Computer Society Press, Los Alamitos, Calif., pp. 207-214.

[16] `http://192.35.232.38/resource/technology/essl.html`

[17] `http://192.35.232.38/resource/technology/spswp1.book_1.html`

[18] `http://192.35.232.38/resource/technology/MASS/index.html`

[19] `http://192.35.232.38/resource/technology/monitor.html`

[20] `http://192.35.232.38/resource/technology/adguide1.html`

[21] `http://192.35.232.38/resource/technology/power2_models.html`

[22] `http://192.35.232.38/resource/technology/p2ppc_tech.html`

An Interactive Parallel Programming Environment Applied in Atmospheric Science*

Gregor von Laszewski
Mathematics and Computer Science Division, Argonne National Laboratory
gregor@mcs.anl.gov

Keywords: Graphical program design, visual programming, metacomputing, Java, data assimilation.

Abstract

This article introduces an *interactive parallel programming environment* (IPPE) that simplifies the generation and execution of parallel programs. One of the tasks of the environment is to generate message-passing parallel programs for homogeneous and heterogeneous computing platforms. The parallel programs are represented by using visual objects. This is accomplished with the help of a graphical programming editor that is implemented in Java and enables portability to a wide variety of computer platforms. In contrast to other graphical programming systems, reusable parts of the programs can be stored in a program library to support rapid prototyping. In addition, runtime performance data on different computing platforms is collected in a database. A selection process determines dynamically the software and the hardware platform to be used to solve the problem in minimal wall-clock time. The environment is currently being tested on a Grand Challenge problem, the NASA *four-dimensional data assimilation* system.

1 Introduction

During the development of programs to solve Grand Challenge problems, many diverse computing environments usually are used.[13,15] These environments include different hardware platforms and software packages and are used to solve various smaller subproblems in order to derive a solution for the overall problem.

*This material is based upon work conducted at Northeast Parallel Architectures Center at Syracuse University, supported by the National Aeronautics and Space Administration under Cooperative Agreement No. NCCW-0027. This work was continued under support by the Mathematical, Information, and Computational Sciences Division subprogram of the Office of Computational and Technology Research, U.S. Department of Energy, under Contract W-31-109-Eng-38.

Because of frequent changes in hardware and software, high overhead costs arise in the program design, maintenance, and execution on the ever-increasing number of computing platforms. Unfortunately, only limited time is available for scientific researchers to obtain detailed knowledge of the diverse software environments and hardware platforms. Therefore, in order to simplify the programming task and to reduce the development costs for the design of parallel programs, a graphical programming environment has been developed. The environment, called the *interactive parallel programming environment* (IPPE), allows scientists to embed their favorite software library while providing enough structural guidance to support parallel programming on the basis of a task-parallel approach.

This article describes some design issues of IPPE and its application on the Grand Challenge problem of *four-dimensional data assimilation* (FDDA) system used in atmospheric science studies.

The next section describes the requirements placed on the design of the environment for parallelizing scientific codes. Section 3 describes the environment developed to fulfill these requirements. Then, the usefulness of the environment is shown while applying it to the Grand Challenge FDDA problem (Sections 4–6). Section 7 points out similar research efforts, and Section 8 summarizes the conclusions and outlines future research directions.

2 Motivating Analysis

Many different approaches for parallelizing sequential programs exist. For example, during the parallelization of the FDDA, a top-down approach is used. A typical top-down approach includes the determination of major program blocks and the decision about which parallel programming paradigm is best suited for the parallelization of the problem.

In this article, a task-parallel programming paradigm is used. Several steps are necessary when converting a sequential program to a task-parallel program:

1. Determine the structure of the sequential program, and divide the program in smaller subproblems.
2. Based on the structure analysis, transfer the program into a task graph and include data dependencies.
3. Maximize the parallelism of the task graph.
4. Determine the tasks that are sequential, and reuse as much of the original sequential code as possible.
5. Determine and rewrite the parts of the task graph that are concurrent.
6. Generate a mapping from the task graph onto a virtual computing platform, which will be mapped onto a real hardware platform.

7. Execute the parallel program, and observe the performance.

To reduce the amount of effort needed to parallelize the serial code, the IPPE supports all steps of the parallel program design. Although the software developer still has to determine which parts should be parallelized, the IPPE supports the visualization of the task graph, its mapping onto a real machine, and the animation of the execution of the final parallel program. During use of the visual representation of the program (a task graph), the final parallel program is augmented automatically with the necessary parallel language constructs. Depending on the available hardware and software, these can be message-passing library calls (e.g., MPI or PVM) or additional language constructs (Fortran M, CC++, etc.).
Besides simplifying the coding effort, the IPPE programming environment takes care of the resource management. Therefore, the scientist can concentrate on the problem to be solved, rather than having to decide what resources are available and on which machine the program should be executed. Resources are selected in such a way that the wall-clock time for solving the problem is minimized, as it is necessary for many Grand Challenge problems.
The next section summarizes some of the requirements arising from a particular application analysis that have influenced the design of the interactive parallel programming environment.

3 Application Analysis

The well-known method of *optimal interpolation* (OI) serves as a test application throughout this article. The OI algorithm, as used in four-dimensional data assimilation, interpolates from irregularly distributed observations into a three-dimensional grid. Later, this grid is used as input for a climate model. A detailed description of the sample application and its parallelization is beyond the scope of this article. More information about the scientific problem can be found in the literature.[5,10,12] The parallelization of different versions of the optimal interpolation algorithm is described elsewhere.[15–17]
Of major importance are the properties of the code, as shown in Table 1, which directly influence the design of the parallel software environment. This code analysis is typical for many scientific codes.
The sequential OI code was developed by many programmers over the past decade. It has been transferred from very early computers to state-of-the-art vector supercomputers. Unfortunately, the documentation and code quality has suffered during this process. Because of the size of the program and its structural complexity, however, a complete redesign of the program is too expensive. Thus, any attempt at parallelizing the code should reuse as many parts of the original program as possible. This restriction also implies that, since the original program was written in Fortran, the core of the program should be kept in this language. Doing so facilitates maintenance by cur-

Table 1: *Code Analysis of the Optimal Interpolation Algorithm as Used in Four-Dimensional Data Assimilation*

	Grand Challenge (NASA)
Legacy code	yes
Number of programmers	very high
Development cost	high
Number of program lines	69000 (large)
Input/Output	not portable
Code quality	poor
Documentation	poor
Programming paradigm	sequential
Programming language	Fortran
Computational load	high
Data volume	high
wall-clock time important	very
Machine precision	high
Parallelization potential	vectorization: medium MIMD: high

rent and future atmospheric scientists who are most likely to know Fortran. Keeping the program in Fortran also provides portability of the code to other machines.

Using the IPPE parallel programming environment improves the program documentation and the overall structure of the program because supporting features for both are embedded in the design of the environment. The environment is able to generate a message-passing parallel code that runs on a wide variety of parallel computing platforms, including heterogeneous computing platforms. The generated programs are executable in batch mode for MIMD supercomputers, as well as interactively on distributed computing platforms performing in time-sharing mode.

Currently, scientists are developing several other modules that are similar in functionality to the optimal interpolation algorithm. With the help of the IPPE parallel computing environment, these modules can be easily incorporated in the overall structure of the data assimilation system. After inclusion of the programs in a module library, an atmospheric scientist can use the module best suited for her/his individual goal (i.e., minimal wall-clock time and/or increased accuracy of the calculation).

Use of IPPE and its expansion with user-defined modules will ultimately lead to a wide variety of predifined computational modules. The availability of these predefined modules will dramatically reduce the programming effort for building new programs. Moreover, the modules can be reused in other projects, thus enabling rapid prototyping.

Many of the decisions dealing with the resource management of the program will be hidden from the user and taken care of by the IPPE. The next section describes part of the actual metacomputing environment in more detail.

4 Parallel Programming Environment

The following list summarizes the principal requirements for the parallel programming environment. The environment should

- be easy to use for experts and nonexperts in the field of parallel programming;
- simplify the parallel program design;
- regulate the software development while providing support for documentation and structure;
- regulate and supervise the program execution;
- be expandable;
- have native language support for Fortran and other high-level languages;
- support heterogeneous computing environments; and
- be expandable to include WWW resources in the environment.

To fulfill some of these requirements, the IPPE environment follows a dataflow programming concept used by other similar approaches.[3] To provide portability and to enable program execution on many MIMD machines, the message-passing paradigm is used to translate the visual program specification of the task graph into a program by using message-passing routines. After the generation of the message-passing program, the executable is executed on a statically or dynamically selected set of computers or processors.

Figure 1 shows the multiple functionality of the IPPE. The IPPE is used to simplify generating parallel programs or transferring a sequential program into a parallel program. Additionally, it can be used to supervise the execution of a sequential as well as a parallel program.

Internally, the IPPE comprises two main layers. First is a *message-passing interface*, intended to enable support of different computing platforms. Currently, the IPPE provides an interface to MPI. On top of this communication library a *parallel support library* provides the necessary functionality to simplify parallel programming. An example is a set of parallel vector routines necessary for the optimal interpolation algorithm or other scientific codes.[15] Code generated with the help of the programming environment can be added to the pool of available modules.

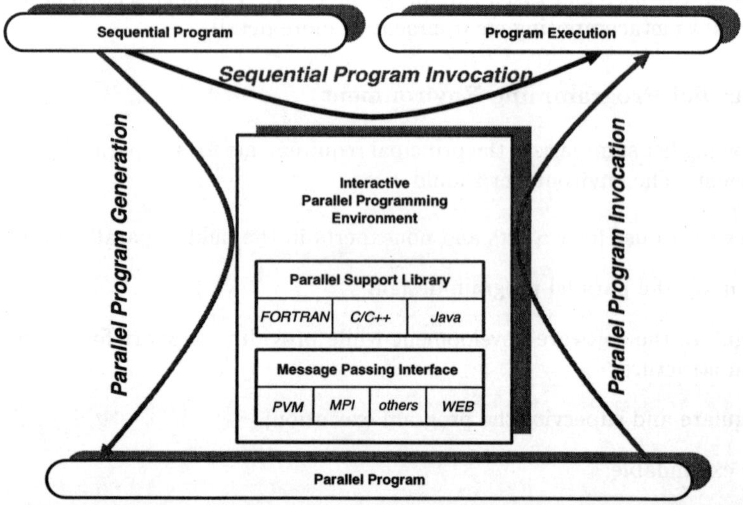

Figure 1: *The multiple functionality of the parallel programming environment while creating and executing parallel programs*

Computationally intensive and time-critical applications are supported, while integrating high-level programming languages often used in scientific programming, for example, Fortran 77, Fortran 90, C, and C++.
While the original prototype of the parallel computing environment was developed in Tcl/Tk, Perl, CGI and Python, the current implementation of the interface is based on Java. This simplifies expansion toward the WWW-driven usage of the interface as suggested in the MetaWeb project.[6] Using Java also reduces the number of software packages involved in the core implementation of the computing environment.

5 Practical Application

After this relatively abstract introduction of the environment, some snapshots will illustrate the usefulness of the parallel programming environment. In Figures 2- 7, snapshots of the software environment are displayed as used during the parallelization and execution of the OI algorithm.
Figure 2 shows the logical division of the sequential program. Data is generally shown in rectangles, while processes working on data are displayed in circles. Dependencies between data and tasks/processes are displayed with the help of directed edges. For better visualization, colors are used in addition to the obvious form distinction.

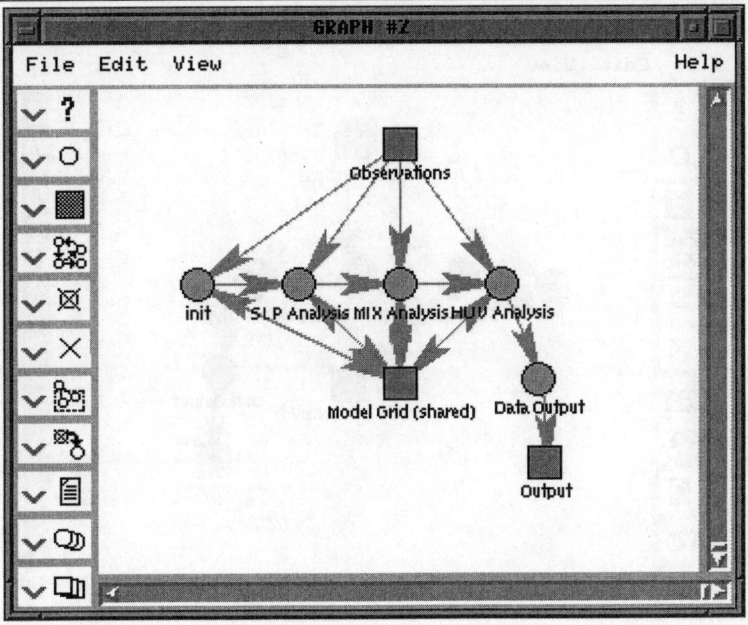

Figure 2: *Window showing the building blocks used in the global program structure of the sequential OI algorithm*

Once the processes with concurrent nature are defined, they are introduced into the process graph as shown in Figure 3. A parallel process is visualized with multiple circles, while distributed data is visualized with multiple rectangles. An example for such distributed data is the block distribution as known from HPF.

Data objects *flowing* between process objects are defined via a simple interface. For example, in the optimal interpolation algorithm, a simplified data structure for the observations and their location is used (Figure 4). The definition of data objects is similar to a record, as known from several Fortran 77 extensions and Fortran 90, and the struct command in C. The definition of the data type generates the necessary message-passing routines, allowing communication between the data objects and the process objects. Figure 5 shows the use of a data flow object in a procedure definition. The indirection of the data flow is marked with the special keywords IN DATA and OUT DATA. Thus, with only a very limited extension to the original sequential programming language (in this case Fortran), task-parallel programs can be defined easily and naturally. More interesting problems can be considered while sending dynamic data structures, as found in irregular problems. For future research, we

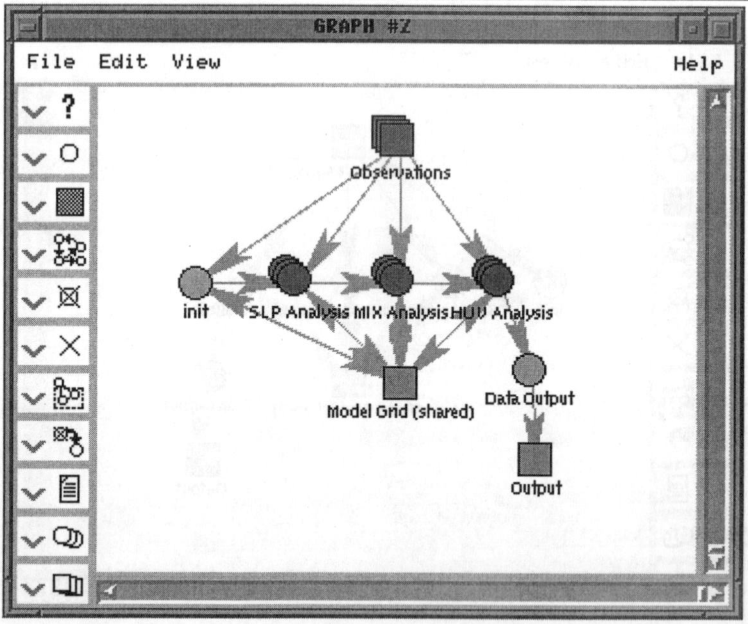

Figure 3: *Window showing how the program is represented after the parallel program blocks have been introduced.*

point out that the extension of the data flow concept, with actual programs as data, will enable the distribution of programs similarly to the distribution of data. Special care has to be taken, however, in order to solve security issues.

After the processes and the data objects are defined, they have to be mapped onto a real computer to be executed (Figure 6). Restrictions during the code development (e.g., the code can be compiled only on one machine and is not portable) may limit the number of choices for the mapping. To minimize the overall wall-clock time of the program execution, dynamic load balancing is used to map the problem on the different processors and/or computers, based on their current load.

To support this strategy, a process monitor keeps track of the status and use of the machines. Figure 7 shows an example of some system variables (here CPU load, load average, and swap load) monitored to support the mapping strategy.

The load monitor helps to display performance bottlenecks of the parallel program during its execution on the machines, while collecting a time-space diagram.

In the example depicted, all processes are mapped to an IBM SP2, and the graphical display is viewed on a SPARC workstation. If the processes are written in a portable

```
DATAFLOW OBJECT OBSERVATIONS

    INTEGER NoOfObservations;
    REAL x(NoOfObservations);
    REAL y(NoOfObservations);

    REAL temp(NoOfObservations);
    REAL pressure(NoOfObservations);

END DATAFLOW OBJECT
```

Figure 4: *Definition of data able to flow between process objects*

way, mapping onto other machines is possible too. Hence, one can execute a parallel program in a heterogeneous (super)computing environment.

6 Dynamic Selection of Software and Compute Resources

The dynamic execution of a program is driven by two factors. These are the software modules and the available computational nodes (hardware). Figure 8 illustrates the process responsible for making the selection of the hardware and software used to solve a problem. Often, a scientific problem is solved many times for similar instances of data. If the runtime is not significantly different between the instances, it can be used for performance prediction. In other cases, a performance prediction function can be used. Once the suspected execution time for a particular machine configuration is stored in the database, the information about the current utilization of the machine is used to predict the real-time performance of the program. If several choices of software and hardware mappings are available, the one with the shortest execution time is chosen. Hence, the selection not only includes a hardware mapping but also can include the usage of completely different algorithms that are best suited for the selected computer to solve the (meta)problem.

7 Related Research

Considerable research has been done in the field of visual programming in respect to parallel computing. The environment described in this article uses an approach similar to that introduced in HeNCE and CODE. A more detailed description of visual programming and their applications in parallel computing can be found in the literature.[3,4] Unlike these other systems, however, the IPPE extends the usage toward a realistic metacomputer while providing a database of performance predictions that guides the selection and mapping of programming tasks to selectable resources.

```
SUBROUTINE slp (IN DATA OBSERVATIONS, IN DATA MODEL, OUT
DATA modified_MODEL)

        !- Quality Control
        do i=1,NoOfObservations
        call buddy_check
        call gross_check
        end do
        !- Matrix Solve
        do i=1,NoOfGridpoints
        call set_up_the_matrix
        call solve_the_equations
        end do


END SUBROUTINE slp
```

Figure 5: *Definition of a process object using a data object*

8 Conclusion and Future Research

This article describes an interactive parallel programming environment. The environment makes it possible to view the available resources as a metacomputer and to reduce the development time for parallel programs. Dynamic process assignment is used to assist the execution of the parallel programs on diverse computing resources. This includes not only the selection of the best suited hardware platform but also the appropriate software for solving the problem.[16]
Many extensions are planned. One of the most striking will be the inclusion of a message-passing layer for the WWW. This will allow access to resources available via the WWW. Integration of a Fortran interpreter or language tools to simplify the distribution of programs is desirable. For mathematical problems, scripting languages like Matlab, Scilab, or Mathematica could be a viable alternative to interpreted Fortran or Java.

9 Acknowledgments

I thank Geoffrey C. Fox and Miloje Macivic for their support an valuable comments. I am grateful to the research team at the Data Assimilation Office, Goddard Space Flight Center, Greenbelt, MD, for they support during the project. I especially thank Peter Lyster, David Lamich, Jim Stobie, and Mike Seablom for their discussions in helping to understand the dark secrets of OI code. I also thank Richard B. Rood for his support and hospitality during several visits at NASA Goddard Space Flight Center as part of the Universities Space Research Association (USRA). Finally, I thank Gail Pieper and Warren Smith from Argonne National Laboratory for their comments on improving the paper.

References

[1] MPI: A Message Passing Interface. Technical report, University of Tennessee, Knoxville, TN, 1994.

[2] A. Beguelin, J. Dongarra, A. Geist, R. Manchek, K. Moore, and V. Sunderam. PVM and HeNCE: Tools for heterogeneous network computing. In J. S. Kowalik and L. Grandinetti, editors, *Software for Parallel Computation: Proceedings of the NATO Advanced Workshop on Software for Parallel Computation, held at Cetraro, Cosenza, Italy, June 22–26, 1992*, volume 106 of *NATO ASI Series F*, pages ix + 363, Berlin, Germany / Heidelberg, Germany / London, UK / etc., 1993. Springer-Verlag.

[3] James C. Browne, Jack Dongarra, Syed I. Hyder, Keith Moore, and Peter Newton. Visual Programming and Parallel Computing. Technical report, University of Texas Austin and University of Tennessee at Knoxville, 1996.

[4] James C. Browne, Syed I. Hyder, Jack Dongarra, Keith Moore, and Peter Newton. Visual programming and debugging for parallel computing. *IEEE parallel and distributed technology: systems and applications*, 3(1):75ff, 1995.

[5] Roger Daley. *Atmospheric Data Analysis.* Cambridge Atmospheric and Space Science Series, Cambridge University Press, 1991.

[6] Geoffrey C. Fox, Wojtek Furmanski, Marina Chen, Claudio Rebbi, and James H. Cowie. WebWork: Integrated Programming Environment Tools for National and Grand Challenges. Joint Boston-CSC-NPAC Project Plan to Develop WebWork, Northeast Parallel Architectures Center at Syracuse University, Syracuse, NY, 1996.

[7] Geoffry C. Fox and Wojtek Furmanski. Neat Tools Overview. Technical Report unpublished, Northeast Parallel Architectures Center at Syracuse University, Syracuse, NY, 1996.

[8] A. A. Khokhar, V. K. Prasanna, and M. E. Shaaban. Heterogenous Computing: Challenges and Opportunities. *IEEE Computer*, pages pp18–27, 1993.

[9] Andrew Lumsdaine, Jeffrey M Squyres, and Brian C. McCandless. Object Oriented MPI (OOMPI): A C++ Class Library for MPI. Technical report, Department of Computer Science and Engeneering, University of Notre Dame, Notre Dame, IN, 1996. http://www.cse.nd.edu/ lsc/research/oompi.

[10] James Pfaendtner, Stephen Bloom, David Lamich, Michael Seablom, Meta Sienkiewicz, James Stobie, and Arlindo da Silva. Documentation of the Goddard Earth Observing System (GEOS), Data Assimilation System - Version 1. NASA Technical Memorandum 104606, Vol.4, NASA GSFC Data Assimilation Office, Greenbelt, Maryland, Jan. 1995. (ftp).

[11] Adrian Simmons. High Performance Computing Requirements for Medium Weather Forecasting. *ECMWF Newsletter, Reading, UK*, (69):8–14, Spring 1995.

[12] Lawrence L. Takacs, Andrea Molod, and Tina Wang. Documentation of the Goddard Earth Observing System (GEOS), General Circulation Model - Version 1. NASA Technical Memorandum 104606, Vol.1, NASA GSFC Data Assimilation Office, Greenbelt, Maryland, Sep. 1994. (ftp).

[13] Gregor von Laszewski. The Parallelization of a Weather Prediction Model. Technical Report SCCS 533, Northeast Parallel Architectures Center at Syracuse University, July 1993.

[14] Gregor von Laszewski. Minimal Requirements for a Graphical User Interface for Parallel Computing Applications. Technical Report SCCS 602, Northeast Parallel Architectures Center at Syracuse University, February 1994.

[15] Gregor von Laszewski. Preliminary Performance of a Parallel Interpolation Algorithm. Technical Report SCCS 713, Northeast Parallel Architectures Center at Syracuse University, December 1995. first edition in June 1995.

[16] Gregor von Laszewski. *A Parallel Data Assimilation System and its Implications on a Metacomputing Environment.* PhD thesis, Syracuse University, December 1996.

[17] Gregor von Laszewski and et al. Design Issues for the Parallelization of an Optimal Interpolation Algorithm. In G-R. Hoffman and N. Kreitz, editors, *Coming of Age: Proceedings of the 4th Workshop on the Use of Parallel Processing in Atmospheric Science, European Centre for Medium Weather Forecast, Reading, UK*, pages 290–302. World Scientific, November 1994. (ftp)

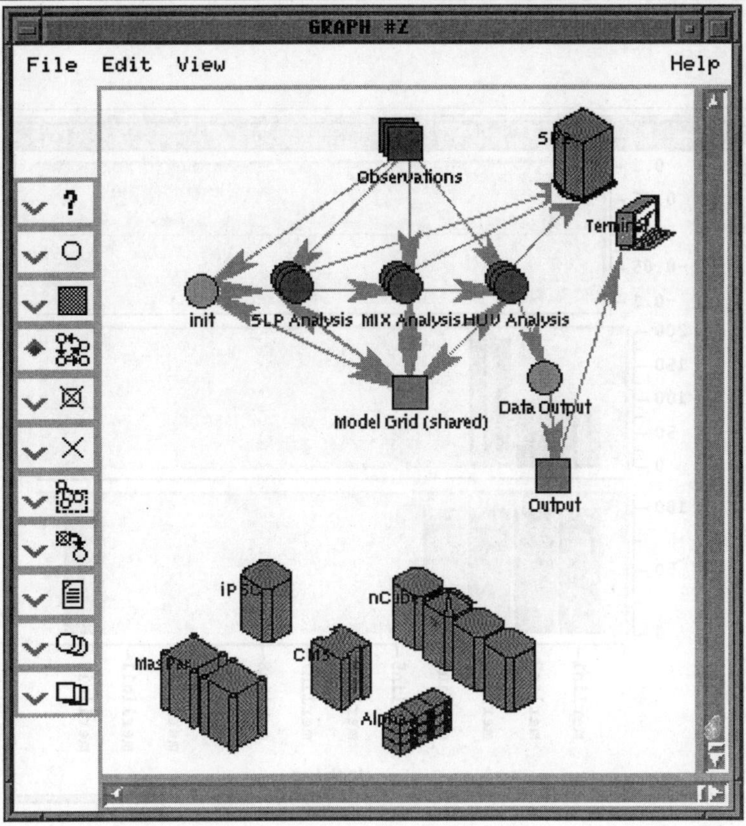

Figure 6: *Window showing the selection of the machines participating in the execution of the program*

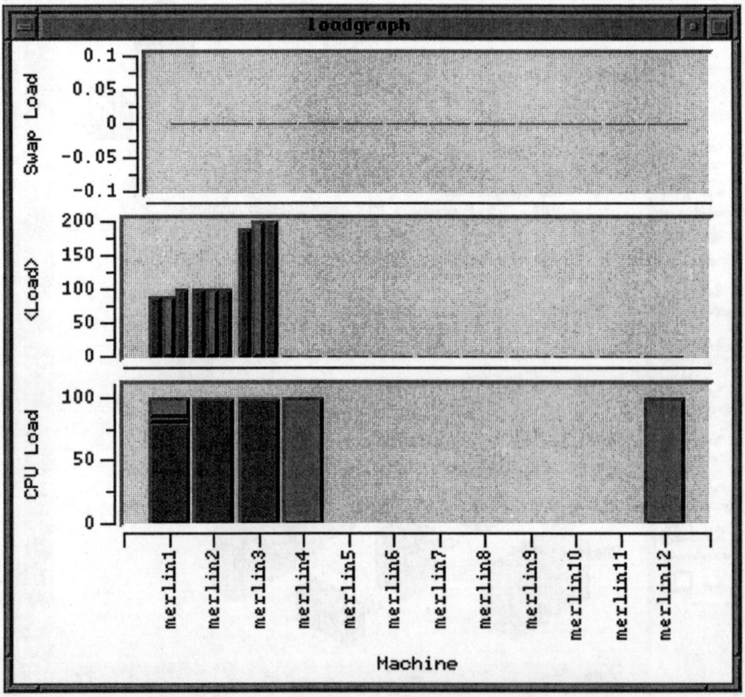

Figure 7: *Window showing the load meter to control dynamic load balancing while executing the code*

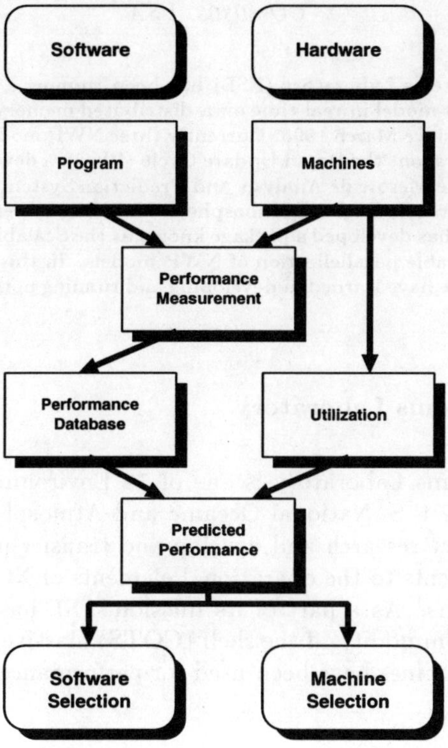

Figure 8: *Dynamic selection process during program execution*

EXPERIENCE DEVELOPING AND RUNNING REAL-TIME WEATHER MODELS ON PARALLEL MACHINES

Leslie Hart[a], Clive Baillie, Jim Edwards[a], Mark Govett, Tom Henderson[b]

NOAA Forecast Systems Laboratory, 325 Broadway, Boulder, CO 80303, USA

The Forecast Systems Laboratory (FSL) has been running a numerical weather prediction (NWP) model in real-time on a distributed memory massively parallel processor (MPP) since March 1995. Currently three NWP models run in real-time on FSL's Intel Paragon: the Rapid Update Cycle (RUC), a development version of RUC known as the Mesoscale Analysis and Prediction System (MAPS), and Colorado State University's Regional Atmospheric Modeling System (RAMS). During this process, FSL has developed a package known as the Scalable Modeling System to aid in the portable parallelization of NWP models. In this paper we describe the lessons that we have learned in developing and running portable parallel NWP models.

1 Forecast Systems Laboratory

The Forecast Systems Laboratory is one of 12 Environmental Research Laboratories within the U.S. National Oceanic and Atmospheric Administration. Its role is to conduct research and develop and transfer new technologies and scientific advancements to the operational elements of NOAA and other operational organizations.[1] As a part of its mission, FSL has been exploring distributed memory commodity off the shelf (COTS) massively parallel processors (MPPs). These machines have been used for performance testing as well as in a real-time testbed.

2 Portable Parallel Software

2.1 The Need for Portability

FSL has chosen to produce portable parallel software for two primary reasons:

[a]Jointly affiliated with the Cooperative Institute for Research in the Atmosphere (CIRA), Colorado State University, Ft. Collins, CO 80523, USA

[b]Formerly with CIRA, currently with Computating Devices International, 6041 South Syracuse Way Suite 300, Greenwood Village, CO 80111, USA

1 FSL's role is as a technology transition *facilitator*.[1] To best accomplish our role, we try to provide as many options to the operational organizations as possible. In other words, we try not to limit procurement choices to the operational organizations.
2 Since the high-performance computing industry has a history of being unstable, we cannot afford to become dependent on one vendor's product line.

2.2 The Scalable Modeling System (SMS) - FSL's Solution

In order to produce portable parallel NWP models, FSL is developing a package known as SMS [7], which currently has three components:

1 The Nearest Neighbor Tool (NNT)
2 The Scalable Runtime System (SRS)
3 The Parallelizing Preprocessor (PPP)

NNT [8] is a subroutine library developed to support the parallelization of finite difference approximation NWP models. NNT provides data decomposition, loop translation, and data movement primitives. Data is decomposed in any combination of the three spatial dimensions resulting in a *decomposition handle* which is the base reference structure for subsequent operations. Loop translation is handled by using the loop begin and end values as indices into lookup tables that are built by utility routines referencing the *decomposition handle*. Data movement is accomplished by associating the arrays that must be updated in their halo region with a *decomposition handle* and then calling the data movement routine. NNT also provides high level-calls that associate *decomposition handles* with arrays for use by the I/O subsystem of SRS.

SRS [3] is a support subsystem that provides name services, process scheduling services, and high-performance I/O. SRS encapsulates procedures that are not yet standardized. We use standards in areas where they are available, such as MPI.[5] Name services are used internally by NNT and SRS to determine the number of processes within a computational group and to detect the presence of different process groups (such as I/O caching groups). Scheduling services allow us to start processes in a platform independent way with our **smsRun**. High-performance input is accomplished by forming large blocks for input and reading collectively, high-performance output is accomplished by caching the output data in *caching processes* that later collectively write large blocks while computations are occurring.

PPP,[2] now in prototype, is a Fortran77 to Fortran77 source code translator that provides a directives based approach to parallelization using NNT. This translator accepts directives in the form of Fortran77 comments for NNT setup calls, loop translation, loop optimization, and data movement.

2.3 Supported Hardware and Models

SMS has been ported to and tested on the following platforms:

1 Intel Paragon
2 IBM SP2
3 SGI Origin 2000
4 SGI Challenge and PowerChallenge
5 Sun SPARC SMP (SunOS and Solaris)
6 HP SMP (HPUX 10.x)
7 Cray PVP

SMS has been used to parallelize the following models:

1 Quasi-nonhydrostatic (QNH) - experimental model within FSL.
2 Mesoscale Analysis and Prediction System (MAPS) - 40KM US National Scale developmental version of RUC.
3 Rapid Update Cycle (RUC) - 60KM US National Scale model operational at the US National Centers for Environmental Prediction (NCEP).
4 Eta - 29KM US National Scale model from NCEP (in progress).
5 Regional Atmospheric Modeling System (RAMS) - Local scale model from Colorado State University.

3 Real-Time Tests

3.1 Reliability

We have found the Intel Paragon to be approximately 98% reliable, meaning only 2% of the attempted runs of real-time models have failed. This is a result of regular preventative maintenance and on-site spares. We believe that newer machines with high-availability features will be able to perform with sufficient reliability in an operational environment.

3.2 Performance

For problems that are large enough, scaling to large numbers of processors is very good with the exception of load imbalanced problems. Larger versions of the MAPS model have been tested on large configurations of Intel Paragons with efficiencies of greater than 75% for 127 processors (40KM) and 80% for 512 processors (20KM)[4].

3.3 Open Issues

Network communication bandwidth on our Intel Paragon is a problem. We believe that some of the newer machines such as the IBM SP2, SGI Origin 2000, or Cray T3E will deal more effectively with this problem. The central issue is that one node of our Paragon is performing network communication, thus there is no parallelism at that point.

An open issue with SMS is that we do not yet deal with load balancing. A significant part of the motivation of PPP is to help hide the complexity of load balancing from the modeler.

4 Development Issues

4.1 Loop Level versus Procedure Level Parallelization

SMS supports both loop level as well as procedure level parallelization. By procedure level parallelization, we mean changing the upper bound for each loop dimension to be the local size rather than the global size and calling each procedure essentially unchanged. This works well if the problem is adequately balanced, but can cause difficulties if load imbalances occur. The current methods of dealing with load imbalances require irregularly shaped regions.[6] This is one reason why we generally use the loop level parallelization method and why PPP is targeted toward loop level parallelization. One of our principal reasons for developing PPP is to allow for irregularly shaped regions while not significantly impacting the look of the code. Another reason for employing loop level parallelization is that it provides us finer control when trading computation for communications.

4.2 *Developing on Different Platforms*

Development of portable parallel codes allows developers to work on a different platform from the real-time compute server. This allows the compute server to be isolated from developmental activities and allows the developers to work wherever they feel most comfortable. At FSL, development and testing takes place on a small SMP or single processor machine and real-time runs occur on the Intel Paragon.

4.3 *Minimizing Code Impact*

One of the driving issues for producing SMS is to minimize code impact. By minimizing code impact, the subject matter experts (meteorologists) can work with a parallelized code with minimal support from parallelization specialists. This allows the code to continue to be an evolving entity.

5 Conclusions

SMS has proven to be an effective tool for the parallelization of NWP models. We have found COTS MPPs to be successful in a real-time environment and believe that high availability techniques will make them even more successful.

Acknowledgments

The authors thank Kevin Brundage for his helpful suggestions. Partial funding for this project is being provided by the FAA/AUA-400, Aviation Weather Research Program Office. This work is supported by the Aviation Weather Development Program.

References

1. N. Fullerton (ed.), FSL in Review Fiscal Year 1995 (NOAA/FSL, Boulder, CO, USA 1995 http://www.fsl.noaa.gov).
2. M. Govett, C. Baillie, L. Hart, A Directive Based Approach to Parallel Programming Using ELI, (submitted to ISC, Vienna, Austria, 1996)

3. L. Hart, T. Henderson, and B. Rodriguez, An MPI Based Scalable Runtime System: I/O Support for a Grid Library, *MPI Developers Conference, U. of Notre Dame, South Bend, IN, USA, electronic proceedings only* http://www.cse.nd.edu/mpidc95/proceedings/papers/html/hart/ (1995).
4. T. Henderson, et. al., High Performance Computing at FSL: Demonstrating Operational Capability of Massively Parallel Processors, *unpublished paper.*
5. Message Passing Interface (MPI) Forum, *International J. of Supercomputing Applications* **8**, Numbers 3 & 4 (1994).
6. J. Michalakes, Implementing the Penn State/NCAR MM5 in Parallel Using RSL, (Technical Report, Mathematics and Computer Science Division, Argonne National Laboratory, 9700 South Cass Ave, Argonne, Il, USA, 1994)
7. B. Rodriguez, L. Hart, and T. Henderson, The Scalable Modeling System Tutorial and User's Guide, FSL Technical Memorandum, in preparation (NOAA/FSL, Boulder, CO, USA, 1996).
8. B. Rodriguez, L. Hart, and T. Henderson, *J. of Parallel and Distributed Computing* **37**, 159-170 (1996).

CUBIC AND OCTAGONAL SPHERICAL GRIDS ON MPPs

MIODRAG RANČIĆ
National Centers for Environmental Predistion
University Corporation for Atmospheric Research
Camp Springs, Maryland, USA
and
CLIVE F. BAILLIE
Cooperative Institute for Research in the Atmosphere
Colorado State University/NOAA Forecast System Laboratory
Fort Collins, Colorado, USA

ABSTRACT

A global shallow water grid-point model, that uses quasi-uniform cubic and octagonal spherical grids, has been run on several different parallel machines. The paper presents the performance results on both parallel vector and massively parallel processors. One of the results is that the model executes about three time faster on the MPP Cray T3D than on the parallel vector Cray C90.

1. Introduction

Among the variety of techniques presently used in numerical models of the atmosphere, the Eulerian grid-point method appears to be the most suitable for a straightforward application of massively parallel processors (MPPs). The most obvious way to parallelize a Eulerian grid-point model is to decompose the computational domain into smaller sub-domains, to assign one processor to each of the sub-domains, and to run all the processors simultaneously with exchange of boundary conditions among the neighboring sub-domains at the end of each time-step. However, in the application over the whole globe, with the standard longitude-latitude grid-geometry, the Eulerian grid-point method has serious problems due to accumulation of grid-points near the poles. A common technique of Fourier filtering of the short-waves in the high-latitudes, apart from gradually distorting numerical solution during the long-term integration, is especially inconvenient on MPPs, where it causes load imbalance, and where its implicit nature additionally increases the expenses of communications.

Recently, [1] (see also [2]) suggested that a so-called 'cubic grid', originally introduced by [3], combined with the Eulerian grid-point technique, may be an optimal choice for global models on MPPs. The approach of [1] solves the problem of the singular lines on the cubic grid via numerical orthogonalization (Fig. 1a) and allows the use of second-order energy/mass conserving finite-differencing schemes. In [4] has been introduced another member of the same family, so-called 'octagonal grid' (Fig. 1b), where two conformal octagons are connected along a circle on the sphere. Most recently, [5] have announced further generalizion of this method by presenting a family of cubic and octagonal grids with a smooth passing of the coordinates across the singular lines and even more homogeneous overall distribution of grid-points. Being square, these grids represent an excellent tool for expansion of operational regional grid-point models to global coverage. Most importantly, these grids do not require application of any area-dependent technique, such as polar filtering. Rather, they naturally decompose the spherical surface into equally treated sub-domains, which is ideal for application of MPPs in modeling of the atmosphere on global scales.

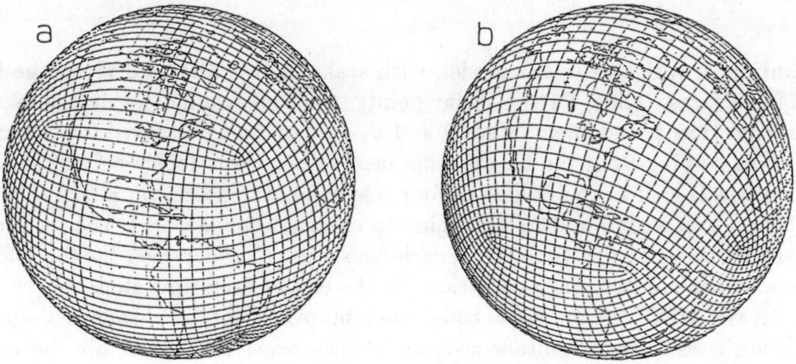

Figure 1: (a) Conformal cubic grid. (b) Octagonal grid.

This paper outlines the parallelization and briefly presents the performance on several different parallel architectures of a shallow water global model (see [1] and [4]) designed on cubic and/or octagonal grid-geometry. More comprehensive comparison, including the developed performance model, will be presented elsewhere.

2. Parallelization

The regular data structures, such as cubic and octagonal grids, are straightforward to parallelize using domain decomposition. Since they naturally decompose the spherical surface in a discrete number of equal basic sub-domains ('faces'), the obvious way to deal with them is to parallelize over these faces. The number of faces, n_f, is 6 for the cube and 14 for the octagon. This provides sufficient parallelism for a vector supercomputer, but for an MPP we can go further and parallelize over each face by splitting them into $i \times i$ portions, where i is an integer. Thus we obtain the amounts of parallelism as shown in Table 1.

The historic variables are the covariant wind components (U and V) and geopotential (H). On the parallel vector processors (PVP) they are defined as 3D arrays dimensioned as $A(0:n_1+1, 0:n_1+1, n_f)$. We have used Arakawa B-grid for

Table 1: Parallelism for the cubic or octagonal grids.

i	cube	octagon
1	6	14
2	24	56
3	54	126
4	96	224
6	216	504
8	384	896

distribution of variables in our model, with scalar points arranged along the boundaries of the faces. Thus, for the scalar points, the first two indices on the faces take values from $[1, n_1]$, and their values 0 and $n_1 + 1$ are being reserved for the external boundaries coming from the neighboring faces. (For the velocity points, located on the Arakawa B-grid in the middle of four scalar points, indices on a face take values from $[1, n_1 - 1]$, with 0 and n_1 belonging to external boundaries.) The computation is organized as a series of time-steps, each consisting of a loop over the faces followed by the exchange of boundary conditions at the end of each time step.

In the MPP version of the code, the loop over the faces disappears since each subdomain is executed simultaneously on its own processor. Therefore the variables are 2D arrays $A(0 : n_i + 1, 0 : n_i + 1)$, where n_i is defined as n/i and i takes values from Table 1. The boundary exchange remains, but now not only between faces but rather between the subdivided pieces of the faces.

Many different choices are available for mapping the processors on the cubic and/or octagonal grid, but on todays MPPs, such as Cray T3D and IBM SP2, the actual choice of topology does not really matter, and the communication time is essentially independent of the processor mapping.

Having a domain decomposed problem, we choose the message passing paradigm, simply because the other two alternatives (work sharing and data parallel) are not yet as efficient. We developed communication subroutines using both one-sided Cray Shmem and two-sided PVM and MPI message passing libraries. The communication subroutines developed and tested using this shallow water model as a prototype will be eventually applied in the full 3D version.

3. Performance Results

The codes have been benchmarked on both PVPs (Cray C90) and on MPPs (Cray T3D and IBM SP2). On T3D we have applied all three developed methods for communications and on SP2 only PVM and MPI.

3.1 Vector Supercomputer Performance

As explained in the previous section, in the vector supercomputer form, both the cube and octagon codes could be parallelized over the faces. Therefore we have benchmarked (in dedicated mode) the cube code on 1 and 6 processors and the octagon code on 1 and 14 processors of the Cray C90 vector supercomputer. We ran the same subdomain size on each face - which means that the octagon code takes roughly 14/6 times longer than the cube code on one processor. The total times and speedups are presented in Table 2. We also list the total number of floating-point operations (in millions) for one processor as reported by the Cray hardware performance monitor. Out of these total number of floating-point operations, 99% are vector floating-point operations, with an average vector length of 97 for cube and 82 for octagon. Dividing the total number of floating-point operations by the total times yields approximately 400 sustained Mflops for both codes - meaning that they are well vectorized. The achieved speedups are quite respectable demonstrating that the parallelizing part of the compiler can efficiently parallelize the outer loop over faces (while still vectorizing the inner loops within a face). The codes were compiled with the "-Zp" flag to the Cray Fortran compiler.

Table 2: Time for 100 iterations of cube and octagon codes on Cray C90.

Code	Processors	Flops	Total time	Speedup
CUBE	1	2878 M	6.74	1
CUBE	6	-	1.30	5.2
OCTAGON	1	6691 M	16.13	1
OCTAGON	14	-	1.67	9.7

3.2 MPP Performance

In order to obtain a fair comparison of the cube and octagon grids we tried to run similar sized physical domains on each, which means that the amount of work per a face should be roughly 6/14 times smaller for the octagon. We chose:

$$n_1 = n \times \nu + 1 \tag{1}$$

where ν is a positive integer and n is equal to 24 for cube a nd 12 for octagonal code. Using the formula for the total number of points:

$$N = n_f \times n_1 \times n_1 \tag{2}$$

yields the values listed in Table 3. Thus, the octagon code ended up doing generally less work than the cube code.

We have done a most comprehensive benchmarking of both the cube and octagon codes for all resolutions on all the possible numbers of PEs listed in Table 1, but just a portion of derived results will be presented here.

Table 3: Total number of grid points versus resolution for cube and octagon MPP codes.

Code	$\nu = 1$	$\nu = 2$	$\nu = 4$	$\nu = 8$
CUBE	3456	13824	55296	221184
OCTAGON	2016	8064	32256	129024

In Fig. 2 we show speedup versus number of PEs for the cube grid code on the T3D with Shmem, PVM and MPI; perfect scaling is shown as a straight line from bottom left corner to top right corner. We see that the Shmem code exhibits super-linear speedup, no doubt due to the fact that as the subdomains on each processor become smaller, the cache behavior, and therefore floating-point performance, improves. We also see that PVM and MPI scale almost identically with MPI being slightly faster.

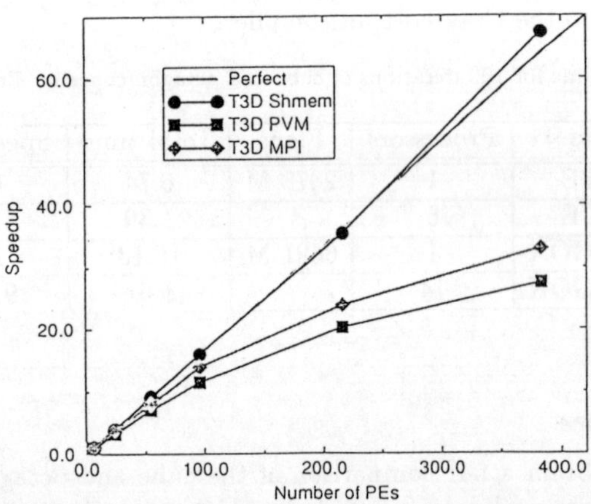

Figure 2: Speed-up versus number of PEs for the cube grid code on the T3D with Shmem, PVM and MPI. We do not show speedups for the SP2 with MPI because we obtained an anomalously high time on the smallest number of processors which leads to unrealistic values. Perfect scaling is shown as a straight line from bottom left corner to top right corner.

Using the 'apprentice' tool on T3D we obtained the total Mflop rates and Mflops per PE shown in Table 4. These Mflops/PE seem rather low, but around 15 Mflops/PE is usually found for a good Fortran code on the T3D (for example, in [5], a multigrid code obtained 3.6 Gflops on 256 PEs, which is 14 Mflops/PE).

In addition to Mflops, the apprentice tells us several other interesting performance numbers which are listed in Table 5. In the second column is listed the percentage of time spent doing useful work, i.e. the total calculation time minus the time spent waiting on data which is not in the cache, which is listed in the third

Table 4: Mflops for cube code with $\nu = 8$ on 54 PEs of T3D.

Message Passing	Total Mflops	Mflops/PE
Shmem	819	15.2
PVM	628	11.6
MPI	713	13.2

column. We see that roughly half of the total run time is spent waiting for data to come from memory. This is happening because the cache on the T3D PE is very small. We expect much better performance on the new version of Cray's MPP - the T3E - which has a much larger cache. In the fourth column we give the twice the % we measured during the main loop of the program because a large amount of communication is done in the set up phase, which we do not measure but which apprentice does. Finally, in the last column is listed the total number (summed over all 54 PEs) of floating-point operations (in millions) which apprentice counts. Note that this number is exactly the same for all three codes since only the message passing library changes. Strangely the number of flops - 2696 M - is less than the number for the vector supercomputer code - 2878 M (in Table 2). This was rather confusing until we realized that it is due to the compiler optimizing the code (all of our benchmarks were done using the highest optimizations possible). When we recompiled with no optimization the number of flops increased to 2872 M.

Table 5: More apprentice performance information for $\nu = 8$ cube code on 54 T3D PEs.

Message Pass.	Work %	Cache Wait %	Commun. %	Flops
Shmem	44	53	3	2696 M
PVM	34	41	25	2696 M
MPI	39	48	13	2696 M

In Fig. 3 we show the communication time as a percentage of total time versus number of PEs for the cube grid code on the T3D with Shmem, PVM and MPI, and on the SP2 with MPI. We see that the time spent in communication with Shmem is almost flat or rising very slowly, whereas the communication times for the two-sided message passing libraries are increasing linearly. Again PVM and MPI on the T3D are very similar with MPI being slightly faster. On the SP2, the percentage communication time is not scaling well - on the largest number of processors we had access to (in dedicated mode) the communication time is more than 50% of the total execution time - twice that of MPI on the T3D and more than ten times that of Shmem. This may be attributed to the fact that the calculation time on the SP2 is decreasing faster than that on the T3D as the number of processors increases.

Fig. 4 shows a log-log plot of total time versus number of PEs for the cube

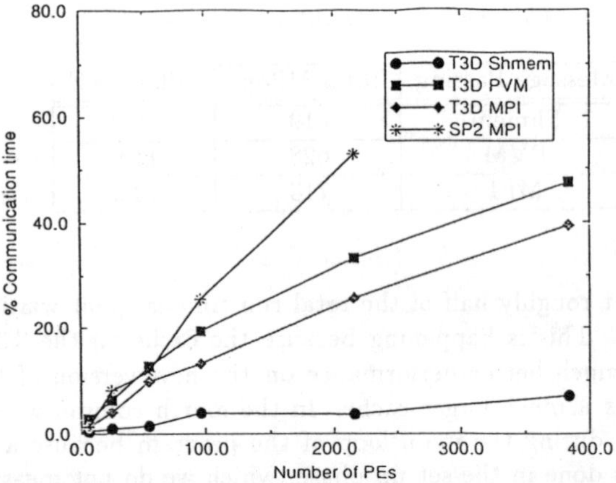

Figure 3: Percentage of total time spent on communication versus number of PEs for the cube grid code on the T3D with Shmem, PVM and MPI, and on the SP2 with MPI.

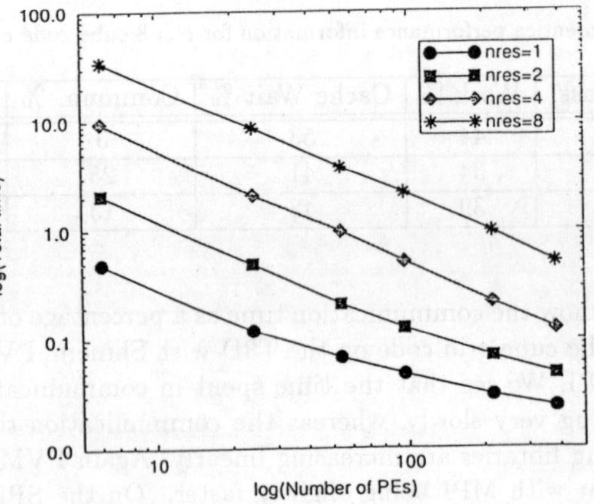

Figure 4: Log-log plot of total time versus number of PEs for the cube grid code on the T3D with Shmem for all four resolutions showing the scaling. As expected, for increasing resolution the lines become straighter with steeper slope.

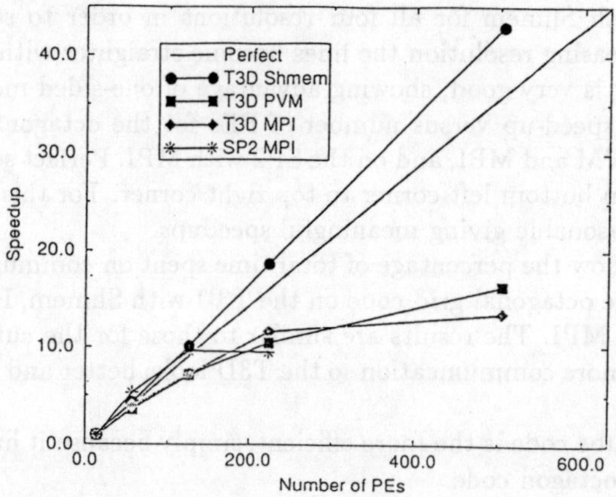

Figure 5: Speed-up versus number of PEs for the octagonal grid code on the T3D with Shmem, PVM and MPI, and on the SP2 with MPI. Perfect scaling is shown as a straight line from bottom left corner to top right corner.

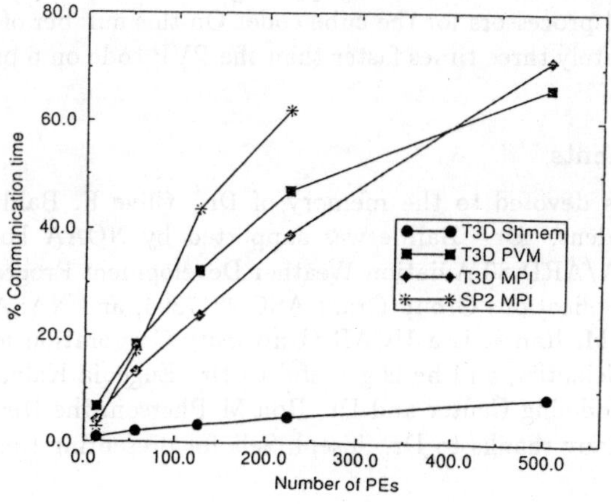

Figure 6: Percentage of total time spent on communication versus number of PEs for the octagonal grid code on the T3D with Shmem, PVM and MPI, and on the SP2 with MPI.

code on the T3D with Shmem for all four resolutions in order to see the scaling. As expected, for increasing resolution the lines become straighter with steeper slope. Moreover, the scaling is very good, showing advantage of one-sided message passing.

Fig. 5 shows speed-up versus number of PEs for the octagonal code on the T3D with Shmem, PVM and MPI, and on the SP2 with MPI. Perfect scaling is shown as a straight line from bottom left corner to top right corner. For this code, the SP2 time on 14 PEs is reasonable giving meaningful speedups.

In Fig. 6 we show the percentage of total time spent on communication versus number of PEs for the octagonal grid code on the T3D with Shmem, PVM and MPI, and on the SP2 with MPI. The results are similar to those for the cube case but for the octagon there is more communication so the T3D looks better and the SP2 comes off worse.

Overall, the cube code is the more efficient, simply because it has less physical boundaries then the octagon code.

4. Summary

We have presented some of the performance results for a global shallow water model using two numerically generated quasi-uniform spherical grids: cubic and octagonal. We implemented these codes on both PVPs and MPPs, using parallel do-loops for the former and both one-sided and two-sided message passing for the latter. The cubic grid proved to be more efficient as it has less communication. On the Cray T3D MPP using Shmem one-sided message passing we obtained super-linear speedup all the way up to 384 processors for the cube code. On this number of processors this code runs approximately three times faster than the PVP code on 6 processors of the Cray C90.

5. Acknowledgements

This paper is devoted to the memory of Dr. Clive F. Baillie who lost his life in a tragic accident. Dr. Baillie was supported by NOAA Forecast Systems Laboratory, the FAA/ARD-80 Aviation Weather Development Program Office, NSF Grand Challenge Applications Group Grant ASC-9217394, and NASA HPCC Group Grant NAG5-2218. M. Rančić is a UCAR (University Corporation for Atmospheric Research) Visiting Scientist, and he is grateful to Dr. Eugenia Kalnay, the Director of Environmental Modeling Center and Dr. Ron McPherson, the Director of NCEP, for their support. Many thanks to Dr. Joseph Sela for presenting this paper.

5. References

1. M. Rančić, R. J. Purser and F. Mesinger, *Quart. J. Roy. Meteor. Soc.* **122** (1996) 959-981.

2. C. Ronchi, R. Iacono and R. S. Paolucci, *J. Comput. Phys.* **124** (1996) 93-114.

3. R. Sadourny, *Mon. Wea. Rev.* **100** (1972) 136-144.

4. R. J. Purser and M. Rančić, *Meteor. Atmos. Phys.* (in press).

5. M. Rančić, R.J. Purser, Z. I. Janjić and C. F. Baillie, in *Preprint Volume of Eleventh Conference on Numerical Weather Prediction,* Norfolk, VA (AMS 1996) 452-453.

6. C. F. Baillie, J. C. McWilliams, J. B. Weiss and I. Yavneh, in *Proceeding of the Supercomputing 95* (ACM Press, New York, 1995)

USE OF A PARALLEL MESOSCALE MODEL IN SUPPORT OF THE 1996 SUMMER OLYMPIC GAMES

JAMES P. EDWARDS, JOHN S. SNOOK
NOAA Forecast Systems Laboratory R/E/FS1
Boulder, CO 80303, USA

ZAPHIRIS CHRISTIDIS
IBM T.J. Watson Research Center, PO Box 218,
Yorktown Heights, New York 10598, USA

The United States National Weather Service (NWS) implemented the Olympic Weather Support System (OWSS) to meet the weather forecast and warning needs of the 1996 Summer Olympic Games. This paper describes the implementation of the local forecast component of that system, the SMS-RAMS parallel model, comprising the FSL developed Scalable Modeling System and the Colorado State University developed Regional Atmospheric Modeling System.

1 Introduction

The United States National Weather Service agreed in 1992 to provide specialized operational weather support to the 1996 Summer Olympic Games in Atlanta.[1] In response to the high-resolution weather forecast and warning requirements of the Olympic Games, the NWS developed the Olympic Weather Support System (OWSS). A key element of the OWSS was the operational implementation of the Forecast Systems Laboratory's Local Analysis and Prediction System (LAPS). A three-dimensional data assimilation system, LAPS incorporates all available data sources to provide meso-β scale analyses and forecasts of the atmosphere covering an area roughly the size of a typical NWS office forecast domain. [2–4] Furthermore, the LAPS system is designed to run in the local forecast office using affordable computer workstation technology.

The demonstration of LAPS within the OWSS was important. For the first time a mesoscale forecast model initialized with comparably high resolution analyses was implemented in an operational environment, using technology representative of that planned for the NWS forecast office in the next several years. The LAPS data assimilation and generation of 8-km grid spacing analyses for the Olympic domain is described in a paper by Stamus and McGinley.[5] The prediction portion of LAPS used a parallel version of the Regional Atmospheric Modeling System (RAMS), [6,7] which was run on a 30-node IBM RS/6000 Scalable Powerparallel (SP2) computer. This paper discusses the operational configuration of the forecast modeling system, the setup of the SP2 hardware, performance of the model as configured, and forecast verification as well as benefits to the local forecast office.

2 Hardware Selection and Configuration

The initial LAPS-RAMS forecast system was installed in May 1995 on an IBM RS6000/590 RISC computer workstation located at FSL.[8] The hardware allowed the generation of one daily 18-hour forecast using an 85 x 85 8-km grid and 25 levels. Although the predictions typically required 10 to 12 hours to complete, the forecast results were encouraging and they provided the basis to pursue more powerful computer platforms for this task. At that time, a parallel version of the model was being developed as a joint collaboration between FSL and Colorado State University. Development took place on a network of IBM RS6000 workstations as well as on an Intel Paragon distributed-memory massively parallel processor. Test results using 81 nodes on the Paragon demonstrated that vast improvements in compute time could be attained at a reasonable cost. During early 1996, IBM, as an official sponsor of the Olympic Games, agreed to provide a 30-node RS6000 Scalable Powerparallel (SP2) system as the operational compute engine for the local-domain mesoscale forecast model. The IBM SP2 was installed in the NWS forecast office at Peachtree City, Georgia, during April 1996.

The Olympic SP Weather system was configured with 28 66 MHz "thin" nodes, each equipped with 2 GB of local disk and 128 MB of memory, and 2 77 MHz "wide" nodes, with 256 MB of memory and 4 GB of local disk each. The wide nodes shared an external 8 GB Serial Storage Architecture (SSA) disk pack, configured with IBM's high availability software (HACMP) for maximum redundancy. User filesystems residing on the SSA disks were physically connected to both wide nodes, while owned by a single preassigned *primary* node. In case of a *primary* node failure, HACMP was designed to take control automatically by reappointing the *secondary* node as *primary*, while physically switching ownership of file systems to the active *primary* node (Figure1). The system was also equipped with two RISC System/6000 39H POWER2 control workstations with another external 8 GB SSA disk pack and HACMP.

User home as well as utility filesystems were owned by the *primary* control workstation, **olcw1**, and they were NFS-mounted across all nodes via the AIX automounter. File-systems containing production software and executables were owned by the *primary* wide node (**ol1n01**), and they were also NFS-mounted across all available nodes. The configured system had a total of 80 GB of storage, 4.2 GB of memory, and it was capable of delivering over 8

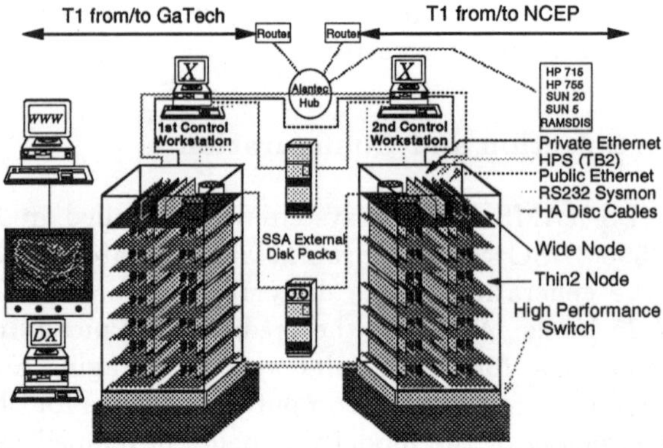

Figure 1: Olympic Weather Forecasting Server Configuration.

GFLOPS of peak performance.

In support of the dissemination of weather forecasts, another IBM RISC System/6000 39H equipped with a GXT1000-2 graphics adapter was configured as an additional node to the total SP configuration. This workstation shared the same filesystems with the rest of the SP nodes, and ran the IBM Visualization Data Explorer (DX) software package. The entire system was assigned its own private Ethernet network, while the two control workstations and the two wide nodes were also connected to the local public network, with direct access to the Internet via a firewall host.

3 Operational Model Design

The forecast system was designed to have as much flexibility and local control as possible. Two model domains were developed: the first had an 8-km horizontal grid spacing and covered the full Olympic domain (672 x 672 km^2) and the second had a horizontal grid spacing of 2 km covering an area of 160 x 160 km^2, and could be centered at any point within the 8 km domain at the forecasters' discretion. The forecasters also had control over model initialization and start time and forecast duration. Typically, data to initialize at the top of any hour was available by 20 minutes after the hour. After experimenting with a range of possibilities the forecasters settled into a schedule of running the 8-km model initialized at 0600 UTC (0200 EDT), then the 2-km model centered over the yachting venue initialized at 0900 UTC, and then updating the 8 km forecast with a new forecast initialized every 3 hours thereafter.

The model uses a finite difference discretization of the primitive dynamical equations which govern atmospheric motion along with a number of parameterizations for turbulent diffusion, solar and terrestrial radiation, moist processes including the microphysical processes responsible for the formation and interaction of clouds, and sensible and latent heat transfers between land and water surfaces, vegetation, and the atmosphere. Lateral boundary conditions were provided at 3 hour intervals using the latest forecast from the 29-km Eta model.

The model is instrumented for parallel operation in a multiple instruction multiple data (MIMD) fashion using a set of layered library calls for ease of programming and understanding. The model was coded with calls to the Scalable Modeling System which was developed by the NOAA High Performance Computing Group (HPCG).[9] SMS is a library of code that handles most of the details of domain decomposition of problems defined on regular grids. SMS in turn calls the Message Passing Interface (MPI) standard.[10] For this project we used the MPI implementation provided by IBM. The problem is decomposed in the two horizontal dimensions only, this causes the majority of the calculations, which are the column based parameterizations, to fall into the catagory of embarrassingly parallel (although not homogeneous) operations. The exceptions are the horizontal turbulent diffusion, advection, and grid nesting. We choose not to use grid nesting for the OWSS project.

4 Performance

Performance of the parallel model is measured using a real forecast initialized at 0600 UTC (0200 EDT) and computed out to 9 hours on 4 August 1996. Most performance tests of previous parallel atmospheric models have used coarser resolution models and much different physical parameterizations. In particular, the authors are aware of no other parallel models in which performance tests have been conducted using an actual forecast and explicit microphysics. This is quite important, since both the single-node performance and the parallel speedup will be affected by the localized nature of convection and the iterative method used to model it. Figure2 is a LAPS analysis of cloud cover within the model domain (primarily Georgia) at 1330 UTC (0930 EDT) showing the scattered cumulus activity normal for the region and the season.

To illustrate the importance of modeling convective processes on model performance, we also tested without the microphysics, which reduces the number of three-dimensional exchanged fields from 24 to 14 and simply treats the

Figure 2: LAPS analysis of cloud volume over the Olympics domain.

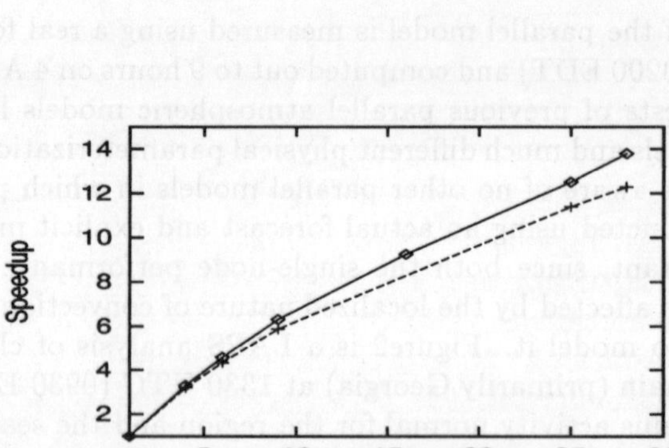

Figure 3: Speedup on the IBM SP2 ◇ = microphysics off, + = microphysics on.

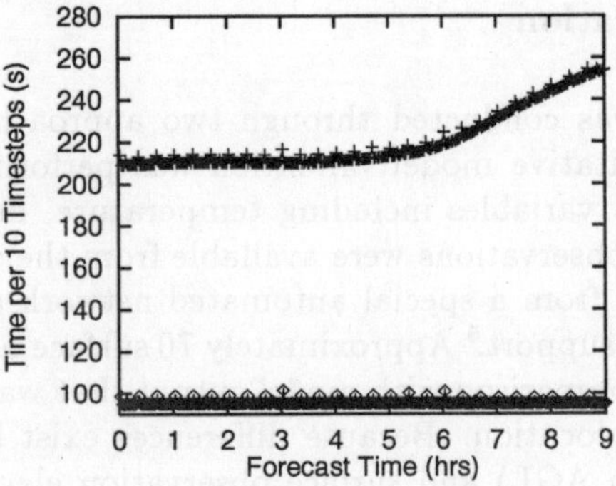

Figure 4: Single node performance on the IBM SP2 ◇ = microphysics off, + = microphysics on.

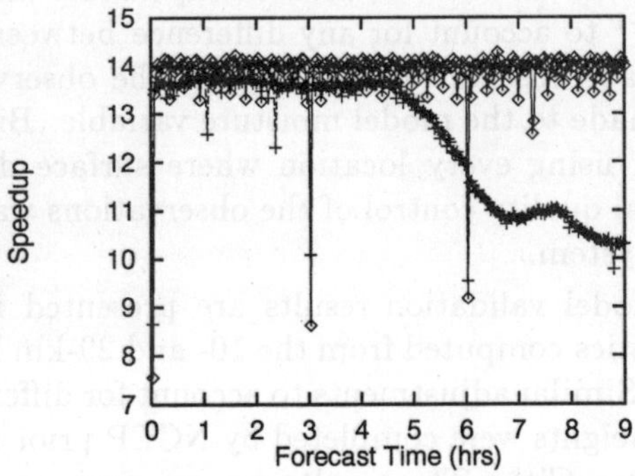

Figure 5: Speedup to 28 nodes as a function of forecast time ◇ = microphysics off, + = microphysics on.

water vapor field as an advective tracer. Figure 3 shows the speedup on up to 28 nodes from a single-node, both with and without microphysics averaged over the entire 9-hour forecast. Figure 4 illustrates the effect of microphysics on single-node performance sampled every 10 timesteps over the 9-hour run. The slower performance at 20-minute intervals is due to the radiation parameterization, and at 3-hour intervals is due to the change of boundary data files. The major slowdown after about 6 hours in the microphysics case is due to the buildup of cumulus activity within the model domain. Finally, the effect of microphysics on speedup is illustrated in Figure 5, which shows the speedup from one node to 28 nodes averaged every ten timesteps over the run.

5 Model Verification

Model validation was conducted through two approaches: quantitative and qualitative. Quantitative model validation was performed automatically on a variety of surface variables including temperature, dewpoint temperature, and wind. Surface observations were available from the standard NWS observation network and from a special automated network assembled specifically for Olympic Games support.[5] Approximately 70 surface observations were typically available for comparison with model output that was interpolated to each surface observation location. Because differences exist between the low-level model height (48 m AGL) and surface observation elevation, several adjustments were made to the interpolated model output. Similarity theory[11] is used to adjust model temperatures and wind speeds to the surface temperature observation level of 1.5 m and the surface wind observation level of 10 m. An additional adjustment is made to the model temperature using a standard lapse rate of -6.5 K km^{-1} to account for any difference between the model terrain height and the surface observation elevation at the observation location. No adjustments were made to the model moisture variable. Bias and RMS statistics were computed using every location where surface data were available. Spatial and temporal quality control of the observations was completed by the LAPS operational system.

Quantitative model validation results are presented in Figure 6 with a comparison to statistics computed from the 10- and 29-km Eta model forecasts provided by NCEP. Similar adjustments to account for differences in model and actual observation heights were completed by NCEP prior to the arrival of the forecasts at Peachtree City. The results are an average of all Eta forecasts initialized at 0300 UTC and all RAMS forecasts initialized at 0600 UTC for the period 2 July – 24 August 1996. The plotted Eta results are displaced by three hours so that the comparisons are displayed at common forecast valid times.

The bias and RMS results indicate a significant improvement with RAMS compared to both 10- and 29-km Eta through 1200 UTC for temperature, dewpoint, and wind. Significant improvements continue to be noted through 1500 UTC for dewpoint and RMS improvements of $0.4 - 1.0$ m s^{-1} are evident through the entire forecast period for wind. Several experimental forecasts were conducted after the Games in an attempt to explain the cool temperature bias after 1200 UTC (0800 LT). These simulations suggested that the model boundary layer was not properly mixed out in the late morning, possibly due to the underestimation of solar radiation at the ground as determined by the model radiation parameterization. The overall improvements over NCEP

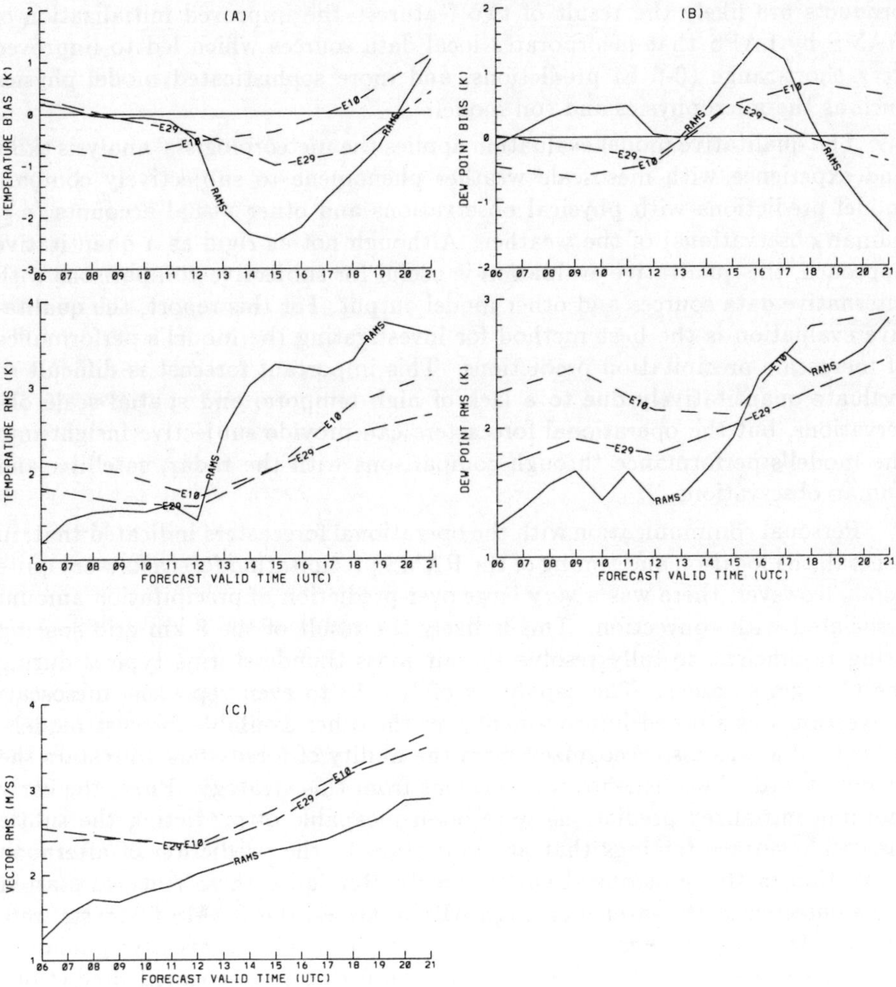

Figure 6: Hourly bias and RMS differences of (A) temperature (K) and (B) dewpoint (K), and hourly RMS differences of (C) wind vector difference (m s^{-1}) averaged for the period 2 July – 24 August 1996. Differences are computed by subtracting surface observations from model forecasts (RAMS – solid, 29-km Eta – long dashed, 10-km Eta – short dashed). Model initialization time was 0600 UTC for RAMS and 0300 UTC for Eta. Plotted Eta results are displaced by three hours so that the comparisons are displayed at common forecast valid times (UTC).

products are likely the result of two features: the improved initialization of RAMS by LAPS that incorporates local data sources which led to improved very short-range (0-6 h) predictions, and more sophisticated model physics such as the microphysics and soil model.

The qualitative model evaluation applies the meteorologists' analysis skills and experience with mesoscale weather phenomena to subjectively compare model predictions with physical observations and other visual accounts (e.g. human observations) of the weather. Although not as rigid as a quantitative approach, the qualitative evaluation is useful for subjective comparisons with alternative data sources and other model output. For this report, the qualitative evaluation is the best method for investigating the model's performance of mesoscale precipitation predictions. This important forecast is difficult to evaluate quantitatively due to a lack of high temporal and spatial scale observations, but the operational forecasters can provide subjective insight into the model's performance through comparisons with the radar, satellite, and human observations.

Personal communication with the operational forecasters indicated that, in general, the location and timing of the RAMS precipitation forecasts were quite good. However, there was a very large over-prediction of precipitation amount associated with convection. This is likely the result of the 8-km grid spacing being insufficient to fully resolve the air mass thunderstorms typical during the Georgia summer. The capability of RAMS to even represent mesoscale convection was a noted improvement over the other available forecast models. Added value was also recognized from the ability of forecasters to restart the model at will. Two benefits were evident from this strategy. First, the early morning initialized predictions were often incapable of predicting the subtle mesoscale surface forcings that are important to the prediction of afternoon convection in the subtropical environment. But, once these features started to be detected in the later morning LAPS analyses, the RAMS forecasts were able to "latch onto" these features and generate reliable convective guidance. Second, as the forecasters observed common features in repeated predictions they became more confident using these particular forecast features.

The 0900 UTC, 2-km RAMS predictions were designed to enhance the sea breeze forecasts required by the yachting venue. Local buoy observations of sea surface temperature were used in the RAMS model initialization. Special point wind forecasts were generated at half-hour increments for the two yachting event locations. The operational yachting forecasters noted that the timing, penetration, and direction of the sea breeze were well forecast by RAMS. A common theme expressed by all the forecasters was that the RAMS forecasts by themselves were generally good, but the predictions were most useful when viewed in combination with all other available guidance.

6 Benefits to the Local Forecast Office

The LAPS-RAMS system has been running quasioperationally at FSL for several years. Results from this system have suggested that the local forecast office could realize many benefits from running a mesoscale analysis and forecasting system on site.[8,12] Now, for the first time, these benefits have been demonstrated in a true operational environment.

Some of the more important demonstrated benefits include the local control of the mesoscale model. The ability to interactively select the model domain, the model grid resolution, the model start time and duration, and the frequency of model predictions allowed the operational forecasters to tailor a strategy that would best meet their needs for the local forecast problem. This is a decision that can only be made in the local forecast office.

Locally produced weather analyses and predictions greatly reduce the amount of required communications, from both a data collection standpoint and a model output dissemination standpoint. Local data sources, which may not be available to the National Center for Environmental Prediction (NCEP) central facility in a timely fashion, can be incorporated into local analyses and prediction. The volume of model output continues to grow exponentially in combination with expanding computer hardware capabilities. Communication of model output over long distances to another computer platform is not necessary when the model runs locally on the same network of computers that controls the operational meteorological workstation. This also eliminates the problem of degrading the frequency and resolution of the model output that frequently occurs when disseminating model data from the NCEP central facility to a local office. Hence, the whole flow of data (from collection into the local analyses, to model initialization, to model computations, to model output visualization) occurs in one location in a timely fashion.

Finally, it is important to understand that the locally produced local-domain numerical weather prediction effort is not intended to replace any guidance that is available from the NCEP central modeling facility. The local-domain forecasting support is designed to provide an *additional* mesoscale forecast tool to the suite of products already available on the meteorological workstation. The experiences at Peachtree City and Savannah have successfully demonstrated this synergy.

Acknowledgments

The authors wish to thank Drs. Roger Pielke and William Cotton of Colorado State University and Dr. Craig Tremback of Mission Research Corporation for their permission to use RAMS for this project. Special thanks also to the IBM corporation for their donation of hardware, software and technical support for this project. Paul Schultz reviewed the article and Nita Fullerton provided technical editing support. The Advanced Computing Group within FSL is acknowledged for their help in parallelizing the RAMS model. RAMS was developed under the support of the National Science Foundation (NSF) and the Army Research Office (ARO). James Edwards is jointly affiliated with NOAA and the Cooperative Institute for Research in the Atmosphere, Colorado State University, Fort Collins, CO 80523 USA.

References

1. L.P. Rothfusz, S.K. Rinard, J.T. Johnson, and M. Valverde, "The Operational Impact of a Comprehensive Collection of Cutting-Edge Forecast, Warning and Observational Tools." *Preprints, 13th International Conference on Interactive Information and Processing Systems,* (American Meteorological Society, Boston, MA, 1997)
2. S.C. Albers, J.A. McGinley, D.L. Birkenheuer and J.R. Smart, "The Local Analysis and Prediction System (LAPS): Analysis of Clouds, Precipitation, and Temperature." *Weather and Forecasting* **11**, 273 (1996)
3. J.A. McGinley, "Opportunities for High Resolution Data Analysis, Prediction, and Product Dissemination Within the Local Weather Office." *Preprints, 14th Conference on Weather Analysis and Forecasting,* (American Meteorological Society, Boston, MA, 1995)
4. J.S. Snook, J.M. Cram, and J.M. Schmidt, "LAPS/RAMS: A nonhydrostatic mesoscale numerical modeling system configured for operational use." *Tellus* **47A**, 864 (1995)
5. P.A. Stamus and J.A. McGinley, "The Local Analysis and Prediction System (LAPS): Providing weather support to the Olympic Games." *Preprints, 13th International Conference on Interactive Information and Processing Systems,* (American Meteorological Society, Boston, MA, 1997)
6. R.A. Pielke, W.R. Cotton, R.L. Walko, C.J. Tremback, W.A. Lyons, L.-D. Grasso, M.E. Nicholls, M.D. Moran, D.A. Wesley, T.J. Lee and J.H. Copeland, "A Comprehensive Meterological Modeling System - RAMS." *Meteorology and Atmospheric Physics* **49**, 69 (1992)

7. R.L. Walko , W.R. Cotton, J.L. Harrington, and M.P. Meyers, "New RAMS Cloud Microphysics Parameterization, Part I: The Single-Moment Scheme." *Atmospheric Research* **38**, 29 (1995)
8. J.S. Snook, "Local domain forecasting support to the 1996 Atlanta Olympic Games." *Preprints, 12th International Conference on Interactive Information and Processing Systems,* (American Meteorological Society, Boston, MA, 1996)
9. B. Rodriguez, L. Hart, and T. Henderson, "NNT 1.0 Users Guide", *In press, Forecast Systems Laboratory Technical Memorandom* (1996).
10. Message Passing Interface Forum, *MPI: A Message-Passing Interface Standard,* University of Tennessee (1995).
11. J.F. Louis, "A parametric model of vertical eddy fluxes in the atmosphere." *Boundary Layer Meteorology* **17**, 187 (1979)
12. J.S. Snook and R.A. Pielke, "Diagnosing a Colorado heavy snow event with a nonhydrostatic mesoscale numerical model structured for operational use." *Weather and Forecasting* **10**, 261 (1995)

Explicit Management of Communication Schedules in HPF+ *

Siegfried Benkner[a] *Piyush Mehrotra*[b] *John Van Rosendale*[b] *Hans Zima*[a]

[a]Institute for Software Technology and Parallel Systems,
University of Vienna, Liechtensteinstr. 22, A-1090 Vienna, Austria
E-Mail: {sigi,zima}@par.univie.ac.at

[b]ICASE, MS 132C, NASA Langley Research Center, Hampton VA. 23681 USA
E-Mail: {jvr,pm}@icase.edu

Abstract

The goal of High Performance Fortran (HPF) is to "address the problems of writing data parallel programs where the distribution of data affects performance", providing the user with a high-level language interface for programming scalable parallel architectures and delegating to the compiler the task of producing an explicitly parallel message-passing program. For some applications, this approach may result in dramatic performance losses. An important example is the inspector/executor paradigm, which HPF uses to support irregular data accesses in parallel loops. In many cases, the compiler does not have sufficient information to decide whether an expector computation is redundant or needs to be repeated. In such cases, the performance of the whole program may be significantly degraded.

In this paper, we describe an approach to solving this problem through the introduction of constructs allowing explicit manipulation of communication schedules at the HPF language level. The goal is to allow the programmer to avoid use of EXTRINSICS to express irregular computation via message-passing primitives, while guaranteeing essentially the same performance. The language features described in this paper are part of the language "HPF+" which is being developed in an ESPRIT IV research project. HPF+ is currently being implemented in the framework of the Vienna Fortran Compilation System (VFCS).

1 Introduction

The High Performance Fortran Forum (HPFF), which first convened during 1992, set itself the task of defining language extensions for Fortran to facilitate data parallel programming on a wide range of parallel architectures, without sacrificing performance. Much of the work on HPF-1 [5] focussed on extending Fortran 90 by directives for specifying alignment and regular data distributions (block and cyclic). Other extensions allow the specification of explicitly parallel loops, in particular with the **FORALL** statement and construct and the **INDEPENDENT** directive, as well as a number of library routines. It soon became apparent that HPF-1 did not provide enough flexibility for an efficient formulation of many advanced algorithms. Such algorithms, for example weather forecasting codes or crash simulations, are often characterized by the need to distribute data in an irregular manner and dynamically balance the computational load of the processors. HPF-2 [6], together with its "approved extensions", is a significant step in supporting such algorithms.

The study of real applications has, however, revealed that even with the enhanced data and work distributions offered by this language, certain "low-level" information that may decisively influence a program's performance cannot be expressed. An important example is conveying to the compiler or runtime system the possibility of reusing the information produced by an inspector computation [9, 7, 8]. In a parallel loop with

*The work described in this paper was partially supported by the ESPRIT IV Long Term Research Project "HPF+" of the European Commission and by the National Aeronautics and Space Administration under NASA Contract No. NAS1-19480, while the authors were in residence at ICASE, NASA Langley Research Center, Hampton, VA 23681.

irregular data accesses, an *inspector* analyzes the data accesses at runtime and, based on the data distributions of arrays and the work distribution of the loop, computes a communication schedule. This schedule is then used in the *executor* to actually perform the required communication for the loop by gathering all nonlocal data to be read in the loop at the beginning, and scattering all nonlocal data that were written in the loop, at the end.

Inspector execution can be very expensive. However, a parallel loop is often enclosed in a sequential loop (for example, a time-step loop), and its communication schedule may be invariant over many or all of the iterations of the sequential loop. The recognition of invariant communication schedules is crucial for obtaining high object code performance. In some cases, a compiler can recognize this invariance automatically. However, in many cases, such loop structures are hidden in a hierarchy of procedure calls, so that even sophisticated interprocedural analysis may be unable to achieve this goal.

In this paper, we describe new language features based on the concept of a *schedule variable*. Communication schedules may be bound to a schedule variable and then applied by mentioning the schedule variable in the context of a parallel loop. A special statement terminates such a binding by resetting the schedule variable. These language features are formulated at the HPF level, without dealing in any way *directly* with the actual communication specified in a schedule. Thus, using these features the programmer is not forced to switch to the message-passing paradigm via the **EXTRINSICS** facility of HPF, while obtaining basically the same performance as with message passing. The features for schedule manipulation are part of the language *HPF+*, which was initially specified in [4] and is currently being developed into a fully-specified Fortran 95-based programming language in the ESPRIT Long Term Research Project "HPF+". The required compiler and runtime support , is being implemented as a part of the *Vienna Fortran Compilation System (VFCS)* [2].

This paper is structured as follows. In Section 2, we give a short overview of the HPF+ language. The subsequent Section 3 introduces the required concepts and the associated notation and terminology. In Section 4, we illustrate the new language features by discussing a number of examples of increasing complexity. The paper ends with a short conclusion in Section 5.

2 HPF+: A Short Overview

A stated goal of the HPF Forum was to "address the problems of writing data parallel programs where the distribution of data affects performance" [5]. HPF-1, defined in 1993, provides the first step towards such a portable programming interface by extending Fortran 90 with a set of directives. However, many advanced algorithms require functionality which is missing from HPF-1. For this purpose, we are developing a language, called *HPF+*, which adds new features to HPF-1, and simultaneously drops a number of the weaker (irrelevant or ill-defined) features of HPF-1. Many of the new features are based upon Vienna Fortran [3, 10] and Vienna Fortran 90 [1]. However, in the interest of standardization, HPF+ uses the syntactic conventions of HPF. In this section, we discuss some requirements of advanced algorithms that are not met by HPF-1 and outline a solution for the problem in HPF+.

In HPF-1, data can be either replicated or distributed across all processors of a distributed-memory architecture via the predefined block, cyclic, and block-cyclic mappings. In addition, linear alignment functions can be specified. These language constructs suffice for the expression of a range of numerical applications operating on regular data structures. However, more complex applications pose serious difficulties. For example, modern codes in such important application areas as aircraft modeling and combustion engine simulation often employ multi-block grids. Even if these grids are to be distributed by block to processors, the constructs of HPF-1 do not permit an efficient data mapping for these programs: this requires distributions to sections of the processor array. HPF-1 is even less equipped to handle advanced algorithms such as particle-in-cell codes, adaptive multi-grid solvers, or sweeps over unstructured meshes. Many of these problems need more complex data distributions if they are to be executed efficiently on a parallel machine. And some of them may require the user to control the execution of major do loops by specifying which processor should perform a specific iteration. These features are not provided by HPF-1. Some programs will require that data be distributed onto processor arrays of different dimensions: the current language definition does not permit the user to prescribe or assume any relationship between such processor arrays.

HPF+ offers a number of language extensions which provide the required functionality, including the following:

- **Distribution to Processor Subsets**
  This feature allows the specification of a distribution target as a subsection of the processors in a processor array, or to an individual processor. This feature can be used, for example, in multi-block and multi-grid algorithms.

- **Processor Views**
  Processor views permit the interpretation of a given processor set according to different declarations of possibly different ranks. The relationship between the processors in such declarations can be established unambiguously using the conventions of the Fortran equivalence statement. This feature is needed whenever different numbers of dimensions of data arrays are distributed to a given set of processors.

- **General Block Distributions**
  General block distributions, while maintaining the contiguity of distribution segments associated with regular block distributions, allow different segment lengths, thus providing a simple and efficient means of dealing with load balancing. If, for instance, the nodes of an unstructured mesh are partitioned prior to execution and then appropriately renumbered, then the resulting distribution can be described in this manner.

- **Indirect Distributions**
  General block distributions cannot deal with every irregular problem: for example, the distributions generated by dynamic partitioners cannot in general be represented in this way. Indirect distributions provide a means for specifying an *arbitrary* replication-free distribution by using a *mapping array* which, for each element of the array to be distributed, points to the corresponding processor. This mechanism is supported by an enhanced version of the PARTI [9, 8] library integrated into the VFCS.

- **User-Defined Mapping Procedures**
  Indirect distributions incur a considerable overhead both at compile time and at runtime. A difficulty with this approach is that when a distribution is described by means of a mapping array, any regularity of structure that may have existed in the distribution is lost; as a consequence, the compiler cannot exploit possible regularities. User-defined mapping procedures provide a facility for extending the set of intrinsic mappings in the language in a structured way, using a notation similar to that of Fortran procedures.

- **On-Clauses**
  On-clauses allow the control of the work distribution in an **INDEPENDENT** loop by explicitly specifying a mapping from iterations to processors. In this way, the the work distribution of irregular algorithms can be adapted to the data distributions.

- **Communication Schedules**
  A facility for the explicit control of communication schedules is included in HPF+. This is the main topic of this paper and discussed in more detail in the subsequent sections.

In parallel to the development of HPF+, the HPF Forum has designed HPF-2. Some, but not all of the above-mentioned features are included in HPF-2 (distribution to processor subsets, general block and indirect distributions, and on clauses). In particular, HPF-2 does not contain high-level language features for schedule manipulation as described in this paper.

A more complete description of HPF+ is given in [4].

3 Schedules and Patterns

In this section, we introduce the concepts and terminology required for a discussion of communication schedules and access patterns. Assume that A is an array with index domain $\mathbf{I}$, and $\mathbf{P}$ denotes a set of processors.

3.1 Data and Work Distributions

A (replication-free) **distribution** for array A with respect to $\mathbf{P}$ is a total function $\delta : \mathbf{I} \to \mathbf{P}$. $\lambda^A(p)$ specifies the set of elements of A owned by processor p under distribution δ. This is called the **distribution segment** of A with respect to p, its elements are the **local variables** of A in processor p.

Let L denote a perfectly nested parallel loop with iteration domain $\mathbf{R}$. A **work distribution** for L is a total function $\omega^L : \mathbf{R} \to \mathbf{P}$. For each $\mathbf{r} \in \mathbf{R}$, $\omega(r)$ specifies the processor in which iteration $\mathbf{r}$ is to be executed. Given a work distribution ω^L, and $p \in \mathbf{P}$, the set of all iterations of L to be executed in p can be determined. This is called the *execution set* of p with respect to L.

3.2 Schedules

A **schedule function** is a total function $\sigma : \mathbf{P} \twoheadrightarrow \mathcal{P}(\mathbf{I})$. It can be associated with an array and a direction indicator. The "direction indicator" determines whether the schedule is being used for scatter or gather communication.

Let A denote an array, $\mathbf{I}$ its index domain, and δ a distribution for A. An **array schedule** for A is a triplet (A, σ, d), where

1. σ is a schedule function with range $\mathcal{P}(\mathbf{I})$
2. $d \in \{G, S\}$ specifies a **direction**: *read (gather)* ("R"), or *write (scatter)*("W").
3. for each $p \in \mathbf{P} : \sigma(p) \cap \lambda^A(p) = \phi$.

Depending on the value of d, we speak either of a *gather schedule* (d= "R"), or of a *scatter schedule* (d= "W"). We discuss now the effect of the **application** of an array schedule.

Assume $\bar{\sigma} = (A, \sigma, R)$ is a gather schedule. Then the application of $\bar{\sigma}$ results for each processor p in the gathering of nonlocal elements of A as prescribed by the schedule. More precisely:

Let $p \in \mathbf{P}$. There exists a one-to-one correspondence from $\sigma(p)$ to a set of local *buffer elements* $\mathbf{B}^p(\mathbf{i})$ in p. The effect of the application of $\bar{\sigma}$ can then be described as:

APPLY($\bar{\sigma}$):
for every $p \in$ **P do**
  for every $\mathbf{i} \in \sigma(p)$ **do**
    communicate $A(\mathbf{i})$ **to** $\mathbf{B}^p(\mathbf{i})$
  endfor
endfor

A complete specification of the communication can be obtained by indicating for each processor pair the sets of data to be respectively sent or received. Let p, p' denote two arbitrary processors. Then *RECEIVE(p,p')* denotes the set of elements of A to be received in p from p', and *SEND(p,p'):=RECEIVE(p',p)* is the set of elements of A to be sent from p to p'. Due to the obvious relationship between send and receive sets, it is sufficient to specify either all receive sets or all send sets. [1]

Given the above gather schedule $\bar{\sigma}$, the receive sets can be immediately specified as follows:

For each $p, p' \in \mathbf{P}$: *RECEIVE(p,p')*$:=\sigma(p) \cap \lambda^A(p')$

The application of a scatter schedule can be described similarly to that of a gather schedule. Note that scatter schedules are only relevant if a work distribution different from the *owner computes* paradigm [11] is used.

A **schedule** is a set of array schedules such that each array is associated with at most one gather schedule and at most one scatter schedule. It can be applied in two ways: a *gather application* applies all gather schedules, and a *scatter application* applies all scatter schedules.

[1] This hides some nontrivial implementation details. For example, an inspector, when scanning the code in a parallel loop for occurrences of non-local references to an array at the right-hand side of assignments, determines first a gather schedule, which makes the receive sets known in each processor. In order to make the corresponding send sets known in each processor, a global communication step is required.

3.3 Patterns

Let L denote a perfectly nested parallel loop with iteration domain $\mathbf{R}$, $\mathbf{I}$ an index domain, A an array with index domain $\mathbf{I}$, and d a *direction*. In this section, we define the concept of a *pattern*. Patterns describe a set of data accesses occurring during the execution of L and can be used to specify communication schedules explicitly, at the HPF language level.

A **pattern function** is a total function $\pi : \mathbf{R} \to \mathcal{P}(\mathbf{I})$. It describes an "abstract" access pattern in iterations of a parallel loop, and can be combined with an array and a direction to obtain an array pattern. An **array pattern** is a triplet $\bar{\pi} = (A, \pi, d)$, where A is an array and d is a direction. An array pattern $\bar{\pi} = (A, \pi, d)$ expresses the fact that for each $\mathbf{r} \in \mathbf{R}$, iteration $\mathbf{r}$ of L accesses the elements $A(\mathbf{i})$, $\mathbf{i} \in \pi(\mathbf{r})$. If d=R, then all these accesses are reads, otherwise all accesses are writes. Correspondingly we respectively speak of a *read pattern* or a *write pattern*.

We assume that each array in an array pattern is associated with a well-defined distribution. A **pattern instance** consists of a parallel loop L, a set of array patterns, and a loop distribution for L. It specifies communication that must be performed before and after execution of the iterations of L in order to satisfy nonlocal accesses in L. More formally, a pattern instance for loop L is a tuple $\Pi := (\omega, \mathbf{A})$, where ω is a loop distribution for L, and $\mathbf{A}$ is a set of array patterns related to L such that each array is associated with at most read pattern and at most one write pattern. It can be easily seen that a pattern instance Π uniquely determines a gather schedule and a scatter schedule for each array involved.

3.4 Schedule Variables

Schedule variables can be bound to pattern instances and the associated schedules. The purpose for introducing schedule variables is to give the user explicit control over construction, application, and reuse of schedules. In HPF+, a schedule variable, S, can be bound to a pattern instance in the following ways:

- by assigning it implicitly, using the results of the analysis of an *inspector*,
- by assigning the value of another schedule variable,
- by assigning a pattern function, together with a loop distribution and an array (distribution), and
- by assigning the pattern instance associated with an array halo.

The binding between a schedule variable S and its schedule can be disconnected by applying a *reset* operation to S. Schedule variables are used by mentioning them in the context of a parallel loop construct.

In this paper, we will only illustrate the first two options. A more complete discussion will be contained in an upcoming Technical Report.

4 Explicit Schedule Manipulation: Examples

In all of the following examples, the declarations of schedule variables are assumed to be included in a Fortran 90 module. We do not explicitly show the module definition and the corresponding use statements which make the module accessible in any program unit that uses the schedule variable.

Example 1 Schedule Reuse

```
!HPF+ SCHEDULE S

      ...
   DO t=1,max_time
       ...
       CALL KFORCE(X,Y,IX,...)
```

```
        ...
        END DO

    SUBROUTINE KFORCE(X,Y,IX,...)
        ...
    !HPF+ INDEPENDENT, GATHER (X::S)
      L: DO I=1,NUMEL, ON HOME(Y(I))
           ...
           ... = X(IX(I)) ....
           ...
           END DO
        ...
    END SUBROUTINE KFORCE
```

We assume that S is not explicitly assigned or reset anywhere in the program. When the independent loop is encountered for the first time, the value of S is undefined. The first execution of L results in the normal execution of the inspector for L; because of the gather-clause, the resulting schedule for X is assigned to S. For all subsequent executions, this value of S is reused, i.e., there is no second activation of the inspector analysis phase.

Example 2 Conditional Schedule Reuse

```
    !HPF+ SCHEDULE S

        INTEGER   RESET_CONTROL
        ...
        RESET_CONTROL=10
        ...
        DO t=1,max_time
        ...
        CALL KFORCE(X,Y,IX,...)
        IF MODULO(t,RESET_CONTROL)=0
    !HPF+      THEN RESET(S)
        ENDIF
        ...
        END DO

    SUBROUTINE KFORCE(X,Y,IX,...)
        ...
    !HPF+ INDEPENDENT, GATHER (X::S)
      L: DO I=1,NUMEL, ON HOME(Y(I))
           ...
           ... = X(IX(I)) ....
           ...
           END DO
        ...
    END SUBROUTINE KFORCE
```

As in the previous example, when the independent loop is encountered first, the value of S is undefined, and the use clause results in the execution of the inspector for loop L and assignment of the resulting schedule to S. For iterations $t = 2, ...10$, the schedule bound to S in the first iteration is reused. At the end of iteration 10, *S is reset, resulting in a disconnection of the existing binding. This sets the value of S to* undefined. *In iteration* 11, *the analysis phase of the inspector is executed again, assigning a possibly new schedule to S. In general, S is assigned (as a result of an execution of the inspector analysis phase) in all iterations $10 * k + 1$ within the loop range, where $k = 0, 1, \ldots$.*

Example 3 Using a Schedule in Different Loops

```
!HPF+ SCHEDULE S1,S2,S3,S4

 ...
    DO t=1,max_time
       ...
       CALL SUB(X,Y,Z,U,V,IX1,IX2)
       ...
    END DO

SUBROUTINE SUB(X,Y,Z,U,V,IX1,IX2)
 ...
!HPF+ INDEPENDENT, GATHER (X::S2,Z::S1,U::S3), SCATTER (X::S1)
   L1: DO I=1,N, ON HOME(Y(I))
      ...
      X(IX1(I)) = X(IX2(I))+Y(I)*Y(I)/(Z(IX1(I))-U(IX1(I),IX2(I)))
      ...
      END DO
      ...
      S4=S1+S2  !union
!HPF+ INDEPENDENT, GATHER (X::S1,V::S4), SCATTER (V::S2)
   L2: DO I=1,N, ON HOME(Y(I))
      ...
      V(IX2(I)) = X(IX1(I))+Y(I)*Y(I)*(V(IX1(I)+V(IX2(I)))
      ...
      END DO
      ...
END SUBROUTINE SUB
```

In this example, the pattern instance associated with $L1$ is given by

$$\Pi_1 = (\omega_1, \{(X, \pi_2, R), (Z, \pi_1, R), (U, \pi_3, R), (X, \pi_1, W)\}$$

where:

- ω_1 is determined by the loop iteration range $[1 : N]$ and the associated on clause **ON** HOME(Y(I))
- the pattern function π_1 maps I to $IX1(I)$ for $I = 1, \ldots N$
- the pattern function π_2 maps I to $IX2(I)$ for $I = 1, \ldots N$
- the pattern function π_3 maps I to $(IX1(I), IX2(I))$ for $I = 1, \ldots N$.

The pattern instance associated with $L2$ is given by $\Pi_2 = (\omega_1, \{(X, \pi_1, R), (V, \pi_4, R), (V, \pi_2, W)\}$, where $\pi_4 = \pi_1 \cup \pi_2$. When $L2$ is entered for the first time, all schedule variables used in this loop already have a defined value, and no inspector analysis is required.

This example is only correct if $\delta^X = \delta^Z = \delta^V$. It relies on the fact that the on clauses specified in $L1$ and $L2$ are identical.

5 Conclusion

In this paper, we have described language features for High Performance Fortran that allow the explicit manipulation of communication schedules for parallel loops containing irregular data accesses. The most important use of these features is the specification of redundant schedules, which can have a significant effect on object code performance.

These language features are part of the language HPF+, which is currently being developed and implemented in the framework of the ESPRIT IV project "HPF+". Initial performance measurements indicate that schedule manipulation can considerably improve the execution time of a range of real codes, including ECMWF's weather forecasting system *IFS*, AVL's combustion code *FIRE*, and the *PAM-CRASH* code developed by ESI. On-going work will extend the work decribed in this paper to include user-defined specifications of irregular halos and access patterns.

Acknowledgement

The authors thank their partners in the ESPRIT project "HPF+", in particular Guy Lonsdale and George Mozdzynski, for many fruitful discussions on this subject.

References

[1] Benkner, S. Vienna Fortran 90 and its Compilation. Ph.D. Thesis. Technical Report TR 94-8, University of Vienna, Institute of Software Technology and Parallel Systems, November 1994.

[2] Benkner, S., Andel, S., Blasko, R., Brezany, P., Celic, A., Chapman, B.M., Egg, M., Fahringer, T., Hulman, J., Hou, Y., Kelc, E., Mehofer, E., Moritsch, H., Paul, M., Sanjari, K., Sipkova, V., Velkov, B., Wender, B., Zima, H.P.: Vienna Fortran Compilation System - Version 1.2 - User's Guide, October 1995.

[3] B. Chapman, P. Mehrotra, and H. Zima. Programming in Vienna Fortran. *Scientific Programming* 1(1):31-50, Fall 1992.

[4] B. Chapman, P. Mehrotra, and H. Zima. Extending HPF for Advanced Data Parallel Applications. IEEE Parallel and Distributed Technology, Fall 1994, pp.59-70.

[5] High Performance Fortran Forum. High Performance Fortran Language Specification Version 1.0. Technical Report, Rice University, Houston, TX, May 3, 1993.

[6] High Performance Fortran Forum. High Performance Fortran Language Specification Version 2.0.Ω Technical Report, Rice University, Houston, TX, January 31, 1997.

[7] C. Koelbel and P. Mehrotra. Compiling global name-space parallel loops for distributed execution. *IEEE Transactions on Parallel and Distributed Systems*, 2(4):440–451, October 1991.

[8] R. Ponnusamy, J. Saltz, A. Choudhary. Runtime Compilation Techniques for Data Partitioning and Communication Schedule Reuse. Technical Report, UMIACS-TR-93-32, University of Maryland, April 1993.

[9] J. Saltz, K. Crowley, R. Mirchandaney, and H. Berryman. Run-time scheduling and execution of loops on message passing machines. *Journal of Parallel and Distributed Computing*, 8(2):303–312, 1990.

[10] H. Zima, P. Brezany, B. Chapman, P. Mehrotra, and A. Schwald. Vienna Fortran – a language specification. ICASE Internal Report 21, ICASE, Hampton, VA, 1992.

[11] H. Zima and B. Chapman. Compiling for Distributed Memory Systems. *Proceedings of the IEEE*, Special Section on Languages and Compilers for Parallel Machines, pp. 264-287, February 1993.

FUJITSU VECTOR PARALLEL COMPUTING
- PRESENT AND FUTURE -

MORIYUKI TAKAMURA
High Performance Computing Group
Fujitsu Limited
4-1-1 Kamikodanaka, Nakahara-ku, Kawasaki 211-88, JAPAN

ABSTRACT

Meteorology, as with other computational sciences, requires vast amounts of computational power in order to model more accurately the physical world. The pairing of state-of-the-art CMOS technology and the vector parallel architecture delivers the needed performance. The high-bandwidth crossbar network and compiler controlled load/store mechanism are key elements of the vector parallel architecture which provide it with an efficiency several times better than that of microprocessor-based systems. A vector processor attains twenty to thirty times the speed of high performance microprocessors and a vector parallel processor performs several times faster than systems based on other architectures. Fujitsu is committed to the vector parallel architecture and to improving performance in order to extend the range of practical large-scale computation.

1. Introduction

Computer design involves the examination of a wide range of technology, architecture, and implementation issues in order to satisfy a certain set of design goals, and hence any one design does not necessarily represent a unique solution. Accordingly, what constitutes a good supercomputer is a question no one can really answer. No one, however, would disagree that a computing system is a combination of semiconductors, packaging, architecture, system software, applications, and operating features. The nature and advance of each of these six necessary technologies and their combination determine the "flavor" of the computer. By using these issues as a starting point, the following describes what Fujitsu firmly believes to be the optimal approach to the fast processing of large-scale scientific and engineering computation. With respect to the VPP Series, we describe in this paper its technology and architecture, present a few performance results, and give an outline of the VPP's history as well as its projected evolution.

2. Semiconductors

For the past twenty years high-end supercomputers have relied on high-speed semiconductor devices and vector processing to achieve high performance. The pursuit of higher performance dictated the exclusive use of ECL devices to increase the clock speed of high-end supercomputers. In 1995, however, we saw the first use of CMOS devices in high-end supercomputers. Despite the reduction in uniprocessor performance compared to their predecessors, this changeover to CMOS resulted from the cost/performance, total system performance, miniaturization, and power consumption advantages that give CMOS-based supercomputers superior characteristics. The requirement of superscalar microprocessors has been a major driving force in the rapid progress of CMOS device technology. While it is well known that current lithography allows a minimum design feature of 0.35μm, laboratory experiments have already verified the feasibility of 0.07 μm CMOS. The progress of CMOS technology in the next ten to fifteen years promises to yield further significant advances. We foresee that for at least the next ten years leading-edge CMOS devices will be the semiconductor of choice in future implementations of Fujitsu's moderately parallel VPP supercomputers.

3. Architecture

3.1. Vector Parallel Architecture

It is widely recognized that massively parallel processing has failed to satisfy the ever increasing performance requirements in meteorology and other computationally demanding fields. The problems associated with MPPs[1] are not likely be solved in the near term. As has been previously asserted[2], recent benchmarks show MPPs have lower performance and exhibit inferior actual-to-peak performance ratios for both kernel and application codes[3] when compared to high-speed vector processing systems.

Among other reasons, we argue that this is due to weaknesses inherent in these architectures and their implementations. Three key aspects characterize current MPPs: low network performance and utilization, dependence on cache, and insufficient processor-memory bandwidth. Efficient communication for some non-nearest-neighbor communication patterns is possible, but often only by restructuring code or switching to a different algorithm. In other cases, such as transposition and FFT, the topology of the network itself prevents efficient implementation of algorithms, especially those that require long distance communication. Also, it is well known that microprocessor performance depends heavily on caches to provide both low latency and high bandwidth memory access. These aspects make necessary a heavy reliance on locality, which is not present in sufficient degree in the great majority of supercomputing applications, and as a consequence prevent high execution rates.

The shared memory vector parallel supercomputer, on the other hand, has provided the highest performance to date for most traditional applications. However, its future depends on the ability to implement memory systems that can supply data at a rate commensurate with processing speed–a considerable and complex problem.

In contrast, the VPP architecture[4,5] takes a unique approach to the difficulties encountered in high performance computing and parallel processing. The architecture assembles a moderate number of the high-performance uniprocessors characteristic of conventional vector supercomputers into a scaleable distributed memory architecture that eliminates both the overhead associated with massively parallel systems and the memory contention found in shared memory vector parallel supercomputers.

The 2.2 GFLOPS vector processor in the VPP700 processing element (PE) minimizes the degree of parallelism needed to attain a given level of performance. The high-speed vector processing element also allows existing software assets to execute at high performance on a single processor by simply recompiling the applications, eliminating the need to parallelize code. By taking advantage of the capability to concurrently run arbitrary combinations of single and multi-processor applications, the VPP can attain increased throughput of existing assets and higher performance of already parallelized applications.

The interprocessor communication mechanism handles communication tasks such as data transfer, remote memory fetch, and barrier synchronization in hardware, in order to avoid much of the overhead and resulting bottlenecks often observed in current parallel machines. The high-bandwidth crossbar allows fast data transfer and eliminates limitations imposed by other network topologies, thereby removing much of the complexity involved in attaining high performance on current MPPs.

The choice of a crossbar for the VPP gives the system a number of advantages over other networks, such as those based on tree, mesh, and torus topologies, mainly:

No network contention. Since communication within a pair of PEs is physically independent of communication between other PEs, the network does not suffer performance degradation due to network contention. Consequently, communication patterns that require long distance communication (as opposed to nearest-neighbor) also do not exhibit performance degradation, regardless of network traffic.

Simpler programming. The programmer is freed from performance considerations made necessary by other distributed memory architectures. The fact that a PE can communicate with all other PEs at the same bandwidth as nearest-neighbors regardless of network traffic simplifies both system and application software. For example, programmers need not employ complicated scheduling schemes to minimize network contention.

Flexible partitioning. The absence of hardware and performance limitations allows arbitrary PE partitions and makes dynamic PE allocation by the operating system simple. Also the interconnection among the PEs in a partition is independent from that of other partitions, so job execution on separate partitions cannot interfere with one another. Furthermore the performance will not vary among repeated runs of a given parallel application, regardless of the allocation of physical PEs in a partition.

Simpler hardware. Since the crossbar topology lends itself easily to a deadlock-free implementation, packet routing schemes such as wormhole routing become unnecessary, simplifying hardware design.

Fault tolerance. By nature the crossbar is also fault tolerant. Any malfunction during data transfer affects, at most, only two processing elements, the sending and receiving PEs.

The only drawback of a crossbar is its cost. However, despite the fact that crossbars have a complexity of $O(n^2)$, the implementation does not cost much more than that for O(n lg n) topologies such as the omega network. The VPP700 implementation requires at most only 16 boards for the crossbar network, which constitute only one-seventeenth of the number of boards needed for a full system (16/(256PEs+16)=1/17). Thus the implemented crossbar comprises a small fraction of the system while providing an efficient interconnect for interprocessor communication.

3.2 Memory Access: Cache vs. Vector Registers

It is generally agreed that memory performance is the critical factor that determines the actual power of a processor. The means to attain fast memory access marks the biggest difference between the high performance microprocessor and the vector processor. In the case of the microprocessor, performance is constrained by the performance of the cache. Since caches provide fast access only to data in memory access patterns with temporal and spatial locality, applications such as large-scale computations that have poor locality utilize only a small fraction of the microprocessor's total capability.

While vector registers take the place of cache, compiler controlled memory access in a vector processor circumvents a cache's dependence on locality. Hence the vector processor spends more of its time performing useful work and less time waiting for memory access than a microprocessor. Figure 1 shows the efficiencies of a single processor VPP500 and an IBM RS6000 workstation on a single loop[6]. Although caches continue to get bigger as technology improves, the shape of the curve shown will remain the same for future generations of microprocessors. Computation on data that fit entirely in cache will execute efficiently, but computation on data that exceeds the cache's capacity will achieve only a small fraction of peak. The VPP500, however, asymptotically approaches its peak performance as the volume of data increases.

Table 1 shows performance data of current microprocessors and the VPP500 on DO loops in a suite of meteorological programs[7]. A single processor VPP500 exhibits roughly five times the efficiency and a sustained performance of over twenty times compared to the microprocessors.

3.3 Progression of Microprocessors Towards Vector Processing

Since the development of the first microprocessor, the design of each successive generation has incorporated additional functionality to increase execution throughput. Additions include the concurrent operation of multiple pipelined functional units and the asynchronous execution of integer, floating point, and memory operations. Thus the trend is for microprocessor development to more closely resemble the capabilities of the vector processor. In other words,

The pursuit of higher performance microprocessors leads asymptotically to vector processing.

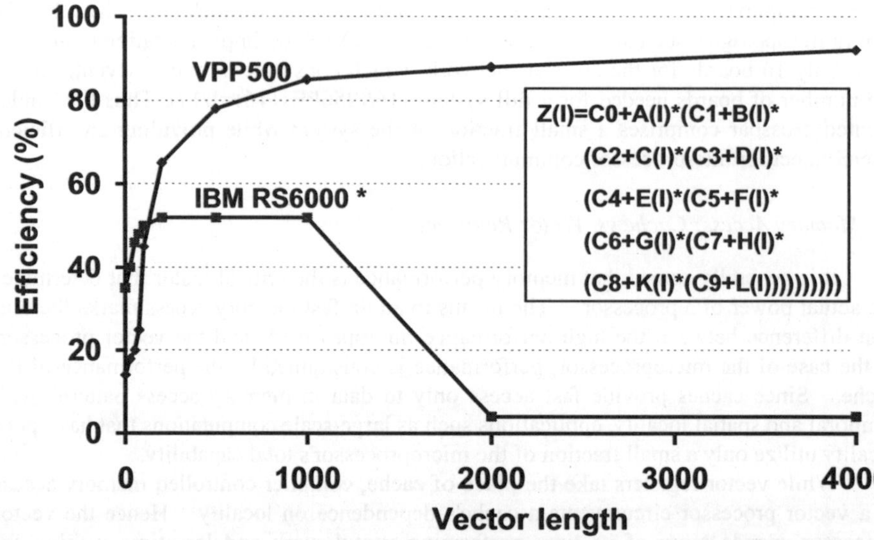

* Measurement condition : RS6000-590(66.6MHz)
FORTRAN77 - 03 - qarch=pwr2 - qtune=pwr2

Figure 1. VPP500 vs. RS6000 on Single Loop

Table 1. Processor Efficiencies and Ratios To VPP500

		RISC Microprocessor						Vector	Ratio
		PA7200	R10000	Power2	Ultra	Alpha	Ave.	VPP500	
Peak MFLOPS		240	400	266	333	600	368	1600	4.3
All DO Loops (64)	Efficiency (%)	7.94	5.25	10.1	7.83	5.63	7.3	38	5.1
	Sustained MFLOPS	19.0	21.0	26.8	26.0	33.7	25	603	23
DO Loops without IF's (40)	Efficiency (%)	8.69	6.31	11.9	9.07	6.34	8.4	47	5.5
	Sustained MFLOPS	20.8	25.2	31.6	30.2	38.0	29	755	25

4. Performance

Existing supercomputers can be classified into one of three categories–distributed-memory vector parallel, shared-memory vector parallel, and microprocessor-based MPP and SMP. The graph in Figure 2 characterizes the performance of each of the three categories of architecture.

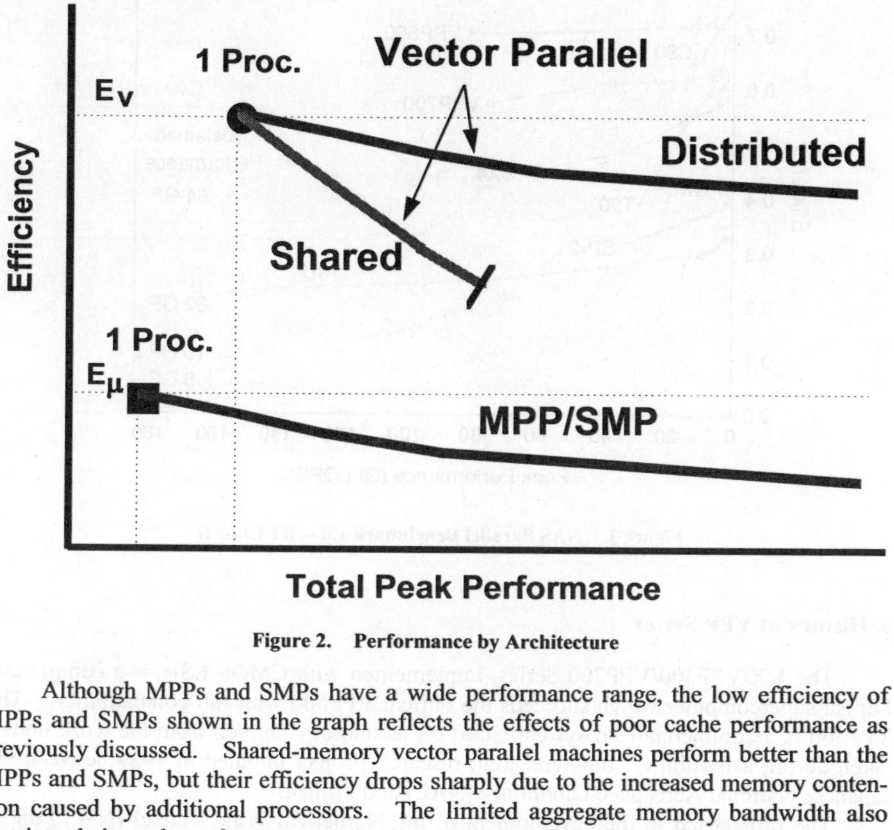

Figure 2. Performance by Architecture

Although MPPs and SMPs have a wide performance range, the low efficiency of MPPs and SMPs shown in the graph reflects the effects of poor cache performance as previously discussed. Shared-memory vector parallel machines perform better than the MPPs and SMPs, but their efficiency drops sharply due to the increased memory contention caused by additional processors. The limited aggregate memory bandwidth also restricts their peak performance to the low range. Distributed-memory vector parallel machines, on the other hand, do not exhibit the limitations of either MPPs/SMPs or shared-memory vector parallel machines. The superiority of the VPP architecture shows in its high efficiency and scalability.

Figure 3 shows the results of the NAS Parallel Benchmark test BT class B for several supercomputer class machines[3]. The Fujitsu VPP500 and VPP700 are distributed-memory parallel vector; the Cray C90 and T90 and NEC SX4 are shared-memory parallel vector; and the SGI Power Challenge XL, IBM SP-2, and Cray T3D are MPP/SMP. Note that the curves for the VPP700, T90, and T3D are similar to the curves for their respective categories as shown Figure 2.

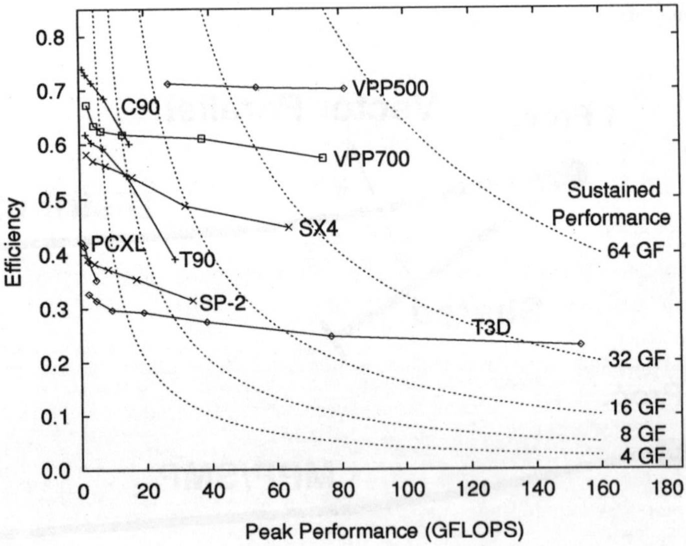

Figure 3. NAS Parallel Benchmark 1.0 -- BT Class B

5. History of VPP Series

The VX/VPP300/VPP700 Series, implemented with CMOS LSIs, is a commercial Fujitsu supercomputer which succeeds the Fujitsu VPP500 with full compatibility. The VPP500 is a commercialized system based on technology derived from the work undertaken during a feasibility study and joint research project initiated in 1989 between the Japanese National Aerospace Laboratory (NAL) and Fujitsu.

The project led to the development of the Numerical Wind Tunnel (NWT) under the strong leadership of Mr. H. Miyoshi of NAL to whom we express our deepest gratitude. In the project, NAL contributed the concept of the vector processing node, the rational positioning of performance, and others; Fujitsu the concept of inter-PE communication with crossbar network, its hardware and software implementations, and others.

6. Future VPP Systems

Fujitsu will extend the vector parallel architecture for future systems while maintaining upward compatibility. Post-VPP700 Series machines will support 64-bit addressing and integer operations. This and other design improvements will make the vector and scalar processors both more efficient and faster. To match the increased execution capability, main memory capacity and performance and the throughput and latency of inter-PE communication will continue to remain commensurate with the speed of the computation throughput.

I/O performance will be greatly enhanced with a faster I/O processing element and parallel distributed file system to meet ever-increasing I/O requirements.

The UXP/VPP operating system and VPP language processing are continually being enhanced. Future system software will integrate additional capabilities and improve performance while being upwardly compatible.

State-of-the-art, in-house CMOS technology used in Fujitsu mainframes and other computer-related areas will continue to form the basis for future hardware.

7. Concluding Remarks

We have discussed the effects of semiconductors and architecture in the design of high performance supercomputers. The vector parallel architecture avoids the performance bottlenecks present in other architectures. Future VPP systems will further enlarge the scope of practical supercomputing by improving cost/performance, efficiency, and computational capacity.

References

1. J. Worlton, "The MPP Bandwagon," HPCwire report, May 5, 1992.
2. Gordon Bell, "Ultracomputer: A Teraflop Before Its Time," *Communications of the ACM*, **Vol. 35, No. 8** Aug. 1992, pp. 27-47.
3. Subhash Saini and David H. Bailey, "NAS Parallel Benchmark Results 12-95," Report NAS-95-021, NASA Ames Research Center, Moffett Field, CA 94035, December, 1995.
4. Moriyuki Takamura and Teruo Utsumi, "Why Vector Parallel?," *The Proceedings of the HPC Conference '94* , Singapore, Sept. 1994, pp. 394-398.
5. Akira Nodomi, Masayuki Ikeda, Moriyuki Takamura, and Kenichi Miura, "Hardware Performance of the VPP500 Parallel Supercomputer," *High Performance Computing : Technology, Methods and Applications* (North-Holland, Amsterdam, 1995), pp. 103-120.
6. Internal memorandum, 1994.
7. Internal memorandum, "Performance Study of VPP500 and Workstations on Meteorological Programs," Aug. 1996. Contents to be made public at a later date.

The Glue That Holds It Together: NCAR's Model Coupling Library

John M. Dennis, Richard D. Loft

National Center for Atmospheric Research
Scientific Computing Division
P.O. Box 3000
Boulder, CO 80307
http://www.scd.ucar.edu/css

Abstract

The Model Coupling Library (MCL) is a robust, adaptable communication library developed at NCAR which supports message passing between coupled climate model components on a variety of platforms. MCL supports message passing over HIPPI IP sockets, PVM, named pipes and System V shared memory calls. MCL also incorporates error handling capabilities including graceful shutdown, timeout and retry. The MCL has been used since May 1996 as an official part of the Climate System Model (CSM).

We present performance results for MCL under a variety of production and experimental configurations. This will include results of coupling PVP's at two centers separated by 2000 km, using the vBNS, an OC3 ATM network, with a sustained data transfer rate of 4 Mbytes/sec.

1 Introduction

The Climate System Model (CSM) project at NCAR has the long-term goal of developing a comprehensive model of the climate system. The interdisciplinary nature of coupled climate modeling presents special obstacles. The component models have been developed autonomously by researchers working in their individual specialties over a number of years. Differences in nomenclature, model grids, data files strucures, and software capabilities all become painfully obvious when a coupled climate model is attempted. It has been suggested by Mechoso[1] Barth[2], and Bryan[3] that the

tasks of coupling earth science models could best be achieved by exchanging coupling fields, such as momentum, mass and energy fluxes, via message passing, while other groups, for example Sekiguchi, et al.. have suggested using Remote Procedure Calls (RPC's)[4]. In either case, the advantage of this approach is that the component models retain a degree of source code autonomy from one another, exchanging only the coupling information at periodic intervals. In this paper, we will focus on the message passing approach, as currently employed by CSM.

The CSM system is made up of three *component models* that each simulate a specific part of the earth's climate system. The atmospheric component, the Community Climate Model Version 3.2 (CCM3.2), also contains the land surface model as a subroutine. CCM3.2 is run in production on a 2.8° (T42) resolution latitude-longitude grid in CSM. The ocean model component is NCOM version 1.2, and is run on a latitude-longitude grid with a resolution of 2.4° in longitude and a stretched latitudinal grid that is 1.2° poleward of 60° in both hemispheres[5]. The NCAR CSM sea ice model, a variant of the Flato-Hibbler model, lies on the same horizontal grid as the ocean[5].

The component models do not exchange information directly with each other. Instead they each pass messages to a centralized piece of CSM software called the *flux-coupler*. The flux-coupler serves both as a "master" process that controls the execution of the overall CSM system as well as a "middleman" that computes the fluxes across the component model interfaces[3].

Early versions of the CSM system, based on PVM 2.4, suffered from poor performance and error handling problems. These problems included difficultly starting up the coupled simulation reliably, as well as occasional hanging of the component processes if any single component failed. While it may be irritating to have a collection of processes hang on a workstation, it is clearly both expensive and unreasonable to lock up a production supercomputers resources in this way.

At the same time that the need of CSM scientists for a production-ready coupling fabric became apparent, NCAR became involved in a wide area distributed computing project, called the Distributed Climate System Labratory (DCSL). The purpose of the DCSL project was to evaluate the potential of using the NSF experimental very high-speed Backbone Network Service (vBNS), a 155Mb/s OC-3 ATM network that connected the National Science Foundation (NSF) CISE centers and NCAR together, to link multiple, distributed high performance computer systems together to form a *metacomputer*. As part of this project NCAR proposed to distribute the NCAR CSM model components between NCAR and one CISE center over vBNS. The center chosen for this project was the San Diego Supercomputing Center (SDSC), located $\approx$ 2000 km away from NCAR.

After examining the status of PVM and MPI provided by the computer industry, we determined that no single message passing environment could simultaneously offer both the error handling capabilities and meet the high performance requirements of coupled climate modeling under all circumstances and on all architectures. For example, a particular concern for DCSL was the ability to use a shared memory data transfer mechanism for those component model residing on the same system, and HIPPI transfers between those distributed on separate systems. Because of these concerns, we recognized that the needs of CSM and the DCSL project could be only be met by the creation of a Fortran subroutine library called the Model Coupling Library (MCL).

The MCL design objectives are ease of use, reliability, adaptibility, and performance. This means that MCL is more than a set of wrappers built around existing message passing libraries, because it provides extensive error handling features to permit graceful shutdown of the entire model, should a component fail. While the MCL does not break any new theoretical ground in computer science, it does fill a real world communication need of coupled climate modeling. MCL is *NOT* designed to replace PVM or MPI. Instead, we *hope* that advances in the sophistication of standard message passing libraries, such as MPI will, in time, render this library obsolete.

In the sections that follow, we first present the library design issues of both performance and reliability. Next we present specifics on a particular section of the library, our high performance TCP/IP implementation, followed by performance from select computation environments.

2 Model Coupling Library Design

The message passing design of MCL is based on the 'star' or 'single hub' organization of CSM, shown in Figure 1. All message passing occur between the central process, called the flux-coupler in CSM, and component models. It was clear from the outset that the functions the MCL had to perform fell into two basic catagories: control and data transfer. The message traffic required to carry out these two functions differ greatly in their characteristics. Control message size tends to be small, on the order of a few bytes, and reliability is the main concern. In contrast, data transfer involves large messages on the order of 1 MByte, and raw bandwidth performance is critical. These tasks were therefore partitioned conceptually in the MCL design to allow a reliable control channel and an independently selectable data channel.

PVM was chosen for the control channel. PVM3.3 has extensive error handling facilities and is widely available on a number of computer platforms. Also, PVM is naturally suited to support a true MIMD application like coupled climate modeling. In contrast, current implementations of MPI seem best suited to SPMD programming. Use of MPI for the control channel was consequently examined and dismissed, principally on the grounds that it did not provide a convenient user interface for

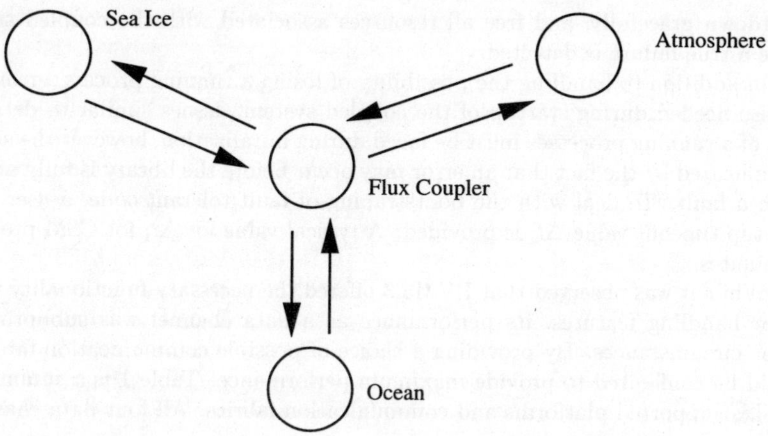

Figure 1: Single Hub Message Passing System

passing messages between separate executables. This decision will be revisited when the "Client/Server Model" feature of MPI-2 is generally avalable.

We next describe the error handling feature implemented in MCL. When a single executable encounters an error, for example a floating point exception, the code will fail, exit, and release its resources back to the operationing system. In the case of multiple processes which exchange messages, one process may fail and exit, leaving the remaining processes waiting indefinitely for a message from their lost group member. A simple solution to this problem is to use a receive with timeout function. This addresses the problem of processes hanging indefinitely, but it may also kill simulations prematurally. In a production environment messages may not be received from perfectly healthy processes because of slow networks, a lost message, or heavily loaded machines. Since MCL is designed to operate on a wide range of production environments, no single choice for a timeout value, Δt, using this simple design would be correct. To solve this problem, a receive with a fixed Δt is used, but is also supplemented with a subsequent polling of the sending process if the timeout value has been exceeded. If the sending process is still alive, then the receive will retry, otherwise the receiving process will return an error code to the user. A failure of any one of the processes, including the flux-coupler itself, will be detected by all other processes after a maximum delay of $2\Delta t$. Each surviving processes can then shutdown gracefully, and free all resources associated with the coupled simulation, once a true failure is detected.

In addition to handling the possibility of losing a running process, error handling is also needed during startup of the coupled system. Issues similar to detecting the loss of a running processes must be faced during initalization, however the situation is complicated by the fact that an error may occur before the library is fully able to deal with a fault. To deal with the bootstraping of fault tolerant code, a user selectable startup timeout value Δt_i is provided. A typical value for Δt_i for CSM production is 6 minutes.

While it was observed that PVM3.3 offered the necessary functionality to provide error handling features, its performance as a data channel was suboptimal under some circumstances. By providing a choice of possible communication fabrics, MCL could be configured to provide maximum performance. Table 1 is a summary of the MCL's supported platforms and communication fabrics. All four data channel types

Table 1: Supported platforms and Communication fabrics

Fabric/Method	Y-MP	SGI	SUN	RS6K
PVM3.3	YES	YES	YES	YES
AF-INET Sockets	YES	YES	YES	YES
Named Pipes	YES	YES	YES	YES
System V Shared Memory	NO	YES	YES	YES

are contained in the same library, and are user selectable through a MCL interface. This allows a coupled simulation to mix and match the most suitable data channel for a particular flux-coupler to component model link. The following is a detailed description of each of the different data channels provided in the MCL.

PVM – This represents the default data channel. This method does not separate the control and data message but insteads packs them together and sends one message.

AF-INET Sockets – The advantage of using this type of data channel is its ability to manipulate the TCP/IP socket parameters. Increasing send and receive buffers, and setting window shifting parameters can result in a considerable performance increases. A detail description of this method will be presented in Section 3.

Named Pipes – This method creates a special FIFO file that two processes can access. Since Named Pipes essentially act as a read/write to/from memory, they can offer considerable performance advantages for processes that reside on the same machine.

System V Shared Memory – The System V Interprocess Communication facility provides the ability to map a piece of memory into the memory space of two separate executables. Passing messages between processes involves the writing/reading of the message into the shared memory space at memory copy bandwidths.

3 A High Performance TCP/IP Implementation

We now examine in detail a specific data channel provided by the Model Coupling Library, AF-INET sockets, and the benefits of the TCP window shifting option. AF-INET sockets form the basis of portable PVM and MPI libraries. TCP window shifting is a technique created in 1992 by the Network Working Group[6], and has been used by both Mechoso[1] and COOP-3D[7] to improve performance in distributed coupling projects. Window shifting is designed to improve bandwidth on Long Fat Networks (LFNs). LFNs are defined as networks in which the product of the bandwidth B and the network latency t_L is greater than 10^6 bits. Transmission Control Protocol (TCP) is a networking protocol developed to provided reliable network communication over an unreliable IP based network. TCP achieves this reliability by breaking a message into fixed size packets and attaching a sequence number to the packets. After a packet is sent, the sender waits to receive either an acknowledgement from the receiver that the packet was received, or a retransmit request from the receiver indicating that a packet was corrupt or lost. This processes continues until the entire message is successfully transmited. Clearly processing the receiver's acknowledgement will reduce sustained bandwidth, especially on a LFN. This degredation can be reduced by increasing the length of the TCP send/receive buffers. While effective, the maximum size is only 2^{16} bytes or 64 Kbytes. Window shifting allows even larger buffer settings, up to 2^{30} bytes or 1 Gbyte. The proper window shift size, w, for a particular network can be determined by calculating the buffer size $b = Bt_L$ and then applying the formula

$$w = \lceil \log_2(b) \rceil - 16.$$

In other words, the proper window shift size is the base 2 logarithm of the ratio of the smallest power of two buffer size which is greater than or equal to b, and 64 Kbytes.

To support these large buffer sizes, kernal memory resources must be specially configured. For this reason, window shifting should not be attempted by the inexperienced, i. e. without consulting a system administrator, since it may have the undesirable side affect of locking a machine up if done incorrectly.

A plot of bandwidth verses buffer sizes for several window shift values for 800 Mbit/sec HIPPI measured between a Cray YMP-8 and a J932 is seen in Figure 2. The relative noiseness of the data is caused by the fact that these tests were

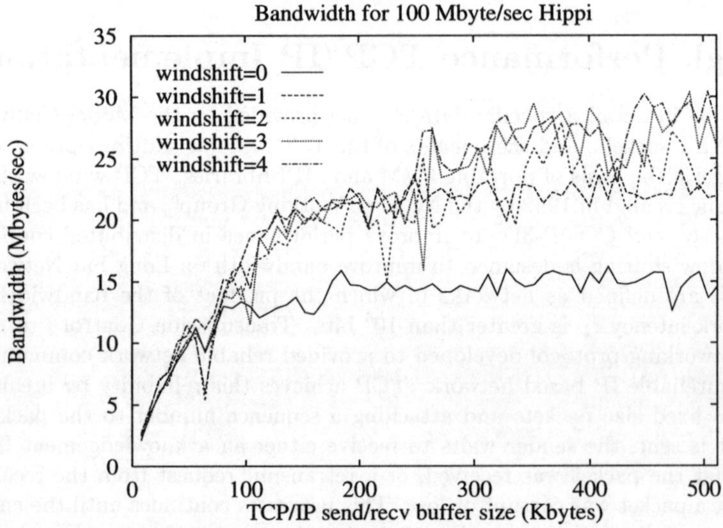

Figure 2: Message passing performance over 800Mbit/sec Hippi

run during regular production hours. As Figure 2 indicates, TCP windshifting can double bandwidth from 15 Mbytes/sec to 30 Mbytes/sec on long messages. These results include all overheads associated with the time to pack the message, perform error checking, send the message, and unpack the message. They are not indicative of theoretical network performance, but rather reflect what an application would achieve in pratice.

4 Performance

The performance of the Model Coupling Library is best understood in the context of several specific projects which use the package, including both CSM and DCSL.

4.1 *Production CSM*

The Model Coupling Library has been used by the CSM models since its public introduction in May 1996. CSM simulations are currently run on single YMP8/64 or a J932/20, using the Named Pipe data channel. A plot of the observed bandwidths for Named Pipes verses PVM is shown in Figure 3. Note that the MCL/Named Pipes data channel has a high latency due to overhead from the PVM -based control channel. In the current CSM system, both the atmosphere and the ice model exchange interface information with the flux-coupler every atmospheric timestep (i.e. every 20 minutes), or 72 times per model day. The ocean exchanges interface information with the flux-coupler once per model day. Thus there are $72 \times 4 + 2 = 290$ exchanges involving buffers approximately 1.26 MBytes in size, for a grand total of 368 Mbytes of data traffic per model day, most of it due to the tight coupling of the atmosphere and ice models. New modifications to the CSM will permit a reduction of the coupling frequency of both models to once per hour.

A comparision of typical CSM message passing performance with percentages of total time for the 20 minute coupling interval is seen in Tables 2 and 3.

Table 2: CSM Simulation on dedicated Y-MP8 142 seconds/day

method	bandwidth (Mbytes/sec)	commun. time (sec)	commun. overhead as percentage of time
PVM3.3	26.0	14.15	10.0%
MCL/Named Pipes	212.0	1.73	1.2%

Table 3: CSM Simulation on dedicated J932/20 160 seconds/day

method	bandwidth (Mbytes/sec)	commun. time (sec)	commun. overhead as percentage of time
PVM3.3	9.5	38.7	24.2%
MCL/Named Pipes	83.1	4.4	2.7%

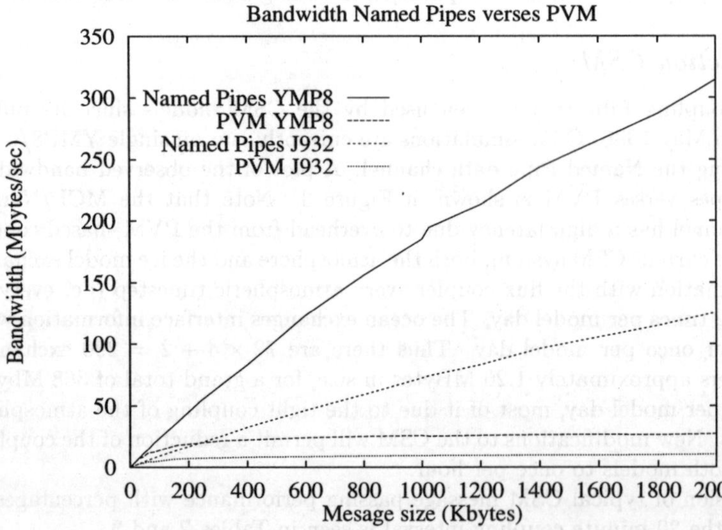

Figure 3: Message passing performance MCL/Named Pipes verses PVM

4.2 *Local Area Coupling*

To test the feasibility of distributing a CSM simulation across multiple machines, dedicated time was arranged on both a YMP8/64 and a J932/20 connected with a 800 Mbit/sec HIPPI network. A ten day CSM simulation was run on both a single system, and distributed across both machines. During these tests, the Cray hardware performance monitors were run to determine the overall Flop count. The results of these tests can be seen in Table 4.

It was determined during the analysis of the test results that an suboptimal windshift and buffer size were used during the simulation. This resulted in messages being passed at 15 MBytes/sec verses the 30 MBytes/sec seen in Figure 2. Corrected values for both the wall clock hours/CSM year and CSM Gflops can be estimate and are in-

Table 4: Dedicated turnaround on CSM Simulation

Models on YMP-8	Models on J932/20	WC hours/CSM year	CSM Gflops
ATM,OCN,ICE,FC	none	14.58	1.08
none	ATM,OCN,ICE,FC	16.63	0.94
OCN,ICE	ATM,FC	9.94(8.9)	1.58(1.65)

cluded in parenthesis in Table 4. While the CSM does not achieve a perfect speedup, due to both load imbalance issues as well as communication costs, the distributed version was able to achieve approximately 1.6 Gflops, which exceeds the performance of any single machine currently avalable at NCAR.

Correction of the defficiency noted above would lead to even better performance. It is noteworthy that the MCL enables the message passing between machines at a higher rate than PVM alone can achieve on a single machine.

4.3 *Wide Area Coupling over vBNS*

The DCSL project was designed to test the feasibility of using the vBNS a 155 Mbit/sec OC-3 ATM network to perform distributed computing on a national scale. A CSM simulation running distributed across the vBNS was demonstrated at the Supercomputing 1995 conference. During this demonstration, the CSM ocean component ran on one processor of a Cray C90 supercomputer at the San Diego Supercomputing Center , while the atmosphere, sea-ice and flux-coupler models were executed on a J932/20 at NCAR . Since the ocean and flux-coupler exchange data only once per model day (about every 30-40 seconds in real time) this configuration reduced vBNS communication overhead to an acceptable level. The one way latency to San Diego measured during this project was 34 ms. With an OC-3 bandwidth of 155 Mbit/sec, the Bt_L product for the vBNS is 5.27×10^6 Mbits, which qualifies the vBNS as a LFN. Using a TCP send/receive buffer of 658750 bytes and a window shift of 4, a sustained bandwidth of 4 MBytes/sec was obtained.

5 Future Work

The Model Coupling Library was designed to fill the needs of a particular application niche that could not be fulfilled by any standard message passing package at the time of its creation. Hopefully in the future advances in message passing standards will make most of functionality provided in MCL obsolete. However until, a single high performance message passing standard with suitable error handling features is supplied by all vendors, development and support of MCL will continue. It is clear from examining Figure 2 that MCL has a very high latency due to its use of PVM as a control channel. As memory bandwidth, and processor performance improves, the latency of the control channel could begin to seriously hinder message passing performance. A natural extention of a user selectable data channel would be the creation of a user selectable control channel. Such a design could allow for use of either PVM or MPI as the control channel.

In addition to improvements in bandwidth and latency, MCL could be enhanced to provided fault tolerance. While MCL currently provides graceful shutdown in response to faults, there has been work by Kohl[8] to exploring the posibility of a truly fault tolerant software system. Such a system could continue to run despite the lost of network connections, or processes, by regenerating itself when a fault is encountered.

6 Conclusions

We presented the motivation and design of an adaptable message passing library design to fullfil a real world computational problem, namely coupled climate modeling. We also have presented specifics on a high performance TCP/IP implentation which has been used in both local and wide area coupling. Finally we present performance figures which reveal that we can achieve up to 7 times the bandwidth by using MCL/Named Pipes verses PVM on a YMP-8, and 4 Mbytes/sec over the vBNS.

Acknowledgements

There are numerous people who assisted in the production of this software package. Special thanks go to: Steve Hammond for his support and ideas behind the philosophy of the package; Brian Kauffman and Mariana Vertenstein for their exhaustive testing and suggestions, of improvements to the package; Dan Anderson, and Chris Fair for their high performance TCP/IP codes; Jeff Kuehn for the idea of using named pipes and System V IPC; Dennis Colarelli, and George Fuentes for their operating system expertise.

References

[1] C. R. Mechoso, C. C. Ma, J. D. Farrara, J. A. Spahr, R. W. Moore. Parallelization and Distribution of a Coupled Atmosphere-Ocean General Circulation Model *Mon. Wea. Rev.*, **121**:2062–2076, 1993

[2] N. Barth, S. Smith. Coupling numerical models of the atmosphere and ocean using the Parallel Virtual Machine(PVM) Package, *Proceedings of the Sixth SIAM Conference on Parallel Processing for Scientific Computation* , R, Sincovec, D, Keyes, M.Leuze, L. Petzold, and D. Redd,eds. SIAM, 71-75

[3] F. O. Bryan, B. G. Kauffman,W. G. Large,P. R. Gent. *The NCAR CSM Flux Coupler* Technical Report NCAR Technical Note/TN-424+STR, Climate and Global Dynamic Division, National Center for Atmospheric Research, PO Box 3000, Boulder, CO 80307, 1996

[4] S. Sekiguchi. Personal communication, Nov. 1996

[5] T. W. Bettge, J. W. Weatherly, W. M. Washington, D. Pollard, B. P. Briegleb, W. G. Strand, Jr. *The NCAR CSM Sea Ice Model,* Technical Report NCAR Technical Note/TN-425+STR, Climate and Global Dynamic Division, National Center for Atmospheric Research, PO Box 3000, Boulder, CO 80307, 1996

[6] V. Jacobson, R. Braden, D, Borman. TCP Extensions for High Performance. *Network Working Group, Request for Comments: 1323*

[7] B. Irwin, C. Ruff. *NCAR/SCD Tech Note regarding TCP_WINSHIFT* Technical Report, Scientific Computing Division, National Center for Atmospheric Research, PO Box 3000, Boulder, CO 80307, July 26,1995

[8] J. Kohl, P. Papadopoulos. Personal communication, Feb. 1996

PROGRESS IN THE PARALLEL IMPLEMENTATION OF THE PHYSICAL-SPACE STATISTICAL ANALYSIS SYSTEM (PSAS)

P. M. Lyster, J. W. Larson, J. Guo, W. Sawyer, A. da Silva, I. Štajner
Data Assimilation Office, NASA/Goddard Space Flight Center, Greenbelt, MD 20706, USA

C. H. Q. Ding[a] and R. Ferraro
Jet Propulsion Laboratory, 4800 Oak Grove Drive, Pasadena, CA 94720, USA

The Physical-space Statistical Analysis System (PSAS) comprises the core analysis scheme for the Goddard Earth Observing System (GEOS) atmospheric data assimilation system. The Data Assimilation Office (DAO) of the National Aeronautics and Space Administration (NASA) developed both GEOS and PSAS, and are currently in the process of constructing a parallel implementation of GEOS. Here we discuss progress in the parallel implementation of PSAS, as well as issues related to the parallelization of an enhanced PSAS capable of direct assimilation of non-state variables (*e.g.*, radiances).

1 Introduction

The Physical-space Statistical Analysis System (PSAS) algorithm is a significant part of the GEOS atmospheric data assimilation system that is used by the Data Assimilation Office (DAO). Apart from the considerable technology surrounding the data input and output, transmission to and from data facilities, and data visualization, the core components of GEOS are: an atmospheric general circulation model (GCM); the data quality control (QC); and PSAS. These are computationally intensive algorithms that, because of the nature of the underlying physical models, are highly coupled.

The DAO is preparing to move its data assimilation system to advanced computing platforms. This will be part of its regular operation, although an important role is expected for the Mission to Planet Earth (MTPE) system in the coming years. Key components of the gridpoint-based GCM [1] [b] have been parallelized, and are expected to be incorporated into the system in the coming year. In 1994, the core components of an early serial version of PSAS were studied by computer scientists at the Jet Propulsion Laboratory (JPL) as part of the High Performance Computing and Communications (HPCC) project. The objective of this study was to devise a parallel version of this

[a]Current Address: Lawrence Berkeley Laboratory, Berkeley, CA 94720
[b]Available online at `http://dao.gsfc.nasa.gov/subpages/tech-reports.html`

developmental version of PSAS; the hope being that this work would lead to the development of techniques that could be applied in the parallelization of later—but already planned—versions of PSAS. The resulting code, *the prototype parallel PSAS*, was parallelized using distributed-memory Single Program Multiple Data (SPMD) message-passing approach [23].

2 The Physical-space Statistical Analysis System (PSAS)

Let w^o be a vector of p observations, and w^f a vector of n forecast variables. The objective is to produce an analyzed state vector $w^a \in \mathbb{R}^n$ (*i.e.*, a vector of physical quantities that describe directly the physical state of the atmosphere). This is accomplished via solution of a pair of equations that comprise the core formulation of PSAS [4]:

The innovation equation

$$(HP^fH^T + R)x = w^o - Hw^f; \qquad (1)$$

and *the analysis equation*

$$w^a = w^f + P^fH^Tx, \qquad (2)$$

where $P^f : \mathbb{R}^n \rightarrow \mathbb{R}^n$ is the specified forecast error covariance matrix, $R : \mathbb{R}^p \rightarrow \mathbb{R}^p$ is the specified observation error covariance matrix, $H : \mathbb{R}^n \rightarrow \mathbb{R}^p$ represents a generalized interpolation from the analysis grid to the observations, and $x \in \mathbb{R}^p$ is an intermediate vector in the observation space. It should be noted here that in the version of PSAS discussed here, both the observation and forecast variables are state variables.

The analysis grid is a 2° latitude by 2.5° longitude spherical grid. The state variables analyzed include: sea-level pressure p_s, sea-level horizontal winds u_s and v_s, and on 18 standard pressure levels the horizontal wind u and v, the geopotential height h, and on standard levels at greater than 300 hPa the specific humidity q.

It is also important to note that PSAS does not use the local approximation as found in optimal interpolation schemes, but instead involves the solution of a global coupled large matrix problem. The innovation matrix $M = HP^fH^T + R$ is dense, although entries associated with locations that are separated by several correlation lengths are negligible (or zero for compactly supported correlation functions) [5 6]. In the present multi-tasked serial version of PSAS, block structure is introduced in M by dividing the observations into 80 equal-area regions via an icosohedral partitioning scheme [7]. In order to introduce some sparseness in $HP^fH^T + R$ and thus save computational effort, the correlations beyond a preset cutoff distance are not included. For PSAS, with a 6,000 kilometer cutoff between centroids of observation regions the innovation, matrix is approximately 26% full, and this uses in excess of 5 gigabytes of storage for a 32-bit representation of the data.

The innovation equation is solved using a preconditioned conjugate gradient (CG) algorithm[8,4]. This is an $\mathcal{O}(N_ip^2)$ operation, where N_i is the number of iterations of a conjugate gradient algorithm. Typically, between eight and twelve iterations are needed to produce a CG solution whose residual has been reduced by at least two orders of magnitude. Experiments in which the residuals are reduced by more than two orders of magnitude resulted in differences much smaller than expected analysis errors[6]. The number of operations in the solution of the analysis equation (2) is $\mathcal{O}(np)$.

3 The Prototype Parallel PSAS

As mentioned in the previous section, a major component of PSAS is the conjugate gradient solution of the innovation equation, which involves a large matrix multiply $(HP^fH^T + R)x'$ where x' is the intermediate conjugate gradient solution for the vector x. The key for message-passing parallelism is to break the matrix into blocks that can be parceled out to different processors (domain decomposition). Each matrix block acts on an appropriate vector segment to form a partial vector. The result of each matrix multiply is a set of partial vectors which must be summed using message passing to form the result. In this way both the total memory and the work in generating the matrix and performing the multiply are divided among the processors; it is a tenet of parallel computing that any data or process that cannot be decomposed is a risk to performance. The large memory that is available on parallel computers, such as the Intel Paragon or the Cray T3D, also affords the possibility of calculating and storing the entire forecast error covariance matrix *once* per analysis cycle. This has the potential for considerable speedup over the serial approach, in which the matrix was reevaluated for each iteration of the conjugate gradient algorithm. This provides strong motivation for creating a parallel implementation of PSAS.

The algorithm of the parallel version of PSAS comprises the five following steps:

- Partition observations using inertial recursive bisection scheme
- Block matrix distribution of the innovation matrix M
- Solve innovation equation using CG method

- Partition of $P^f H^T$
- Solve the analysis equation.

Each of these steps in the algorithm will be examined in greater detail in subsequent sections.

3.1 Partition of the Observations

This algorithm divides p observations among N_r regions using bisection on the surface of a sphere. The data are read in as innovations ($w^o - Hw^f$) and distributed in random order (but in equal numbers) on N_p processors. In the first iteration, the observations are divided along the orthogonal cut of their combined principal axis moment of inertia. Successive iterations are performed in a tree structure that repeats the decomposition on half the remaining processors with approximately half the data. At each stage, the calculation of the principal axis is performed in parallel. As regions replicate, data are moved between different processors using split MPI communicators. It should be noted that only the ***geographical locations*** of the observations, along with the processor on which the data associated with the observation itself resides is communicated during the partitioning process. It is only ***after*** the partitioning is complete that the observations themselves are relocated to the processors corresponding to their respective observation regions. This two-step process minimizes communications costs.

The inertial division guarantees some degree of compactness of the resulting decomposition, and results in equal numbers of observations in each region. This algorithm requires that the number of processors is a power of 2, and less than or equal to the number of regions; $N_r \geq N_p$. Typically, $N_p = 256$, or 512, and $N_r = 512$. It is possible to modify the power-of-two restriction on numbers of regions and processors by moving the location of the orthogonal cut to some point(s) along the principal axis to give a division of observations other than halving, and maintain approximately equal numbers of observations per region. However if the number of regions on each processor is not fixed for all processors, there may be some load imbalance in the analysis equation.

3.2 Matrix Block Partition

In this part of PSAS the parallel decomposition of observations is used to generate a list of matrix blocks. This is used to determine on which processor a matrix block will reside. Particular attention must be paid to this process in order to ensure load balance in the partitioning of memory, and work in generating the matrix elements (from a look-up table), and the work involved in the matrix-vector multiply found in the conjugate gradient solver.

For the prototype parallel PSAS, there are typically 512 observation regions containing approximately $p \sim 10^5$ observations. The 6,000 kilometer correlation cutoff condition gives rise to a forecast error covariance matrix that is approximately 26% full. Therefore there are about $512^2 \times 0.26 \times 0.5 \sim 35,000$ blocks (the factor of 0.5 accounts for the symmetry of the matrix).

Recall that each matrix block in the innovation matrix is associated with one observation region in the case of diagonal blocks, or two regions in the case of off-diagonal blocks. Each block is stored with the associated vector segment of x' that must be used to perform the matrix-vector multiply. Diagonal segmentss are stored on the processor that stores the observation for that region. For the off-diagonal blocks, the number of vector segments on each processor is minimized by storing the block on one of the two processors that owns the observations corresponding to one of the dimensions of the block.

The blocks are approximately of equal size, since there are approximately the same number of observations per region. Therefore, an equal number of blocks per processor provides a good guess for the load balanced distribution. From the initial state, the load-balancing scheme proceeds towards this goal by first refining the initial distribution to produce a new initial state using the following scheme:

1. The load on each processor L_i to perform work for the conjugate gradient solver is estimated, along with the *average load* $\bar{L}$. This leads to a set of imbalances $\Delta_i \equiv \bar{L} - L_i$
2. For each pair (i, j) of processors, an exchange probability $P_{ij} = \frac{L_j}{L_i + L_j}$ is calculated.
3. Use a random number $r \in [0, 1]$ to decide if processor i or processor j is assigned the block. Processor i is assigned the block if $r \leq P_i$, while processor j is assigned the block if $r > P_i$. This amounts to a *weighted coin toss.*

From this refined initial state, the block distribution is improved iteratively through the following adjustment process:

1. For iteration $m \leq N_{\text{iter}}$, for each pair (i, j) of processors deviations from the average load balance Δ_i and Δ_j are calculated, along with the exchange probability P_{ij} described above. The number of blocks on each processor, $\mathcal{N}_{\mathrm{i}}$ and $\mathcal{N}_{\mathrm{j}}$, respectively, are counted along with the average number of blocks on each processor $\bar{\mathcal{N}}$.

2. If $\Delta_i > 0$ and $\Delta_j < 0$, the block is transferred from processor i to processor j. If $\Delta_i < 0$ and $\Delta_j > 0$, the block is transferred from processor j to processor i.
3. If both Δ_i and Δ_j have the same sign, then if $\mathcal{N}_{\mathrm{i}} - \mathcal{N}_{\mathrm{j}} > -\delta\bar{\mathcal{N}}$, then the block is moved from processor i to processor j.
4. Finally, for all other cases not covered above, the block is assigned to a processor based on the weighted coin toss described in the refinement of the initial state.

Following this scheme for $N_{\mathrm{iter}} = 10$ and $\delta = 1/20$ results in a load imbalance of approximately 10%. The messages that must be exchanged during the matrix block partition are matrix indices and geometrical data. No complex structures, such as in the parallel partition, are exchanged until the block partition is complete.

3.3 Solution of the Innovation Equation

The conjugate gradient algorithm uses block preconditioning, and is described in Golub and van Loan[8] and da Silva and Guo[4]. Computationally, the core is the large matrix-vector multiply. In the parallel partitioner the innovations are separated into regions that are distributed among the processors. These provide the initial condition for the vector iterate x'. The vector is actually composed of vector segments. Depending on how the off-diagonal blocks are distributed, some of these vector segments must be replicated on multiple processors. At each step in the iteration, the intermediate solution vector x' must be assembled using global communications. Since only about a third of the processors contain data needed to calculate a particular set of vector segments of x', it is more efficient to do explicit message passing to achieve this end rather than use a global summation such as `MPI_REDUCE`.

For the prototype parallel PSAS, an older and considerably less efficient algorithm was used to perform the table lookup (it has been made faster in the present development PSAS by an order of magnitude). Hence, the generation of the matrix elements was a dominant part of the cost to produce the CG solution. The savings in evaluating the matrix only once per analysis cycle was considerable; effectively N_i (the number of CG iterations) multiplied by the cost of generating the matrix.

3.4 Partition of $P^f H^T$ and Solution of the Analysis Equation

The analysis increment is $P^f H^T x$. The operator $P^f H^T$ represents the forecast error covariance between the observation grid and the analysis grid. The domain decomposition for this must account for the unstructured distribution of the observations and the structured analysis grid. The solution vector x from the conjugate gradient algorithm is decomposed in the same manner as the observations. The decomposition for the analysis grid is based on the fact that gridpoints within the 6,000 mile cutoff of the location of an observation are affected by the value of the corresponding coordinate of the vector x.

The first step in the process is the generation of an equal-area rectangular partition of the analysis grid. These regions are called ***analysis regions***. Next, analysis gridpoints are then thinned longitudinally, resulting in an equal number of analysis gridpoints in each analysis region. In this manner the matrix $P^f H^T$ may be generated as a number of equal-sized blocks such that the centroids of their associated regions are within the 6,000 km cutoff distance of the centroids of each region of x. Since there is a fixed number of regions per processor (i.e., usually 1 or 2) the number and size of the corresponding matrix blocks and the work that is performed in the partial matrix-vector multiply is approximately the same on all processors.

4 Summary of Timings for the Prototype Parallel PSAS

Table 1. shows the timing for the prototype parallel PSAS for 80,000 observations (model resolution $2.5^o \times 2^o \times 14$ levels) on 512 processors of the Intel Paragon at the California Institute of Technology. The solver achieves 18.3 gigaflop/s (77 megaflop/s per processor), which is 36% of peak performance for 512 processors. The parallel partitioner takes 3.1 seconds. This is a relatively small cost; the communications are complex but because of the strategy of sending minimal data and buffering there is little overhead. The calculation of block distribution lists uses a small amount of communications to pass simple lists between processors. During the replication of observation regions, observation vector segments are sent between processors. These are relatively long messages and hence have little communications overhead. The calculation of matrix entries is very time consuming (23.8 seconds), but is performed only once per analysis. The CG solver uses BLAS level 2 library calls and messages to send vector segments of x'. This is iterated $N_i \sim 100$ times[c], taking 36.4 seconds. The dominant cost of the analysis equation is the time spent generating the matrix elements. Since one dimension of $P^f H^T$ is n which is larger than p this takes longer than the generation of the innovation matrix. The

[c]This number of iterations differs from the value of N_i cited earlier because the version of PSAS that was parallelized had a more stringent convergence criterion than subsequent versions of PSAS. As was mentioned in Section 2, the condition cited here is more stringent than was later learned to be necessary.

communication (1.5 seconds) involved in reassembling the analysis grid vector is similar to the assembly of vector segments in the innovation equation.

For the innovation equation:

Task	*Time(sec)*
Read input data	14.6
Partition observations	3.1
Calculate block distribution lists	3.8
Replicate observation regions	3.3
Calculate matrix entries	23.8
Solve parallel CG	36.4
Miscellaneous	1.7
Total	87.0

Secondly, for the analysis equation:

Task	*Time(sec)*
Create grid partition	0.4
Assemble and evaluate $P^f H^T x$	67.6
Reassemble model vector	1.5
Total	69.5

Table 1. Timings for the prototype parallel PSAS on the 512 processor Intel Paragon at the California Institute of Technology.

Table 2 shows the comparison between wall-clock times for three versions of PSAS: the early serial code that was parallelized to create the prototype parallel PSAS as run on a single processor Cray C90, a recent optimized serial PSAS also run on a Cray C90; and the prototype parallel PSAS run on 512 processor of the Intel Paragon. The older serial code is slower than prototype parallel PSAS by two orders of magnitude. The main reasons for this are the higher net flop rate of the parallel processor (approximately 10 gigaflop/s to 1 gigaflop/s) and the calculation of the matrix elements only once per conjugate gradient solve. The more recent, optimized, PSAS runs faster on the C90 because the algorithm for table lookup has been improved (making more use of redundancy and using nearest-neighbor lookup).

The prototype parallel PSAS has also been implemented on a 512 processor

Cray T3D at the NASA Goddard Space Flight Center, and the scaling behavior of the code on this platform is presented in Figure 1. The dots at 128 and 256 processors depict performance for a run with 51,990 observations, while the diamonds at 256 and 512 nodes are for a run using 79,938 observations. The two different numbers of observations were used because there was not sufficient combined memory on 128 nodes to accommodate the larger number of observations.

Version	*solve conj grad.*	*analysis equation*
1994 PSAS	9120. (sec)	9000.
1996 PSAS	750.	750.
Parallel PSAS	87.	71.

Table 2. Comparison of timings for two versions of the serial PSAS and parallel PSAS.

5 Conclusions and Future Work

The prototyping exercise applied to the parallelization of the Physical-space Statistical Analysis System (PSAS) has been sufficiently successful to justify the parallelization of the full GEOS data assimilation system. The next major step in this effort is to learn how applicable the parallel technology developed in the prototype parallel PSAS is to the new operator formulation of PSAS. It is likely these techniques will have to be abstracted and extended if they are to be of use in the solution of this problem. The immediate challenges include:

- The early version of PSAS from which the parallel prototype was built was written in FORTRAN 77. The prototype parallel PSAS is now predominately written in C with some FORTRAN 77 subroutines retained. The version of PSAS currently under development is written in fortran 90, and following its strict typing requirements, fortran 90 derived datatypes, with some portions of FORTRAN 77 code that will be subsequently converted to fortran 90. This change has resulted in a code that is more object-oriented, as well as safe and stable. Finally, the fortran 90 interface to MPI is still under development [9].
- The formulation of PSAS has matured beyond its statement in equations (1) and (2) to include direct assimilation of non-state variables. This presents the major complication in the process of building the next parallel version of PSAS, due to the need to generate a new set of data structures to represent observations that are non-state variables, and

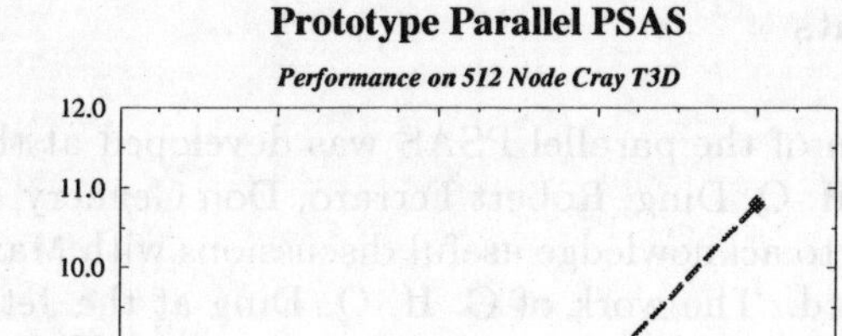

Figure 1: Scaling behavior for the prototype parallel PSAS on a 512 node Cray T3D.

their corresponding variables in state space. These changes will render the load balance problems conquered in the prototype parallel algorithm more complex, and it will also be necessary to extend some of the concepts regarding load-balancing in sections 3.2 and 3.4 to accommodate these enhancements.

Once the above challenges are met, a fully functional parallel version of PSAS capable of assimilating non-state data will result. When this is combined with the parallel version of the GEOS GCM, and lessons learned concerning parallelization of the data quality control system[10], the realization of a fully parallel data assimilation system will be possible.

Acknowledgments

The original design of the parallel PSAS was developed at the Jet Propulsion Laboratory by C. H. Q. Ding, Robert Ferraro, Don Gennery, and Clara Chan.. We would also like to acknowledge useful discussions with Max Suarez and Dan Schaffer at Goddard. The work of C. H. Q. Ding at the Jet Propulsion Laboratory and Peter Lyster and Jay Larson at the Data Assimilation Office was funded by the High Performance Computing and Communications Initiative (HPCC) Earth and Space Science (ESS) program.

References

1. L. L. Takacs, A. Molod, and T. Wang. Documentation of the Goddard Earth Observing System (GEOS) general circulation model–version 1. Technical Report NASA Techinical Memorandum 104606, NASA Goddard Space Flight Center, Greenbelt, Maryland, 1994.
2. C. Ding and R. D. Ferraro. An 18 GFLOPS parallel data assimilation PSAS package. In *Proceedings of the Intel Supercomputer Users Group Conference*, page 70, 1995.
3. C. Ding and R. D. Ferraro. Climate data assimilation on a massively parallel computer. In *Proceedings of Supercomputing, 96*, 1996.
4. A. da Silva and J. Guo. Documentation of the physical-space statistical analysis system (PSAS). part I : The conjugate gradient solver, version PSAS-1.00. Technical Report DAO Office Note No. 96-02, NASA Goddard Space Flight Center, Greenbelt, Maryland, 1996.
5. G. Gaspari and S. E. Cohn. Construction of correlation functions in two and three dimensions. Technical Report DAO Office Note 96-03, NASA Goddard Space Flight Center, Greenbelt, Maryland, 1996.
6. D. A. O. Staff. Algorithm theoretical basis document, version 1.01. Technical report, NASA Goddard Space Flight Center, Greenbelt, Maryland, 1996.
7. J. W. Pfaendtner. Notes on the icosohedral domain decompostion in PSAS. Technical Report DAO Office Note 96-04, NASA Goddard Space Flight Center, Greenbelt, Maryland, 1996.
8. G. H. Golub and C. F. van Loan. *Matrix Computations.* The John Hopkins University Press, Baltimore, second edition, 1989.
9. M. Hennecke. A fortran 90 interface to MPI version 1.1. Technical Report Internal Report 63/96, RZ Universitat Karlsruhe, Karlsruhe, Germany, 1996.

10. G. von Laszewski. *The Parallel Data Assimilation System and its Implications on a Metacomputing Environment.* PhD thesis, Syracuse University, Syracuse, New York, 1996.

NEW CAPABILITIES AND PLANS AT THE FLEET NUMERICAL METEOROLOGY AND OCEANOGRAPHY CENTER

PAUL F. MOERSDORF
Scientific and Technical Director
7 Grace Hopper Avenue
Fleet Numerical Meteorology and Oceanography Center
Monterey, CA 93943-5501
U.S.A.

ABSTRACT

The U.S. Navy's Fleet Numerical Meteorology and Oceanography Center (FNMOC) is taking delivery of some new and exciting models as well as adding hardware to form the basis for its C-90 replacement. The capabilities described are those associated with the delivery of the new relocatable atmospheric model and the update to the global model. Additionally, the status of the new global ocean model, higher resolution wave model, and new ice model will be given. Finally, FNMOC has improved its communication connectivity with other Centers and is augmenting its hardware suite in preparation for its C-90 replacement.

1. Introduction

The Fleet Numerical Meteorology and Oceanography Center (FNMOC) has the U.S. Department of Defense mission to support worldwide meteorological and oceanographic numerical modeling. This mission is accomplished using nested suites of global, regional, and mesoscale models. The suite includes atmospheric, oceanic, ice and wave models run twice a day on a twelve hour watch. Products are numerical gridded fields, satellite images, and gif/tif images. The basis of FNMOC's mission is the safe and smart conduct of Naval aviation and Naval sea operations globally. This is done through not only the day-to-day atmospheric and oceanographic products, but with tailored optimum path aircraft and ship routing services. Additionally, special, tailored products are provided in support of sensor and weapon systems.

Distribution, to over 70 Naval Meteorology and Oceanography Command activities located worldwide, is made through a variety of communication circuits, ranging from high volume, asynchronous transfer mode (ATM) to low volume, slow dial-up connections. Additional users include many other segments of the U.S. government, including the single biggest customer of FNMOC products, the Air Force Global Weatehr Center . Network connectivity is the fastest growing means by which our customers worldwide access FNMOC's products. This growth is reflected in the number of active customers directly serviced by FNMOC, up from roughly a dozen six years ago to over 2500 today.

2. Models

2.1 *Current Model Suite*

The support described above is routinely provided by a combination of a suite of nested, regional and mesoscale atmospheric, oceanic, wave and ice models and the global models that drive them. For special cases, FNMOC can also run even higher resolution atmospheric planetary boundary models; one for use over water and one for use in the land/water littoral environment. Table 1 shows the capabilities of our global data analysis system and the models that use the MVOI output.

Table 1. Global Capabilities (1996)

Data	Multi-variate Optimum Interpolation
Atmospheric	0-144 hrs / 80 km / 18 level / twice daily Eight member, 10 day ensemble / daily
Oceanographic	100 km wave / daily 100 km thermal / daily
Ice	25 km coupled ocean & ice / daily

The data sources used in the multi-variate optimum interpolation (MVOI) scheme include surface, ship, RAOBS, aircraft, XBT, buoys, and satellite observations (SSM/I winds and water vapor; TOVS and SSM/T sounders; TIROS multi-channel sea surface temperature; scatterometry; altimetry; and geostationary derived winds). These are combined in the MVOI analysis six times during each 12 hour watch.

In order to provide support to operating forces in various, changing locations around the world, FNMOC has a suite of relocatable regional or mesoscale models which can be run to provide continency and exercise support. Because the location and duration of the operations to be supported may not be known much in advance, this suite of models must be easily relocatable. The capabilities of these relocatable models are shown in Table 2.

Table 2. Relocatable Capabilities (1996)

Data	Multi-variate Optimum Interpolation
Atmospheric	5 areas / 4000 x 4000 km 0-48 hrs / 45, or 45 & 15 km / 36 level / twice daily
Oceanographic	20 km wave for 4 areas / 1000 km avg area
Tropical Cyclone	2 storms / twice daily

Even though NOGAPS gives excellent results on both tropical system development and movement, FNMOC, with the help of NOAA's Geophysical Fluid Dynamics Laboratory (GFDL), has implemented a modified version of the GFDL Hurricane Model at FNMOC for use in the Pacific and Indian Oceans. This has quickly become the main typhoon prediction tool in support of the U.S. Joint Typhoon Warning Center in Guam. The majority of the modifications were to enable the model to run within the FNMOC computational environment and to run for multiple typhoons. The improvement in skill of our tropical forecasts after this Navy implementation was noticeable.

2.2 *Future Model Suite*

Future improvements include the move to three, and then four, dimension variational assimilation (3,4-DVAR) techniques; increasing the vertical resolution of the global model; the addition of a global ocean model and the improvement in horizontal resolution of both the ocean analysis and wave models. These improvements in capability are shown in Table 3.

Table 3. Global Capabilities (2000)

Data	3-D variational data assimilation
Atmospheric	0-48 hrs / 50 km / 50 level / twice daily 48-168 hrs / 80 km / 50 level / twice daily 24 member. 14 day ensemble / daily
Oceanographic	12 km (5.5 layer) ocean circulation / daily 50 km wave / daily 25 km thermal / daily
Ice	25 km ocean & ice / daily

In addition to the new assimilation techniques being used for the relocatable models, the most significant improvements will be to the regional and mesoscale model nested within the global model. FNMOC has taken delivery of the atmospheric portion of a new, **c**oupled **o**cean-**a**tmosphere **m**esoscale **p**rediction **s**ystem, referred to as COAMPS, from the Naval Research Laboratory. This component has shown a significant improvement in forecast capability over that presently available from FNMOC. This improvement is attributed to both the higher resolution and the better physics packages within the model. In mid-1997, after operational testing, FNMOC anticipates the output of this new model will be available to its customers. Toward the end of the decade it is anticipated that this atmospheric model will be able to drive an underlying mesoscale ocean model. FNMOC will then implement the total coupled capability for those areas and missions where it is appropriate and necessary to run the coupled air/ocean model. At that point FNMOC will have a suite of models that includes the coupled air/ocean global and regional/mesoscale models as well as stand-alone mesoscale atmospheric, ocean circulation, wave, and planetary boundary layer models. In the latter case, there will be two version available, one for over coastal oceans only and one for over both the coastal ocean and littoral regimes. These boundary layer models will be used in support of specific customer requirements and will be highly relocatable. The improvements anticipated in the relocatable mesoscale capabilities at FNMOC can be seen in Table 4.

Table 4. Relocatable Capabilities (2000)

Data	3-D variational data assimilation
Atmospheric	4 areas / 4000 x 4000 km 0-36 hrs / 9 km / 36 level / twice daily
Oceanographic	4 areas / 1000 km box atm / wave coupled 0-36 hrs / 9 km / 40 level / twice daily 4 areas / 2000 km box ocean only 0-36 hrs / 9 km / 20 level / daily
Tropical Cyclone	3 doubly nested storms / twice daily
Boundary Layer	10 areas / 1000 km box 0-48 hrs / 1 km surface wind over land 0-48 hrs / 5 km surface wind over water

3. Hardware Suite

The current FNMOC hardware suite is composed of an eight processor SGI/Cray C-916/8256, YMP-2E, and J-916/6; two StorageTek silos; over 70 Sun Sparc workstations; all interconnected by high speed networks. The hardware architecture can be seen in Figure 1. The upgrades planned for 1997 include adding eight more processors to the C-90, procuring another J-916 and a first J-932, and upgrading to gigaring technology connectivity within the hardware suite. The replacement of the current platforms will begin in 2000, and will extend over three years. It is anticipated that FNMOC will migrate its processing load to distributed systems during that technology modernization effort.

4. Summary

The United States is involved in more international support efforts than ever before, and FNMOC continues to support its customer base worldwide. At the same time as the number of FNMOC's users is increasing and so are the dramatic differences in the methods by which those users receive the data; from both the high end (high volume, "push" distribution) and low end (PC based, low volume, "pull" distribution). Communications pathways range from T-3 (16MBS) to dial-in over ship-to-shore circuits. The demand for products from our relocatable mesoscale models is the fastest growing segment of FNMOC's

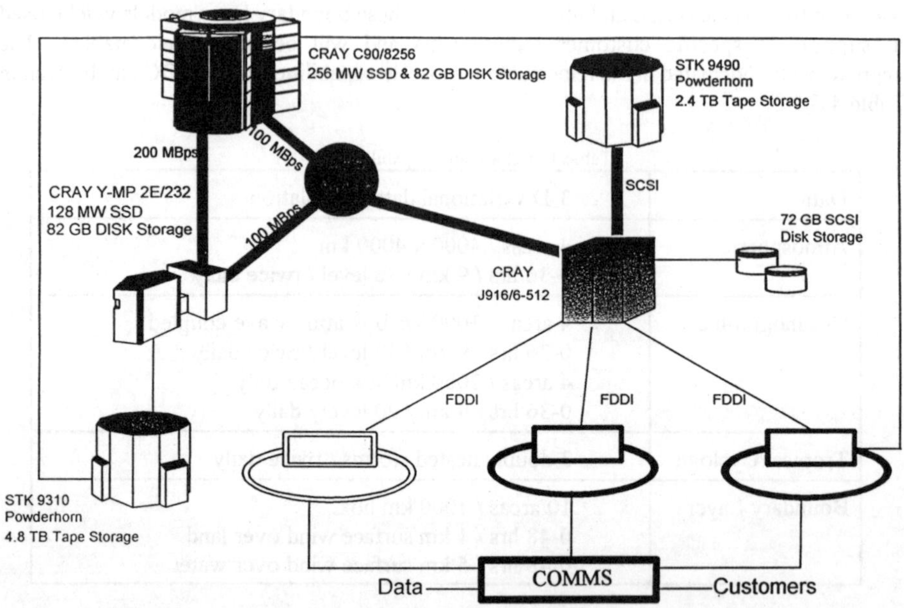

Figure 1. Hardware configuration (as of December, 1996).

business. Contingencies, Fleet exercises, and humanitarian efforts have become a normal routine as the U.S. has emerged from the one adversary, Cold War mentality into today's reality. During 1996 alone, FNMOC brought on 13 new and separate mesoscale areas in addition to the four regional areas (continental U.S., Europe; southwest Asia; northeast Asia) that are routinely run. Both the regional and mesoscale geographic areas supported during 1996 can be seen in Figure 2. The models run in these areas are initialized by the output of the global atmospheric model which is run six times per 12 hour watch. FNMOC nominally does a separate data collection and analysis for each run. The forecast period for each of these global runs is of varying length depending on the regional atmospheric and wave model suite to be initialized. The products are then sent to the customers in the various geographic areas of interest. The near term improvements to the hardware architecture discussed in section 3 are designed to facilitate the addition of more areas as well as speed up dissemination of products to the areas currently supported.

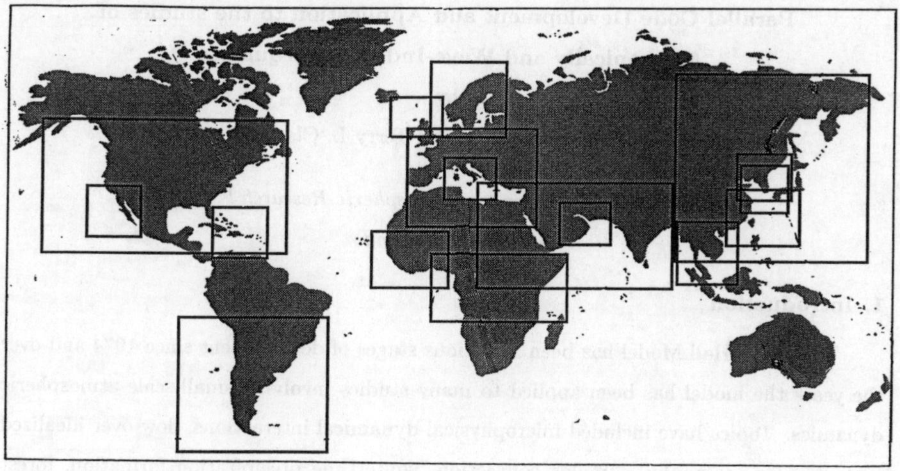

Figure 2. Contingency and Exercise Areas Supported in 1996.

5. Acknowledgments

This author and the Fleet Numerical Meteorology, and Oceanography Center are supported through the operational Naval Meteorology and Oceanography Program under the sponsorship of the Commander, Naval Meteorology and Oceanography Command, Stennis Space Center, Mississippi, USA.

Parallel Code Development and Application to the studies of Mechanically and Wave-Induced Turbulence

by

William D. Hall and Terry L. Clark

National Center for Atmospheric Research [1]

Boulder, Colorado

1. Introduction

The Clark-Hall Model has been in various stages of development since 1974 and over the years the model has been applied to many studies involving small-scale atmospheric dynamics. Topics have included microphysical-dynamical interactions, flow over idealized and complex orography, cumulus convection, wintertime precipitation formation, forest fire dynamics, wave-induced turbulence and mechanically generated turbulence. This paper will describe recent developments and plans for the Clark small scale nonhydrostatic anelastic three-dimensional model with fully two-way interactive grid nesting that includes new parallel processing capabilities on shared and distributive memory computer platforms. Two recent applications of the parallelized model to the study of mechanical generated turbulence within mountainous surface layer (Hong Kong Project) and wave-generated turbulence (during a down slope windstorm) will be illustrated.

2. Model Description

The anelastic nonhydrostatic model developed by Clark (1977), Clark and Farley (1984), Clark and Hall (1991), and Clark and Hall (1996) includes terrain following coordinates with fully two-way interacting grid nesting. A brief description of the model's physical and numerical formulation will be presented in order to clarify the procedures

[1] The National Center for Atmospheric Research is sponsored by the National Science Foundation

required to implement the code in a shared memory parallel environment and to describe how the current code will be converted to run on a distributive parallel environment. A complete description can be found in Clark, Hall, and Coen (1996).

a) Model equations

The model equations have been cast in geo-spherical terrain following coordinates. The terrain following coordinate utilizes the transformation

$$z = F(\zeta)(1 - h/H) + h \tag{1}$$

where ζ is the vertical coordinate. The model is formulated using $\Delta\zeta$ = constant for any particular domain and $F(\zeta)$ is chosen as a monotonically increasing function of ζ where

$$F(0) = 0 \quad \text{and} \quad \mathrm{F(H) = H} \tag{2}$$

so that the surface of the model follows the terrain ($z = h(x, y)$ at $\zeta = 0$) and the model top is at the constant level of $z = H$ or $r = R + H$, where R is the radius of the earth at sea level.

Using the chain rule it is easily shown that

$$\sqrt{J}\frac{\partial}{\partial x^*}\bigg|_z = \frac{\partial}{\partial x^*}\sqrt{J}\bigg|_\zeta + \frac{\partial}{\partial \zeta}\sqrt{J}J^{13} \tag{3a}$$

$$\sqrt{J}\frac{\partial}{\partial y^*}\bigg|_z = \frac{\partial}{\partial y^*}\sqrt{J}\bigg|_\zeta + \frac{\partial}{\partial \zeta}\sqrt{J}J^{23} \tag{3b}$$

$$\sqrt{J}\frac{\partial}{\partial z} = \frac{\partial}{\partial \zeta} \tag{3c}$$

where

$$\sqrt{J} = F_\zeta(1 - h/H) = F_\zeta\sqrt{G} \tag{4a}$$

$$\sqrt{J}J^{13} = \left(\frac{F(\zeta)}{H} - 1\right) h_{x^*} \tag{4b}$$

$$\sqrt{J}J^{23} = \left(\frac{F(\zeta)}{H} - 1\right) h_{y^*} \tag{4c}$$

Defining ρ as

$$\rho = \sqrt{J}\rho^* = \sqrt{J}\Gamma^2 \cos\phi\tilde{\rho}, \tag{5}$$

The notation will be that δ_x and $\overline{}^x$ refer to a finite difference derivative and an average in the x direction, respectively, and similarly in the other coordinate directions. The formulation adopted for mass conservation is

$$\delta_x(\overline{\rho}^x U) + \delta_y(\overline{\rho}^y V) + \delta_\zeta(\overline{\rho}^\zeta \omega) = 0. \tag{6}$$

and for the relation between ω, w, U and V is

$$\overline{\sqrt{J}}^\zeta\,\overline{\rho}^\zeta \omega = \overline{\rho}^\zeta w + \overline{\overline{\sqrt{J}J^{13}\overline{\rho}^x U}^\zeta}^x + \overline{\overline{\sqrt{J}J^{23}\overline{\rho}^y V}^\zeta}^y. \tag{7}$$

The three momentum equations are approximated with second order finite differences using the Arakawa (1966) quadratically conservative form as

$$\delta_t\overline{(\overline{\rho}^x u)}^t + \delta_x\left(\overline{\overline{\rho}^x U}^x\,\overline{u}^x\right) + \delta_y\left(\overline{\overline{\rho}^y V}^x\,\overline{u}^y\right) + \delta_\zeta\left(\overline{\overline{\rho}^\zeta \omega}^x\,\overline{u}^\zeta\right) - \frac{tg\phi}{\Gamma R}\overline{\rho}^x u\overline{v}^{xy} + \frac{1}{\Gamma R}\overline{\rho}^x u\overline{w}^{x\zeta} =$$

$$-\Gamma\left(\delta_x\left(\sqrt{J}p\right)+\delta_\zeta\left(\sqrt{J}J^{13}\overline{p}^{x\zeta}\right)\right)+\overline{\overline{\rho}^y v}^{xy} f-\overline{\overline{\rho}^\zeta w}^{x\zeta} f^*+\frac{1}{\Gamma\cos\phi}\delta_x\left(\sqrt{J}\tau_{11}\right)+\frac{1}{\Gamma}\delta_y\left(\overline{\sqrt{J}}^{xy}\tau_{12}\right)$$

$$+\delta_\zeta\tau_{13} + \frac{1}{\Gamma}\delta_\zeta\left(\sqrt{J}J^{13}\overline{\tau_{11}}^{x\zeta}\Big/\cos\phi + \overline{\overline{\sqrt{J}J^{23}}^x\,\overline{\tau_{12}}^\zeta}^y\right) - \frac{tg\phi}{\Gamma R}\overline{\sqrt{J}}^x\,\overline{\tau_{12}}^y + \frac{\overline{\sqrt{J}}^{x\zeta}}{\Gamma R}\overline{\tau_{13}}^\zeta \tag{8}$$

$$\delta_t\overline{(\overline{\rho}^y v)}^t + \delta_x\left(\overline{\overline{\rho}^x U}^y\,\overline{v}^x\right) + \delta_y\left(\overline{\overline{\rho}^y V}^y\,\overline{v}^y\right) + \delta_\zeta\left(\overline{\overline{\rho}^\zeta \omega}^y\,\overline{v}^\zeta\right) + \overline{\rho}^y\frac{tg\phi}{\Gamma R}\overline{u^2}^{xy} + \overline{\rho}^y\frac{1}{\Gamma R}v\overline{w}^{y\zeta} =$$

$$-\Gamma\overline{\cos\phi}^y\left(\delta_y\left(\sqrt{J}p\right)+\delta_\zeta\left(\sqrt{J}J^{23}\overline{p}^{y\zeta}\right)\right)-\overline{\overline{\rho}^x U}^{xy}\,\overline{\cos\phi}^y\,\overline{f}^y+\frac{1}{\Gamma\overline{\cos\phi}^y}\delta_x\left(\overline{\sqrt{J}}^{xy}\tau_{12}\right)+\frac{1}{\Gamma}\delta_y\left(\sqrt{J}\tau_{22}\right)$$

$$+\delta_\zeta \tau_{23} + \frac{1}{\Gamma}\delta_\zeta\left(\overline{\overline{\overline{\sqrt{J}J^{13}}^{y}\,\overline{\tau_{12}}^{\zeta}}^{x} \Big/ \overline{\cos\phi}^{y}} + \sqrt{J}J^{23}\overline{\tau_{22}}^{y\zeta}\right) + \frac{\overline{tg\phi}^{y}}{\Gamma R}\overline{\sqrt{J}\tau_{11}}^{y} + \frac{\overline{\sqrt{J}}^{y\zeta}}{\Gamma R}\overline{\tau_{23}}^{\zeta} \quad (9)$$

$$\delta_t\overline{(\overline{\rho}^\zeta w)}^{t} + \delta_x\left(\overline{\rho}^x \overline{U}^\zeta \overline{w}^x\right) + \delta_y\left(\overline{\rho}^y \overline{V}^\zeta \overline{w}^y\right) + \delta_\zeta\left(\overline{\overline{\rho}^\zeta \omega}^{\zeta}\,\overline{w}^\zeta\right) - \frac{\overline{\rho}^\zeta}{\overline{\Gamma}^\zeta R}\overline{\left(\overline{u^2}^x + \overline{v^2}^y\right)}^{\zeta} =$$

$$-\overline{\Gamma^2}^\zeta \cos\phi\left[\delta_\zeta p - \overline{\left(\frac{p\Gamma^2 F_\zeta}{\rho}\right)}^{\zeta}\delta_\zeta\left(\frac{\rho}{\Gamma^2 F_\zeta}\right)\right] + \overline{\overline{\rho}^x U}^{x\zeta}\cos\phi f^* + g\overline{\rho}^\zeta \overline{B}^\zeta + \frac{1}{\overline{\Gamma}^\zeta \cos\phi}\delta_x\left(\overline{\sqrt{J}}^{x\zeta}\tau_{13}\right)$$

$$+\frac{1}{\Gamma}\delta_y\left(\overline{\sqrt{J}}^{y\zeta}\tau_{23}\right) + \delta_\zeta\tau_{33} + \frac{1}{\Gamma}\delta_\zeta\left(\overline{\overline{\sqrt{J}J^{13}}^{\zeta}\,\overline{\tau_{13}}^{\zeta}}^{x} \Big/ \cos\phi + \overline{\overline{\sqrt{J}J^{23}}^{\zeta}\,\overline{\tau_{23}}^{\zeta}}^{y}\right) - \frac{\sqrt{J}}{\overline{\Gamma}^\zeta R}\left(\overline{\tau_{11}}^\zeta + \overline{\tau_{22}}^\zeta\right). \quad (10)$$

In the above equations

$$U = \frac{u}{\Gamma\cos\phi} \quad (11a)$$

$$V = \frac{v}{\Gamma} \quad (11b)$$

$$\rho = \left(\sqrt{J}\Gamma^2\cos\phi\right)\rho_0(z) \quad (11c)$$

with $\cos\phi$ and Γ colocated at the grid centered positions with ρ, p and θ.

The Reynolds stress terms τ_{ij} are defined as

$$\tau_{ij} = \rho_0 K_m D_{ij} \quad (12)$$

where D_{ij} is the deformation tensor and K_m is the coefficient of eddy mixing. The subgrid closure is taken from Smagorinsky (1963) and Lilly (1962) where

$$K_m = \begin{cases} (\Delta_\ell)^2|Def|\sqrt{1 - Pr\cdot Ri} & \text{for } Ri < Ri_c \text{ and } z > \Delta_\infty \\ 0 & \text{for } Ri > Ri_c \text{ and } z > \Delta_\infty \\ (\Delta_\ell)^2|Def| & z < \Delta_\infty\ , \end{cases} \quad (13)$$

where Ri is the Richardson number and Ri_c is a critical value (0.1). The grid scale, Δ_l, used the Blackadar (1962) formulation

$$\Delta_\ell = \kappa\frac{(z+z_0)}{(1+\kappa(z+z_0)/\Delta_\infty)} \tag{14}$$

with $\Delta_\infty = .2\Delta_h$ where Δ_h is the horizontal grid resolution and $k = .4$ is the von Karman constant. The deformation, Def, is given as

$$Def^2 = \frac{1}{2}\sum_i\sum_j D_{ij}^2. \tag{15}$$

The potential temperature, θ, and other scalar fields, $q_v, q_c, q_R, q_{IA}, N_{IA}, q_{IB}$, and $\mathrm{N_{IB}}$ are treated using the equation advection scheme of Smolarkiewicz (1983, 1984) where one corrective iteration is typically employed. This results in fully second-order spatial and temporal numerics.

$$\bar\rho\frac{\partial\psi}{\partial t} + \nabla\cdot(\bar\rho\vec V\psi) = +\frac{\partial}{\partial x}(\frac{\bar\rho K_\psi}{\sqrt{J}}\frac{\partial}{\partial x}\sqrt{J}\psi) + \frac{\partial}{\partial y}(\frac{\bar\rho K_\psi}{\sqrt{J}}\frac{\partial}{\partial y}\sqrt{J}\psi)$$

$$+\frac{\partial}{\partial\zeta}(\frac{\bar\rho K_\psi}{\sqrt{J}^2}\frac{\partial}{\partial\zeta}\psi) + \frac{\partial}{\partial x}(\frac{\bar\rho K_\psi}{\sqrt{J}}\frac{\partial}{\partial\zeta}\sqrt{J}J^{13}\psi) + \frac{\partial}{\partial y}(\frac{\bar\rho K_\psi}{\sqrt{J}}\frac{\partial}{\partial\zeta}\sqrt{J}J^{23}\psi) \tag{16}$$

/noindent Here , ψ, represents an arbitrary scalar field variable.

The momentum equations (1–34) through (1–36) can be written in the abbreviated form

$$\delta_t\overline{(\overline{\rho}^x U)}^t + FX = pfx = -\frac{1}{\cos\phi}\left(\delta_x\left(\sqrt{J}p\right) + \delta_\zeta\left(\sqrt{J}J^{13}\overline{p}^{x\zeta}\right)\right) \tag{17}$$

$$\delta_t\overline{(\overline{\rho}^y V)}^t + FY = pfy = -\overline{\cos\phi}^y\left(\delta_y\left(\sqrt{J}p\right) + \delta_\zeta\left(\sqrt{J}J^{23}\overline{p}^{y\zeta}\right)\right) \tag{18}$$

$$\delta_t\overline{(\overline{\rho}^\zeta w)}^t + FZ = pfz = -\overline{\Gamma^2}^\zeta\cos\phi\left[\delta_\zeta p - \overline{\left(\frac{p\Gamma^2F_\zeta}{\rho}\right)}^\zeta\delta_\zeta\left(\frac{\rho}{\Gamma^2F_\zeta}\right)\right] \tag{19}$$

where only terms involving p remain on the right-hand side. FX, FY and FZ represent the advective, Coriolis, buoyancy and mixing terms.

Combining the momentum equations with the anelestic divergence equation

$$\delta_t \overline{Div}^t + Q = -\frac{1}{\cos\phi}\left(\delta_{xx}\left(\sqrt{J}p\right) + \delta_{x\zeta}\left(\sqrt{J}J^{13}\overline{p}^{x\zeta}\right)\right)$$

$$-\left[\delta_y\left(\overline{\cos\phi}^y \delta_y\left(\sqrt{J}p\right)\right) + \delta_y\left(\overline{\cos\phi}^y \delta_\zeta\left(\sqrt{J}J^{23}\overline{p}^{y\zeta}\right)\right)\right]$$

$$-\cos\phi\;\delta_\zeta\left[\frac{\overline{\Gamma^2}^\zeta}{\overline{\sqrt{J}}^\zeta}\left(\delta_\zeta p - \overline{\left(\frac{p\Gamma^2 F_\zeta}{\rho}\right)}^\zeta \delta_\zeta\left(\frac{\rho}{\Gamma^2 F_\zeta}\right)\right)\right]$$

$$-\delta_\zeta\left[\frac{1}{\overline{\sqrt{J}}^\zeta}\overline{\sqrt{J}J^{13}\frac{1}{\cos\phi}\overline{\left(\delta_x\left(\sqrt{J}p\right) + \delta_\zeta\left(\sqrt{J}J^{13}\overline{p}^{x\zeta}\right)\right)}}^{\zeta x}\right]$$

$$-\delta_\zeta\left[\frac{1}{\overline{\sqrt{J}}^\zeta}\overline{\sqrt{J}J^{23}\overline{\cos\phi}^y\left(\delta_y\left(\sqrt{J}p\right) + \delta_\zeta\left(\sqrt{J}J^{23}\overline{p}^{y\zeta}\right)\right)}^{\zeta y}\right] \tag{20}$$

and putting the future value of the divergence (Div) equal to zero results in our diagnostic elliptic equation in p.

Equation (20) is solved in a "block" sense where that portion of (20) to be solved directly is taken over to the left-hand side, whereas the explicit forcing terms and remaining pressure terms are taken to the right-hand side. The right hand side pressure terms are treated using the most recently available iteration level of p. The reduced elliptic equation (or block) is taken as

$$\frac{1}{\cos\phi}\delta_{xx}\left(\sqrt{J}p\right) + \delta_y\left(\overline{\cos\phi}^y\delta_y\left(\sqrt{J}p\right)\right)$$
$$+\cos\phi\;\delta_\zeta\left[\frac{\overline{\Gamma^2}^\zeta}{\overline{\sqrt{J}}^\zeta}\left(\delta_\zeta p - \overline{\left(\frac{p\Gamma^2 F_\zeta}{\rho}\right)}^\zeta\delta_\zeta\left(\frac{\rho}{\Gamma^2 F_\zeta}\right)\right)\right] = R \tag{21}$$

and since $\sqrt{J} = \sqrt{G}F_\zeta$ we can rearrange (1–47) as

$$\sqrt{G}^2\left[\delta_{xx}\left(\sqrt{G}p\right) + \cos\phi\delta_y\left(\overline{\cos\phi}^y\delta_y\left(\sqrt{G}p\right)\right)\right]$$
$$+\frac{\cos^2\phi}{F_\zeta}\delta_\zeta\left(\frac{\overline{\Gamma^2}^\zeta}{\overline{F_\zeta}^\zeta}\delta_\zeta\left(\sqrt{G}p\right)\right) - \frac{\cos^2\phi}{F_\zeta}\delta_\zeta\left(\frac{\overline{\Gamma^2}^\zeta}{\overline{F_\zeta}^\zeta}\overline{\left(\frac{\sqrt{G}p}{\rho_0}\right)}^\zeta\delta_\zeta\rho_0\right) = \frac{\sqrt{G}^2\cos\phi}{F_\zeta}R. \tag{22}$$

Note that only ρ_0 has been retained on the left hand side of (22) so that

$$\frac{\cos^2\phi}{F_\zeta}\delta_\zeta\left[\frac{\overline{\Gamma^2}^\zeta}{\overline{F_\zeta}^\zeta}\left(\overline{\left(\frac{\sqrt{G}p\Gamma^2F_\zeta}{\rho}\right)}^\zeta\delta_\zeta\left(\frac{\rho}{\Gamma^2F_\zeta}\right)-\overline{\left(\frac{\sqrt{G}p}{\rho_0}\right)}^\zeta\delta_\zeta\rho_0\right)\right]$$

appears on the right-hand side and is treated explicitly. Now (22) can be written in matrix form where $\hat{\gamma}$ represents the vertical column vector of $\sqrt{G}p$ components at each (x,y) position. The right-hand side will be written simply as $\hat{R}$. The matrix equation then becomes:

$$\sqrt{G}^2\left(\delta_{xx}\hat{\gamma}+\cos\phi\delta_y\left(\overline{\cos\phi}^y\delta_y\hat{\gamma}\right)\right)+\cos^2\phi\ \mathbf{A}\hat{\gamma}=\hat{R} \tag{23}$$

where $\mathbf{A}$ is the matrix (of order NZ–2) representing the vertical numerical operators on $\sqrt{G}p$. The upper and lower boundary conditions have to be considered to derive the components of $\mathbf{A}$. It is important to note that $\mathbf{A}$ has been formulated in such a manner that all of its components are functions of ζ only.

Given that we have derived the coefficients of $\mathbf{A}$, we simply replace $\hat{\gamma}$ in (23) with $\mathbf{T}^{-1}\hat{\alpha}$ where $\mathbf{T}^{-1}$ is the eigenvector matrix of $\mathbf{A}^T$. We then multiply the equation by $\mathbf{T}$ where $\mathbf{T}$ is the eigenvector matrix of $\mathbf{A}$. This results in the equation

$$\sqrt{G}^2\left(\delta_{xx}\hat{\alpha}+\cos\phi\delta_y\left(\overline{\cos\phi}^y\delta_y\hat{\alpha}\right)\right)+\cos^2\phi\ \mathbf{\Lambda}\hat{\alpha}=\mathbf{T}\hat{R} \tag{24}$$

where $\mathbf{\Lambda}=\mathbf{TAT}^{-1}$ is the diagonal eigenvalue matrix. Equation (24) then represents the decoupled two–dimensional Helmholtz equations. We solve these Helmholtz equations for $\hat{\alpha}$ and then reconstruct $\hat{\gamma}$ and, of course, p.

In the past two or three block iterations were required for convergence with a measure of convergence being the resulting r.m.s. divergence error in the updated velocities. A new approach only performing one iteration is now being practiced. This approach is to use both iteration values of pressure in formulating the pressure forcing term in the momentum equation as the pressure was solved. In the new approach the root means square divergence error is useful as a diagnostic only in determining how well (21) has been solved. Normally,

$\sqrt{G}$ is close to unity and (21) is easily solved to near roundoff error with little difficulty in two or three spatial dimensions. Clearly, other measures of accuracy are required. Two that might be considered are first, the monitoring of the erroneous sources and sinks of kinetic energy due to the pressure and secondly, the comparison of solutions with time dependent benchmark solutions. Comparisons with known steady state solutions of flow over orography are of little use in assessing errors here as both methods converge in the limit of steady state. This last point in itself suggests that the new approach should be useful in the type of problems where one is attempting to approach a steady state. This covers a large range of numerical simulations involving flow over two- and three-dimensional obstacles.

b). Grid nesting procedure and Vertical Grid

The averaging operators and interpolation formula are derived from the matching of the continuity equation for $\overline{\rho}^x u$, $\overline{\rho}^y v$, and $\overline{\rho}^\zeta \omega$, and from the requirement of complete reversibility for the dynamic and thermodynamic variables.

Consider the fine mesh continuity equation in two spatial dimensions for purposes of demonstration only. The extension to three dimensions is straightforward. The continuity equation is

$$\delta_x \overline{\rho}^x u + \delta_\zeta \overline{\rho}^\zeta \omega = 0 \tag{25}$$

in the x and ζ two-dimensional case. Now we sum (25) over an area defined by an interacting coarser mesh. There is a grid ratio of RX:1 in the $x-$direction and RZ:1 in the $\zeta-$direction. This summation results in

$$\frac{1}{\mathrm{RX} \times \mathrm{RZ} \times \Delta x} \sum_k \overline{\rho}^x u \bigg|_{x=0}^{1} + \frac{1}{\mathrm{RX} \times \mathrm{RZ} \times \Delta \zeta} \sum_i \overline{\rho}^\zeta \omega \bigg|_{\zeta=0}^{1} = 0 \tag{26}$$

where we have divided by the total number (=RX×RZ) of summed fine mesh points. The limits of $\zeta = 0$, 1 and $x = 0$, 1 correspond to the boundaries of the coarse mesh grid. Now the relationship between the grid sizes is

$$\begin{aligned} \mathrm{RX} \times \Delta x &= \mathrm{DX} \\ \mathrm{RZ} \times \Delta\zeta &= \mathrm{DZ} \end{aligned} \tag{27}$$

between the fine mesh grid sizes $(\Delta x, \Delta\zeta)$ and the coarse grid sizes (DX,DZ). Now if we define

$$\begin{aligned} < \overline{\rho}^x u > &= \frac{1}{\mathrm{RZ}} \sum_k \overline{\rho}^x u \\ < \overline{\rho}^\zeta \omega > &= \frac{1}{\mathrm{RX}} \sum_i \overline{\rho}^\zeta \omega \end{aligned} \tag{28}$$

as the averaging operator from fine to coarse grids where $< >$ refers to a coarse grid variable, then (26) through (28) result in

$$\delta_X < \overline{\rho}^x u > + \delta_Z < \overline{\rho}^\zeta \omega > = 0 \tag{29}$$

where δ_X is the coarse grid equivalent of the fine mesh δ_x operator. Note that definition (28) applies at common lines (2-d) or common surfaces (3-d) that correspond to common collocation lines (or surfaces in three dimensions) of the appropriate staggered velocity component for both the fine and coarse grid domains. The full averaging operators in three spatial dimensions become

$$\begin{aligned} < \overline{\rho}^x u > &= \frac{1}{RY \times RZ} \sum_j \sum_k \overline{\rho}^x u \\ < \overline{\rho}^y v > &= \frac{1}{RX \times RZ} \sum_i \sum_k \overline{\rho}^y v \\ < \overline{\rho}^\zeta \omega > &= \frac{1}{RX \times RY} \sum_i \sum_j \overline{\rho}^\zeta \omega \end{aligned} \tag{30}$$

where the grid ratios are RX:1, RY:1, and RZ:1 and the summation in i, j, and k refers to sums over fine meshes in the $x-$, $y-$, and $\zeta-$directions, respectively.

The averaging operator for a thermodynamic variable such as θ, q_v, q_c, q_R or for a tracer is given by

$$< \rho\phi >= \frac{1}{\mathrm{RX} \times \mathrm{RY} \times \mathrm{RZ}} \sum_i \sum_j \sum_k \rho\phi \tag{31}$$

where ϕ refers to any of the above mentioned field variables. Equations (30) and (31) are used to supply information from the fine mesh to the coarse mesh after both domains have attained identical time levels. This procedure can be redundant for the velocity components when a pressure deficit correction is used as discussed in Clark and Hall (1991).

We also need to supply information from the coarse grid domain to the fine grid domain, both to initialize fine grid data using coarse grid data and to supply boundary conditions to the fine grid domain boundaries using coarse grid data. The interpolation formula are based upon a least squares fit to quadratic interpolation using a variational procedure. The constraint in the variational calculation is that (30) and (31) are satisfied, *i.e.*, the interpolation is consistent with the averaging operators. The derivation of the interpolation formula is described in Clark and Farley (1984) and will be shown at the end of this section. The formula used in the model is

$$\phi = E_- \Phi_- + E_0 \Phi_0 + E_+ \Phi_+ \tag{32}$$

for application along a line *i.e.*, the case of, say, $\overline{\rho}^x u$, interpolation in two spatial dimensions, where ϕ is the fine grid variable, Φ is the corresponding coarse grid variable, Φ_0 corresponds to the coarse grid data point that occupies some common space with the fine mesh point, and $\Phi_\pm$ corresponds to those points lying on either side of Φ_0. The formulae for the weighting coefficients are

$$\begin{aligned} E_- &= \epsilon(\epsilon - 1)/2 + \alpha \\ E_0 &= (1 - \epsilon^2) - 2\alpha \\ E_+ &= \epsilon(\epsilon + 1)/2 + \alpha \end{aligned} \tag{33}$$

where

$$\alpha = \left[(\Delta \ell/\Delta L)^2 - 1\right]/24 \tag{34}$$

$\Delta \ell$ and ΔL correspond to the grid sizes of the fine mesh and coarse mesh in the chosen direction. $\epsilon = 0$ corresponds to the position of $\Phi_0, \epsilon = 1$ corresponds to the position of Φ_+ and $\epsilon = -1$ corresponds to the position of Φ_-. The value of ϵ used in (33) is that which exactly corresponds to the position of the variable ϕ. The reason for having α in the formulae (33) is to ensure that

$$\frac{1}{\mathrm{RX}} \sum_{\ell} \phi = \Phi_0 \tag{35}$$

which is derived from the variational calculation.

For interpolation in two or three dimensions, (32) is either bilinearly or trilinearly applied. Since (32) is a linear equation the order of application does not matter.

These averaging and interpolation formulae guarantee that the continuity equations are consistent between the two interacting domains. The formulae also ensure that there is reversibility of the field variables between the domains. Initialization of fine mesh data and immediate averaging back to the coarse grid will not affect the values of the coarse grid.

Equations (32) through (34) are derived by minimizing Λ with respect to a_0, a_1, and a_2 where

$$\Lambda = \sum \left[\widetilde{\phi} - \phi(\epsilon)\right]^2 \Delta \ell + \lambda \left[\sum \widetilde{\phi} \Delta \ell - \Phi_0 \Delta L\right] \tag{36}$$

and

$$\widetilde{\phi} = a_0 + a_1 \epsilon + a_2 \epsilon^2. \tag{37}$$

λ is a Lagrange multiplier and ϵ varies between ± 1. $\epsilon = -1$ is at the exact collocation position of Φ_- and $\epsilon = 0$ at the position of Φ_0 and $\epsilon = +1$ at the position of Φ_+. The $\sum$

in (36) ranges then between $\pm\frac{1}{2}$, which exactly corresponds to the Φ_0 grid box. $\phi(\epsilon)$ in (36) is the quadratic fit to (32) obtained by putting $\alpha = 0$ in (33). Thus, (36) states that we are determining $\widetilde{\phi}$ to be a least squares fit to $\phi(\epsilon)$ but demanding exact conservation of ϕ on the fine mesh with Φ_0 on the coarse mesh.

The final topic to be discussed in this subsection is the method of choosing finer mesh vertical coordinates. In the vertically stretched coordinates

$$z = F(\zeta)(1 - h/H) + h \tag{38}$$

and ρ is weighted with the Jacobian, F_ζ. The Jacobian also includes $\cos\phi$ and $1 - h/H$, which will be brought into the discussion later. For the present discussion we define the large mesh Jacobian as F_ζ and the fine mesh Jacobian as f_ζ. Given a coarse mesh grid structure F and F_ζ then

$$\overline{\rho}^\zeta u = E_- \left(\overline{R}^\zeta U\right)_- + E_0 \left(\overline{R}^\zeta U\right)_0 + E_+ \left(\overline{R}^\zeta U\right)_+ \tag{39}$$

would be used to initially specify u throughout the fine mesh domain or to specify u at a boundary at all time levels. Now if U is a constant, then we require u to be the same constant. This means that

$$\overline{\rho}^\zeta = E_- \overline{R}^\zeta_- + E_0 \overline{R}^\zeta_0 + E_+ \overline{R}^\zeta_+. \tag{44}$$

This formula is in fact used to consistently define the Jacobian-weighted fine mesh density using coarse mesh data. To define the actual fine mesh grid positions relative to the coarse mesh we simply put $\rho_0(z) = 1$ in (40) resulting in

$$f_\zeta = E_- (F_\zeta)_- + E_0 (F_\zeta)_0 + E_+ (F_\zeta)_+ \tag{41}$$

Equations (40) and (41) are necessary to suppress noise generation through domain interactions. With the inclusion of orography we also must match ρ in the $x-$ and

$y-$directions as well as in the ζ. This makes calculating ρ for an inner domain extremely complicated. For this reason ρ has been made into a precalculated three-dimensional array.

From the above discussion, the vertical grid structure is determined by the outermost domain and the vertical grid, ratio between the nested and parent grid since equation (41) must be used to establish inner domain grid positions. Clark and Hall (1996) illustrate that considerable care needs to be exercised when designing the vertical grid. The above procedure can lead to extremely noisy grids when this vertical refinement is employed. High amplitude oscillations in $f_{\zeta\zeta}$, for example, can cause noisy and possibly unstable structures in the velocity components. The vorticity components will, of course, be much worse. One should design the stretching for the outer domain so as to minimize such high amplitude oscillations in $f_{\zeta\zeta}$.

3. Virtual nesting and parallel processing

To parallelize the model a coarse grain or virtual nesting method was chosen. The virtual nesting method of solution is designed for parallel processing on both shared memory and distributed memory systems. Currently this feature has been implemented only on Cray and SGI shared memory systems. Virtual nesting is defined as a solution method where conventional two-way interactive nesting software is used to decompose one large domain into numerous similarly sized subdomains with identical resolution. The subdomains are then treated simultaneously by parallel processors. By retaining sufficient boundary condition information between subdomains, one obtains identical results to those using a single large domain. Clark and Hall (1991) first developed this solution method to limit the central memory requirements needed for large calculations with a single processor. Dividing the computational work among processors is a natural extension of the method. The model's present grid refinement architecture allows for multiple domains of different resolution where any of these domains can be subdivided into subdomains (virtual domains of identical resolution) for parallel computing.

A given grid is subdivided into similarly sized subdomains with identical resolution in the horizontal plane by keeping the vertical depth intact. The term layer refers to the entire domain represented by a set of virtual subdomains. A layer may be nested within a coarser resolution (parent) grid and/or contain a finer resolution (child) grid. A general rule when subdividing any grid layer is that each child subdomain must have only one parent.

Implementing virtual nesting allowed the model the additional capability to use more than one central processing unit (CPU). The specific problem that had to be overcome when using the nested technology of the model was to make all finite difference operators transparent at virtual boundaries. This was accomplished by saving various planes of data throughout the code to be used in a later time. Because all the planes of data for a given boundary condition within a grid layer need to be defined before any are applied, a loop structure was designed where, within a given loop, all the subdomains of a given layer that either establish new or use previously defined boundary information are applied in parallel. The planes of data that are saved during a given small loop are used in future loops. This algorithm will be possible on a massively parallel distributed memory computer by adding a passing system where the messages are passed between the loops. In order to minimize the work effort to convert to a distributed system, this data exchange has been incorporated into the model's data storage system. This should allow all the required modifications to implement any message passing procedures to a small area of the code.

The pressure solution requires some special handling due to its elliptic nature. The solution method is to first perform a vertical transformation on each subdomain of a given layer to reduce the three-dimensional problem to a set of decoupled horizontal Helmholtz equations that can be treated independently. These decoupled equations are then solved either by direct or relaxation methods. The solutions of the horizontal Helmholtz equations are then transformed back to physical space establishing the new pressure solution. Each of these steps are done in parallel.

Converting the present memory shared code to a distributed memory system should not require any redesign of the solution algorithm. The required changes involve adding data communications between processors, where three general types of communications will be needed. These include passing boundary condition information (virtual or real) to each subdomain, injecting data from an averaged child subdomain data to its parent which does not generally share the same processor, and adding a transpose procedure to reorder pressure data to treat the decoupled horizontal Helmholtz equations. Boundary condition data can be passed in a straightforward manner by introducing message passing procedures or using virtual memory. The data injection of properly averaged child subdomain data can be achieved by message passing. The last transpose problem, where the pressure is first transformed by a vertical operator on each subdomain and then reordered into decoupled horizontal Helmholtz equations that are sent to independent processors for solution, requires a good gather/scatter procedure. An efficiency problem exists with the present algorithm when the number of processors exceeds the number of Helmholtz equations to be solved. Under such circumstances, one may use a fine grain parallelism to treat these equations or use an alternative pressure solver, such as the congugate residule method Smolarkiewic and Margolin (1996).

4. Mechanical generated turbulence

Numerical simulations of terrain–induced turbulence associated with airflow over Lantau Island of Hong Kong were performed. Lantau is a relatively small island with three narrow peaks rising to between 700 and 950 m MSL. This research was undertaken as part of a project to better understand and predict the nature of turbulence and shear at the new airport site on the island at Chek Lap Kok, which is located to the lee of Lantau. Intensive ground and aerial observations were taken from May through June 1994, during the LANTau EXperiment (LANTEX). Severe turbulence was observed during the passage of Tropical Storm Russ on 7 June 1994.

The nature of the environmental and topographic forcing on 7 June 1994 resulted in the turbulence and shear being dominated by the combination of topographic effects and surface friction. High resolution numerical simulations, initialized using local sounding data, were performed using the Clark-Hall model. The simulation results indicate that gravity wave dynamics played a very minor role in the flow distortion and generation of turbulence. As a result of this flow regime, relatively high vertical and horizontal resolution was required to simulate the mechanically generated turbulence associated with Tropical Storm Russ.

Results are presented using a vertical resolution of 10 m near the surface and with horizontal resolutions of both 125 and 62.5 m over local, nested domains of about 13 to 24 km on a side. The 125 m model resolution simulated highly distorted flow in the lee of Lantau, with streaks emanating downstream from regions of sharp orographic gradients. At this resolution the streaks were nearly steady in time. At the higher horizontal resolution of 62.5 m the streaks became unstable, resulting in eddies advecting downstream within a distorted streaky mean flow similar to the 125 m resolution simulation. The temporally averaged fields changed little with the increase in resolution; however, there was a 3 to 4 fold increase in the temporal variability of the flow as indicated by the standard deviation of the wind from a 10 min temporal average. Overall, the higher resolution simulations compared quite well with the observations, whereas the lower resolution cases did not. The high resolution experiments also showed a much broader horizontal and vertical extent for the transient eddies. The depth of orographic influence increased from about 200 m to over 600 m with the increase in resolution. The nature of the flow separation is analyzed using Bernoulli's energy form to display the geometry of the separation bubbles. The height of the 80 $m^2\ s^{-2}$ energy surface shows eddies forming in regions of large orographic gradients and advecting downstream.

Tests using both buoyancy excitation and stochastic backscatter to parameterize the under-resolved dynamics at the 125 m resolution were performed, as well as one experiment testing whether static stability suppressed turbulence development. All these tests showed no significant effect. This illustrates the need to address the effect of topographic gradients normal to the direction of the flow as these appear to represent the source of flow distortion and subsequent turbulence.

Fig. 1 shows the orography, $h(x, y)$, and outlines of all domains. Inner domains were activated only after the flow in the parent domain had sufficient time to adjust to initialization. To show the transient nature of the eddies and the separation bubble we estimate the envelope of the separation bubble using Bernoulli's energy form.

$$Q_B = q^2/2 + p/\rho - gz\theta^* \tag{42}$$

Fig. 2, plates a) and b), show two vertical cross sections from HRUSSB and HR4M, respectively, that are closely aligned with the mean wind direction. Fig. 2(a) shows that for HRUSSB the $Q_B = 80$ contour stays reasonably close to the mountain and departs significantly from the surface only over the water. Animation of this cross section shows this pattern is relatively steady. On the other hand, Fig. 2(b) shows significant departures of the $Q_B = 80$ contour, both over the mountain and downstream. Animation of these results shows a highly transient character. Using a fixed value of Q_B to display the separation bubble in three dimensions assumes we are looking at regions of rapid distortion where viscous mixing has had little time to smooth out flow structures in the vertical. Otherwise, Q_B would not approximate a conservative variable. We are also looking at values of Q_B close enough to the surface so as to avoid multiple value problems in our selection of Q_B.

Having demonstrated the utility of near surface values of Q_B giving an approximate representation of the separation bubble, the analysis is easily modified to show the full three–dimensional pattern of the flow separation. Fig. 3 shows the height AGL of $Q_B = 80$ for four times in experiment HR4M. The plot shows regularly spaced eddies propagating downstream of the mountain. The speed of propagation is about 17 m s^{-1}, which corresponds to the mean flow speed. An animation of these fields shows regular patterns forming mostly in the lee of the peaks, as well as at the outflows of some of the valleys. After forming, the eddies propagate downstream and reform, repeating the cycle. The maximum height of the separation bubbles regularly exceeds the full depth of the fourth domain, which was 600 m. The penetrations shown are then to some extent affected by the change in resolution occurring at this height level. The maximum 600 m penetration depth shown in Fig. 3 should therefore be viewed as an underestimate. For a complete description of these experiments see Clark et. al. (1996).

5. Wave-induced turbulence

The present modeling study is motivated by the extensive Doppler lidar, radar wind profiler, and RASS (radio acoustic sounding systems) observations taken by NOAA/ERL/Environmental Technology Laboratory that document the temporal and spatial variations of the mountain wave and downslope wind structure during a frontal passage. The lidar observations are unique and have captured elevated regions of wave breaking in the lower stratosphere associated with clear air turbulence that damaged a DC-8 cargo jet.

Many characteristics of mountain wave and downslope winds have been numerically studied during the past two decades. These studies include Clark and Peltier (1979), Klemp and Lilly (1978), Peltier and Clark (1979), Durran and Klemp (1983), Clark and Farley (1984), Pierrehumbert and Wyman (1985), Bacmeister and Schoeberl (1989), Scinocca and Peltier (1989), among others. While these studies have predominately used idealized orography and atmospheric stratification they have contributed much to our physical understanding. These studies have demonstrated many characteristics of flow over mountains including the important role that breaking gravity waves play in establishing an intense downslope wind. Aloso they show how the competing mechanisms of orographic wave excitation and wave breaking causing turbulent breakdown lead to a surface gustiness signature.

In a three dimensional study, Clark el al. (1994) used realistic orography and time-dependent inflow conditions based upon upstream observations during a severe windstorm event at Boulder Colorado. They found that a propagating lee wave resulting from strong transience of the wave breaking region resulted in a highly time dependent effective mountain shape. This study also found that as the resolution of the model increased the flow tended to be two-dimensional structure which was attributed to the scale selection determined by the internal fluid dynamics responding to the mean north-south orographic forcing becoming more prominent than the modulating effects of the 3 dimensional forcing. This study also found as did Richard et al. (1989) the position of the jump region lee of the mountain to be sensitive to the surface friction.

The present study is an extension of the previous study of Clark et al. (1994). The model is the nonhydrostatic interactive grid nesting small-scale dynamic Clark-Hall model with terrain following coordinates. Two new features of the model have been introduced. The first feature allows the incorporation of large-scale datasets to initialize the model domain and update the outer domain boundary conditions during the simulation. The second is the incorporation of a vertically stretched grid where resolution is finer near the ground was chosen to improve the treatment of surface friction.

Large-scale analysis datasets were provided by the NOAA Forecast Systems Laboratory Mesoscale and Analysis Prediction System (MAPS) to initialize and update the outer domain boundary conditions.

The MAPS datasets have a spatial resolution of approximately 60 km (horizontal) and the vertical 25 levels are spaced along boundary layer and isotropic surfaces. The data sets are given at 3-hour intervals. The initial time data is projected onto the model surfaces and processed using a potential flow adjustment scheme to satisfy mass continuity to the accuracy of the pressure solver. The model then integrated forward in time updating the outer boundary conditions by linear interpolation of the MAPS data. We are currently investigating using a time digital filter similar to that proposed by Lynch and Huang (1992) to filter out any spurious oscillations introduced by the initial data or data projection onto the model domain.

The observations of the case study have been summarized by Ralph et al. 1995. On December 9, 1992, a moderate downslope windstorm began in Boulder at 12 UTC with surface wind gusting between 20 and 30 m/sec for several hours preceding a weak frontal passage. During this day numerous aircraft reported clear-air turbulence near and along the eastern slope of the northern Colorado mountains including a DC-8 cargo jet that made an emergency landing after losing an engine and part of a wing in extreme turbulence at 1520 UTC at 9.5 km MSL. From 14 UTC December 9 to 01 UTC December 10 NOAA ETL Doppler lidar was scanning at a site 3 km east of the steep eastern edge of the Colorado Rockies just north of Boulder. The main features that were observed were a flow reversal region in the lower stratosphere at 11.5 and 16.5 km MSL indicating a wave breaking region along with the temporal evolution of a low altitude wind maximum and the eastward progression of the surface windspeed maximum.

The model employed five domains with increasing horizontal resolutions 24.3,8.1,2.7,.9,and .3 km. Fig. 4 shows an east-west composite cross-section plot of the potential temperature and east-west velocity of the highest resolution domain (300 m) for several times just prior to the accident. This figure shows a significant shear zone begining in the main wave overturning region above the continental divide and extending downstream to lower levels. The time and location of this shear zone corresponds closely to the time and location of the accident. Model comparisons with NOAA's lidar data are currently in progress.

6. Parallel performance

The current version of the code has been in production for about one year. While the first applications were performed on a Silicon Graphics Power Challenge, most all of the parallel applications have been performed on the Cray YMP or Cray J-90 series shared memory machines. On the YMP we have been able to sustain up to 7.86 of 8.0 processors for long runs on a dedicated machine. Fig. 5 shows a typical Cray system analysis of a typical parallel loop within the code plotting the model speed up vs. number of CPUs. This report was obtained while running a production code in a dedicated mode on a 16 processor J-90 Cray. The dotted diagonal line shows the theoretical speed up. The solid line shows the actual speed up of the parallel region. The step behavior of the later curve is a result of the coarse grain parallelism employed. Each layer of the model was divided into 16 subdomains. The results indicate that when the number of processors divides evenly into 16, better than 99 percent parallelism is achieved.

References

Arakawa, A., 1966: Computational design for long term integration of the equations of fluid motion: two-dimensional incompressible flow. Part I. *J. Comput. Phys.*, **1**, 119 – 143.

Bacmeister, J.T. and M.R. Schoeberl, 1989: Breakdown of Vertically Propagating Two-Dimensional Gravity Waves Forced by Orography. *J. Atmos. Sci.*, **46**, 2109–2134.

Blackadar, A.K., 1962: The Vertical Distribution of Wind and Turbulent Exchange in a Neutral Atmosphere. *J. Geophys. Res.*, **67**, 3095– 3102.

Clark, T.L., 1977: A small scale numerical model using a terrain following coordinate transformation. *J. Comput. Phys.*, **24**, 186 – 215.

Clark, T.L. and R.D. Farley, 1984: Severe downslope windstorm calculations in two and three spatial dimensions using anelastic interactive grid nesting: a possible mechanism for gustiness. *J. Atmos. Sci.*, **41**, 329 – 350.

Clark, T.L. and W. D. Hall, 1991: Multi-domain simulations of the time dependent Navier Stokes equation: Benchmark Error Analysis of Some Nesting Procedures. *J. Comp. Phys.*, **92**, 456–481.

Clark, Terry L. and W.D. Hall, 1996: On the design of smooth, conservative vertical grids for interactive grid nesting with vertical stretching. *J. Appl. Meteor.*, **35**, 1040 –1046.

Clark, Terry L., and W. R. Peltier, 1977: On the evolution and stability of finite–amplitude mountain waves. *J. Atmos. Sci.*, **34**, 1715–1730.

Clark, T. L., and W. R. Peltier, 1984: Critical level reflection and the resonant growth of nonlinear mountain waves. *J. Atmos. Sci.*, **41**, 3122–3134.

Clark,T. L. , T. Keller, JCoen,P. Neilley, H Hsu, and W. Hall, 1996: Terrain-induced Turbulence over Lantau Island: 7 June 1994 Tropical Storm Russ Case Study. *J. Atmos. Sci., in press.*

Klemp, J.B., and D. K. Lilly, 1978: Numerical simulation of hydrostatic mountain waves. *J. Atmos. Sci.*, **35**, 78–107.

Lilly, D.K., 1962: On the numerical simulation of buoyant convection. *Tellus*, **14**, 145 – 172.

Peltier, W. R., and T. L. Clark, 1983: Non–linear mountain waves in two and three spatial dimensions. *Quart. J. Roy. M. S.*, **109**, pp. 527–548.

Ralph M.,P.J. Neimann and D. Levinson, A Large-Amplitude Mountian Wave with Wave breaking aloft, clear-air turbulence and a downslope windstorm, Part 1 :Observational overview. Procedings seventh conference on mountain meteorology, Breckenridge,Colorado, A.M.S.

Richard,E., P.M. Mascart and E.C. Nickerson, 1989: The role of surface friction in downslope windstorms. *J. Appl. Meteor.*,**28**, 241–251.

Scinocca, J.F. and W.R. Peltier, 1991: On the Richardson Number Dependence of Nonlinear Critical-Layer Flow over Localized Topography. *J. Atmos. Sci.*, **48**, 1560–1572.

Smagorinsky, J., 1963: General circulation experiments with the primitive equations. *Proc. Int. Symp. on Numerical Weather Prediction* Tokyo, Meteor. Soc. Japan, 85 – 107.

Smolarkiewicz, P.K., 1983: A simple positive definite advection scheme with small implicit diffusion. *Mon. Wea. Rev.*, **111**, 479 – 486.

Smolarkiewicz, P.K., 1984: A fully multidimensional positive definite advection transport algorithm with small implicit diffusion. *J. Comput. Phys.*, **54**, 325 – 362.

Smolarkiewicz, P.K. and L.G. Margolin, 1996: On Forward-in-Time Differencing for Fluids: an Eulerian/ Semi-Lagrangian NonHydrostatic Model for Stratified Flows. *Atmos-Ocean* **in press**

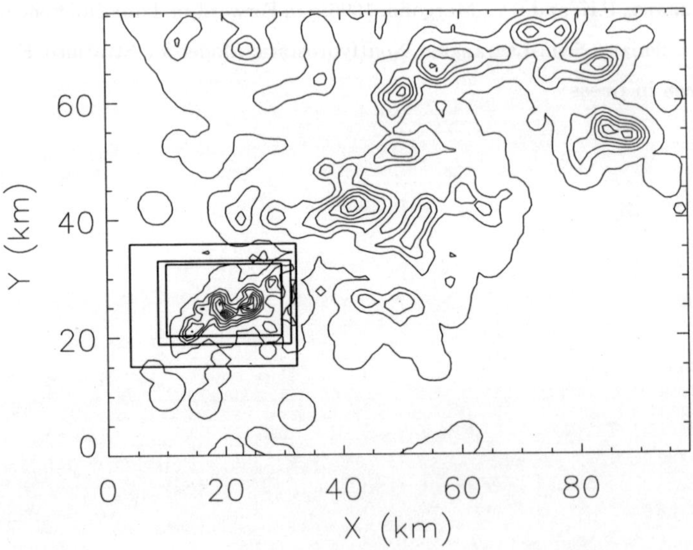

Figure 1. Domain of integration. The full domain orography at a horizontal resolution of 1 km is shown in orography. The contour interval is 90 m.

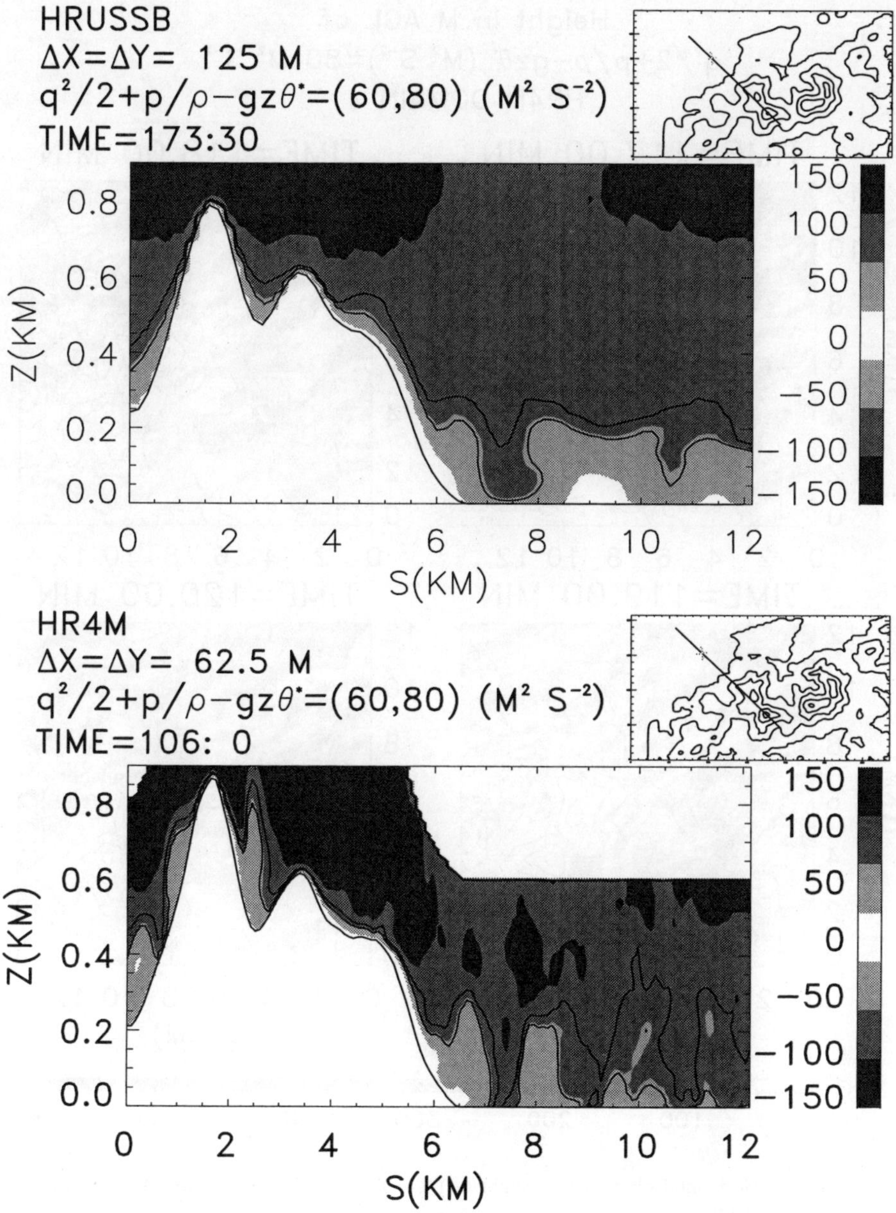

Figure 2. Vertical cross sections of Q_B for (a) HRUSSB at t = 173.5 min and (b) HR4M at t = 105.75 min. Contours of Q_B = 60 and 80 $m^2\ s^{-2}$ are shown.

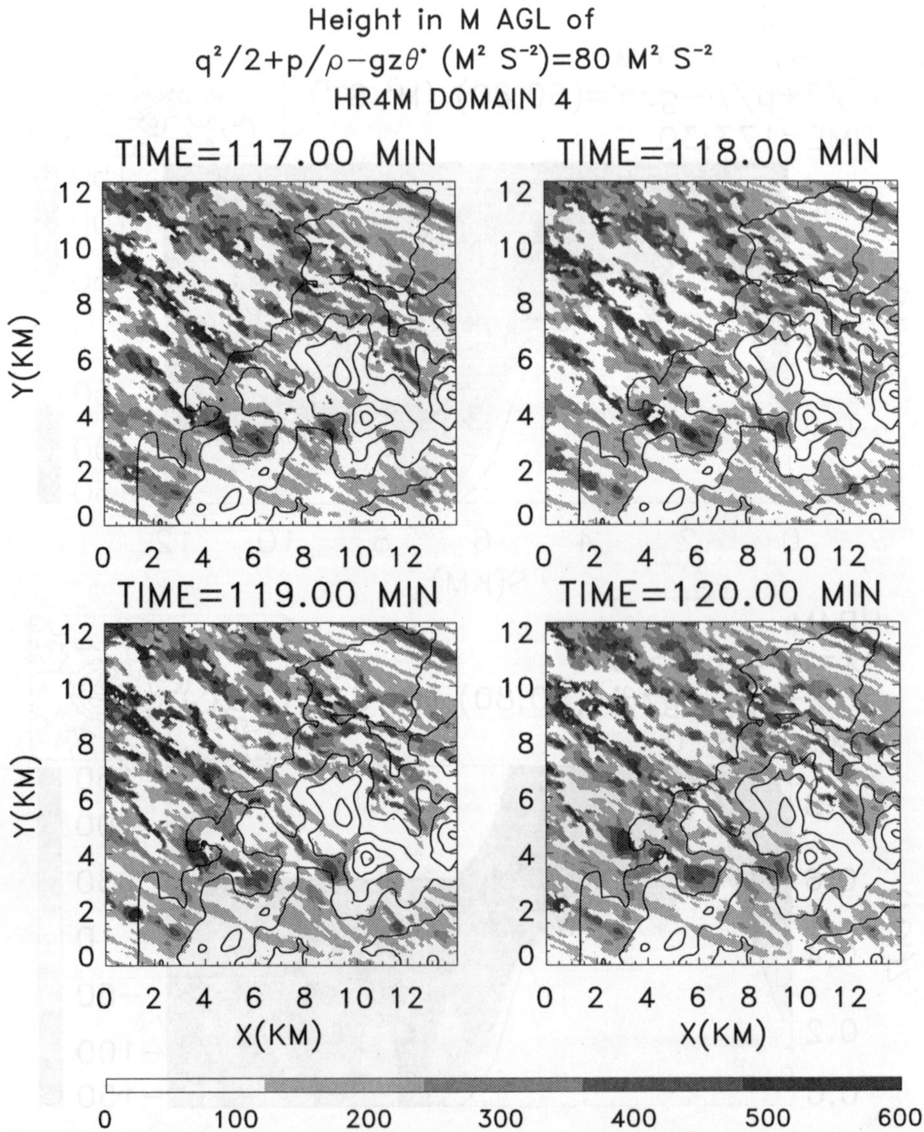

Figure 3. Horizontal plots of the height of $Q_B = 80$ m^2 s^{-2} at four times for experiment HR4M.

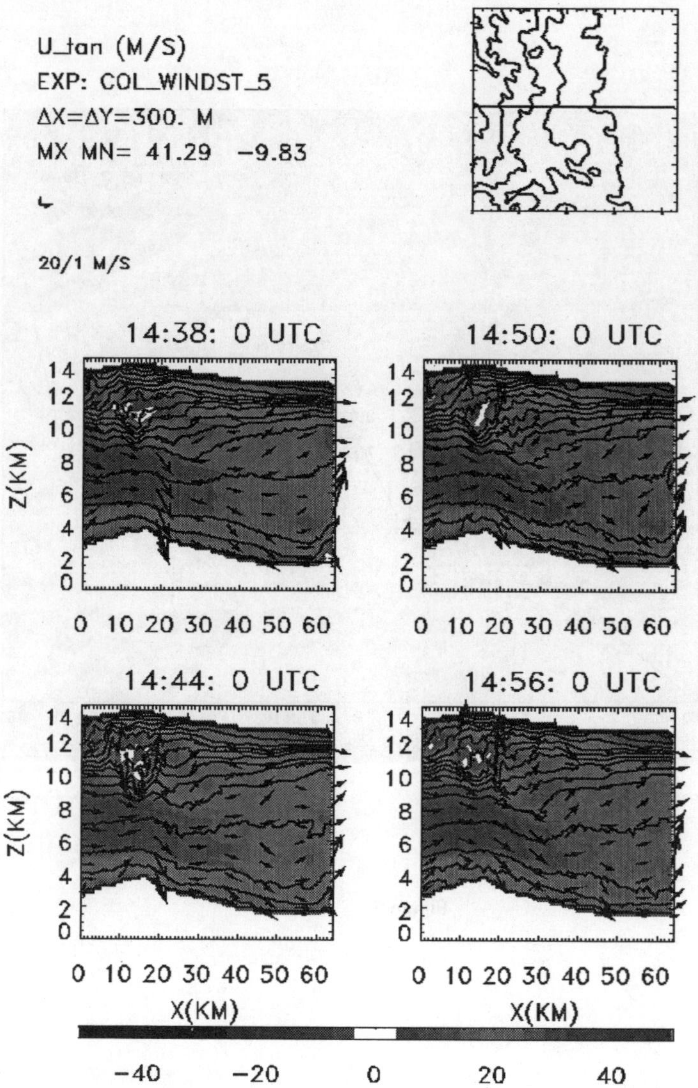

Figure 4. December 9, 1992, Colorado windstorm simulation high-resolution 3 wm domain

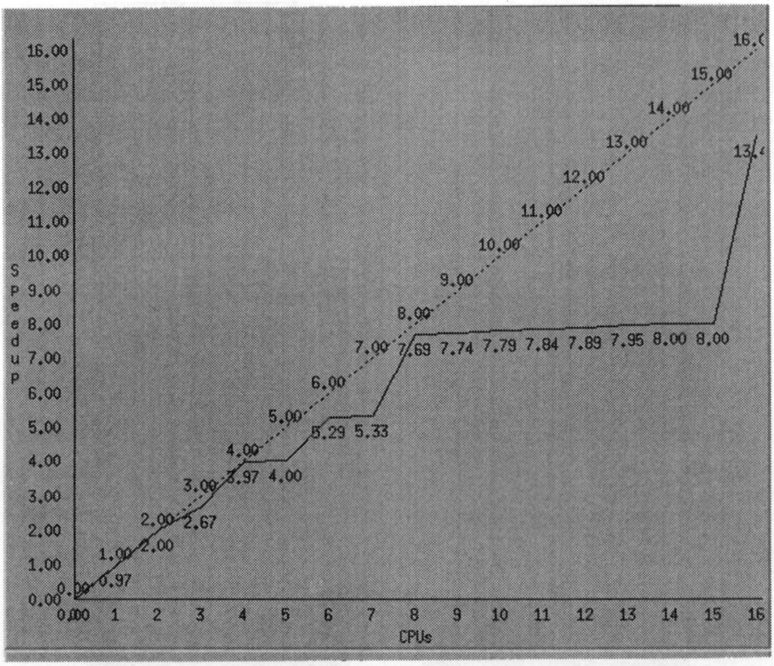
Speedup
CPUs
0.00 1.00 2.00 3.00 4.00 5.00 6.00 7.00 8.00 9.00 10.00 11.00 12.00 13.00 14.00 15.00 16.00
0 1 2 3 4 5 6 7 8 9 10 11 12 13 14 15 16
0.97 2.00 2.67 3.97 4.00 5.29 5.33 7.69 7.74 7.79 7.84 7.89 7.95 8.00 8.00

Figure 5.

Load Balancing of a Comprehensive Air Quality Model

Hendrik Elbern
Institute for Geophysics and Meteorology, EURAD,
University of Cologne,
Aachener Str. 201–209, D–50931 Köln, Germany
e–mail: he@eurad.Uni–Koeln.de

ABSTRACT

Comprehensive mesoscale tropospheric chemistry–transport models (CTMs) must combine numerous physical, chemical, and other processes to simulate realistically transformations and transport of manmade and natural trace species: Starting with anthropogenic emission, meteorological transport and diffusion processes disperse the material, while radiation induced photolysis and gas phase chemistry act on the chemical composition. Aqueous phase and heterogeneous chemistry processes must also be considered in clouds and on aerosols, respectively.

Chemistry–transport models are generally claimed to be well suited for massively parallel processing since the arithmetic–to–communication ratio is usually high. However this observation proofs insufficient to account for an efficient parallel performance with increasing complexity of the model. The local state of the atmosphere ensues very different branches of the modules' code and greater differences in the computational work load and consequently runtime of individual processors occur to a much larger extend during a time step than reported for meteorological models. Variable emissions, changes in actinic fluxes, and all processes associated with cloud modeling are highly variable in time and space and are identified to induce large load imbalances which severely affect the parallel efficiency.

Based on a horizontal grid partitioning approach a method is proposed where the integration domain of the individual processors is locally adjusted to accomodate for load imbalances. This ensures a minimal communication volume and data exchange only with next neighbours. The interior boundary adjustments of the processors are combined with routine boundary exchange which is required each time step anyway. Two dynamic load balancing schemes were implemented and compared against a conventional equal area partition and a static load balancing scheme on an Intel Paragon with 136 compute nodes. A midsummer episode of highly elevated ozone concentrations was taken as test case. The dynamic load balancing approaches were found to perform significantly better and reduce idle times of the processors considerably. The efficiency was raised from 36% to 69% for a 128 processor configuration.

1. Introduction

There is now a growing effort of parallelizing existing AQM codes. Examples for early implementations on different computational platforms are given in studies of Carmiachael et al. [3], Dabdub and Seinfeld [5], or Shin and Carmichael [7]. These studies, and more general studies on the future perspectives of comprehensive air quality modeling corroborate the suitability of MPP systems for atmospheric chemistry calculations [9]. This is mainly due to the fact that the chemistry component of the simulation, which is the most consumptive part of computational resources, must be calculated on each grid point but does not require any intermediate communication. It will also hold true for expected model extensions mainly awaited in the areas of heterogeneous chemistry and aerosol phase processes. Particularly high computational demands will result from future needs to simulate aqueous phase chemistry of binary and ternary solutions, particulate formation processes, advanced dynamics of clouds, and their interaction with radiation schemes. Further to this, cloud, precipitation, and radiative transfer simulations are often designed for calculations on a vertical grid point column, which can be easily arranged on an individual compute node. This also strengthens confidence that more sohisticated physics and chemistry packages of AQMs can be implemented without or with low communication overhead. Hence, it may be concluded that the favourable arithmetic–to–communication ratio of CTMs can be expected to persist or even increase in the future, and the prospects for parallel performance appear to be promising.

However, in recompense, further sophistication of many model components inevitably leads to different computational load of the grid points and columns. For example, many processes related to the existence of clouds, perform with a run time behaviour strongly dependent on the state of atmosphere and the numerical treatment of the individual components, the most important of which are aqueous chemistry conversions, rainout of aqueous phase and particle phase species, modification of the actinic flux, lateral entrainment, and vertical mass distribution of pollutants. Therefore cloud formation as single cause impinges on numerical work load by multiple effects in an accumulative manner. Dependent on the parallelization strategy, the consequence will be the occurrence of severe computational load imbalances or extremely extended and frequent communication. That these and similar effects distorting the parallel efficiency will be accounted for by automated parallel compilers, must not be expected for the next future [8].

Up to now studies on load balancing of parallel atmospheric models are rare, and those with explicit focus on chemistry–transport models not known to the author.

Unlike meteorological models the dynamics part of a CTM is mainly limited to the transport and diffusion modules. Therefore this computational portion which is equally distributed over the integration domain by algorithm can be estimated to be small compared to the work load imposed by the chemistry and cloud modules for which the required computational efforts are dependent on the state of the atmosphere. Load imbalances of limited area AQMs must be expected to be of pronounced spatially and temporally irregular type, especially in the presence of clouds. It is therefore the objective of this study (i) to quantify the specific contributions of dynamics, physics, and chemistry modules to work load imbalances and the subsequent degradation of parallel performance for a comprehensive Eulerian air quality model, and (ii) to introduce and examine a family of robust load balancing schemes to equally distribute the computational work with validity for a wide range of parallel platforms.

The paper is divided into four parts. Section 2 introduces the model and its principal modules which have a major impact on parallel implementation and performance and presents the relative time portions of the modules. In Section 3 the parallelization strategy is described and the load balancing algorithms are presented. Section 4 gives performance results of the different load balancing approaches. The last section presents some conclusions and discusses further issues on load balancing strategies.

2. The chemistry transport model

The EURAD CTM is based on a modified version of the Regional Acid Deposition Model RADM II [4], operating on the mesoscale–α, i.e. an extended synoptic scale with integration domains of the order of 5000 km boundary length. In standard configuration the model domain covers the area from the eastern North Atlantic to the Black Sea and from northern Africa to central Scandinavia. The model's horizontal grid structure is defined by the "Arakawa C" grid stencil, where the locations of the concentrations C_n and the horizontal wind components u and v are at different positions each. The chemistry transport model calculates the transport, diffusion, chemical transformation, dry and wet deposition of about 60 chemical species with 158 reactions. CTM2 is supplemented by a special emission model which provides contributions of anthropogenic and biogenic emission sources with diurnal and weekly variations. Meteorological input data is provided by the mesoscale meteorological model MM5 [1]. The present study is based on a $67 \times 55 \times 15$ grid with a horizontal grid spacing of 50 km. The parallel version of CTM2 is implemented on a massively parallel processor Intel Paragon XP/S as target machine with in total 136 compute nodes.

The model is designed from an Eulerian viewpoint with the principal tendency equation

$$\frac{\partial C_n}{\partial t} + \frac{\partial}{\partial \mathrm{x_i}}(\overline{U_i C_n} + \overline{U_i' C_n'}) = G_n + H_n + A_n + W_n + S_n - V_n, \qquad (1)$$

where C_n is the concentration of the nth species, $\overline{U_i}$ is a mean wind component, U_i' the turbulent part, G_n, H_n, A_n are the contributions due to gas phase, aqueous phase, and aerosol phase chemistry, respectively. S_n denotes the source strength of the emissions, W_n changes due to cloud effects like vertical redistribution and rainout, and V_n the deposition to the surface.

Computational limitations and modularity requirements dictate that the processes are not calculated simultaneoulsy but sequentially by an symmetric operator splitting technique in the following sequence, when stepping from t to $t + \Delta t$

$$C^{t+\Delta t} = (T_h D_h T_z D_z AWHGD_z T_z D_h T_h)^t \circ C^t \qquad (2)$$

Here T, D denote transport and diffusion operators in horizontal (h) resp. vertical (z) direction. In the model version presented here, the dynamic time step Δt of the advection and diffusion operators is 10 minutes.

The assessment of the relative computation times of the individual processes and their variabilities provides elementary features, about which considerations of the subsequent parallelization strategy must pivot. The case study taken for the present paper is a midsummer episode with elevated surface ozone concentrations during the first week of July 1990.

For this case Fig. 1 displays the relative average portions of cpu time during a dynamic time step used for the following principal modules: horizontal advection, vertical advection and diffusion, and a combined chemistry including gas phase, aerosol phase, together with clouds and aqueous phase conversions.

It is a common feature of AQMs that the chemistry fraction of the code requires the highest computational demand relative to other portions. Averaged over the episode CTM2 is spending about 56% of its time running the ODE solver of the gas phase chemistry. The cloud module with the aqueous phase solver requires about 30% of cpu time of a grid column, strongly depending on the extension of the cloud and other state variables. There are also indirect effects of local cloud simulation acting on the gas phase solver's step size adaption, such as altered photolysis rates and additional chemical imbalances, generally enhancing computational demands. The horizontal advection as well as the vertical advection and diffusion both account for 13 and 30% of the CPU time, respectively.

In contrast to the dynamic solvers for advection and diffusion, chemistry time steps of the stiff ordinary differential equation solvers and the cloud module are highly variable in time. This is due to the adaptive time step technique of the algorithm and the variety of branches in the cloud code. Since these parts do not require any communication, the use of massively parallel machines promises to be very efficient. However, the impact of solar radiation change at dawn and dusk or as a consequence of changes in cloud cover and thickness, mixing of polluted and unpolluted air, as well as sources of pollutant emissions enforces the ODE solver to sometimes drastically reduce the time step length on a limited number of mostly neighbouring grid points. This is the principal cause for severe work load imbalances.

Altogether, three major reasons for spatial and temporal variability of the compu-

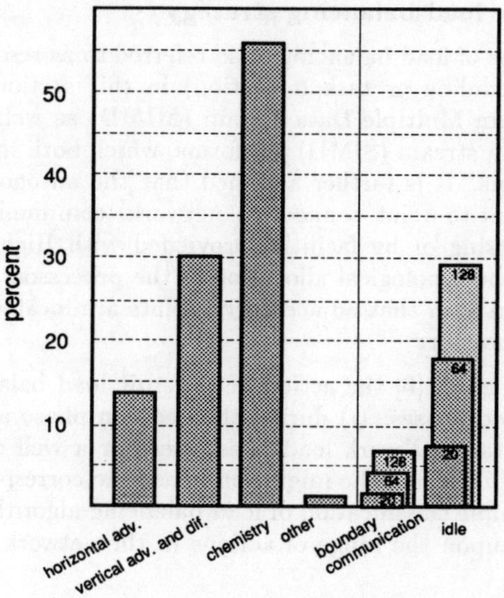

Figure 1: Fractions of CPU usage for the major modules of CTM2. Columns left of the vertical line indicate the seuential partitions and sum up to 100%. Values given for communication and idle times relate to the total of the sequential fractions. Figures indicate the number of processors used.

tational load at a grid point or a grid column can be identified:

(i) A variance of time step length by the chemistry ODE solver due to the diurnal cycle of anthropogenic and biogenic emissions and mixing of polluted and unpolluted air. The regional patterns of areas with enhanced industrial emissions determine the spatial variance. The temporal variances of both man-made and natural emissions is mainly dictated by the diurnal cycle, thereby imposing a computational load variance of a large time scale with respect to the dynamic time step of 10 minutes.

(ii) A drastic decrease of time step length in the chemistry ODE solver due to changing photolysis rates during sunrise and sunset. It takes about three hours for the dawn or dusk area to cross the model domain, which accounts for a sudden increase of time steps at the cloud free grid points of the model domain for a time scale similar to the dynamic time step.

(iii) Finally, variable work load is imposed by cloud effects and aqueous phase chemistry. As in (i) the time scale of the temporal variance is large compared to the dynamic time step.

3. The dynamic load balancing strategy

The considerations of load balancing (also referred to as resource sharing, resource scheduling, job scheduling or task migration) in this section are valid for Multiple Instruction stream Multiple Data stream (MIMD) as well as Single Instruction stream Multiple Data stream (SIMD) platforms, which both imply distributed memory processor systems. It is further assumed that the autonomous processor nodes are linked by a point-to-point connection network, communication is executed by explicit message passing or by facilities provieded with High Performance Fortran Versions, and that the topological allocation of the processors complies with the design of the algorithm such that adjacent gridpoints are located on the same or on a direct neighbour processor.

Following Lüling et al. [6] the action of dynamic load balancing algorithms may be separated into two phases: (i) during the decision phase an optimized redistribution of the computational work load is assessed for a well defined portion of due computations, and (ii) during the migration phase the corresponding data fluxes are accomplished. A simple classification of load balancing algorithms can then be introduced, which relies upon the range of actions in the network during phases (i) and (ii):

1. A local decision base (LD) is applied if the actual work load of a selected processor is compared only with adjacent nodes, whereas a global decision base (GD) is applied if the individual load situations of all processors is considered.
2. A local migration base (LM) is applied if the load units of computational work are carried forward to neighbour processors, whereas a global migration space (GM) allows for the selection of an arbitrary processor of the logically connected network with suitably low work load.

Table 1: Basic classification of dynamic load balancing strategies: Two phases (Decision and Migration) and range of action (Global and Local) can be combined.

Decision base	G	GD LM	GD GM
	L	LD LM	LD GM
		L	G
		Migration base	

Load balancing considerations then focus on the estimation of two break-even points where the benefits of idle time reductions must not be be traded by global communication overhead of
(i) generally small messages for control parameters of work load (decision base),
(ii) and usually larger messages of process data (migration base).
Since global communication overhead increases rapidly with the network size, and

the choice of the pertinent method depends on the number of available processors. Nevertheless, the extent of load imbalances to be levelled, the skill to predict imbalances for each processor, and the properties of the algorithm which dominates the compute costs are further important criteria. It is therefore obvious that global decision or migration operations become increasingly attractive for high arithmetic–to–communication ratios per grid point.

In sequel we restrict our load balancing considerations on Eulerian grid point models, where grids with horizontal grid point numbers of order $\mathcal{O}(NX \cdot NY) = 100^2$ are widely used. NX and NY give the numbers of grid columns in x– and y–direction, respectively. Applying an horizontal grid partitioning approach the network configurations which can be kept busy may not exceed about 1000 compute nodes.

In a time stepping partial differential equation solver as used in the CTM the two principal migration strategies read as follows:

- circular dissemination of due subtasks to idle compute nodes (control decomposition),
- assignment of subdomains of the integration domain to individual processors (grid partitioning).

In the case of control decomposition the principal advantage is the levelling of computational work already during the actual time step. However, the global control of busy and idle processors, the identification of the remaining subtasks to be distributed, the submission of the subtasks and recollection of their results have considerable costs of communication and administrative overhead associated with it. Further to this, today's comprehensive parallel model codes mostly evolved from sequential developments, the data structures of which generally were not designed to accomodate to comfortable data exchange and ensue a comprehensive revision of the code.

In contrast with this, the principal advantage of grid partitioning is based on the general retention of local relations. During a time step the code's coarse grained or high level parallelism given by the chemistry and cloud calculations, can be at best exploited by dividing the integration domain into subdomains which then will be assigned to individual processors. Communication patterns remain local since adjacent areas of the integration domain can be easily mapped on neighbouring processors. Because of its limitation to interface data exchange and the area–to–perimeter effect this method is superior in saving communication volume. Further to this, the single hop required for adjacent processors limits competition in network resources considerably.

Except input/output operations, data dependency such as with finite difference algorithms only occur with next neighbour processors. The overlapping interaction zones (or halo regions) which encompass the individual processor's subdomain are delineated by the order of accuracy of the algorithm. Since the update of the boundary data via next neighbour communication is necessary before the horizontally dependent algorithm is executed, boundary data exchanges take place in preparation of two calls of the horizontal advection and the horizontal diffusion module per time step. This is visualized in Fig. 3, where the required communication between the compute nodes is denoted by arrows. It is indicated that during time stepping data exchange is restricted to neighbours, whereas input and output data dissemination is managed globally.

This exchange process can be generalized to a twofold task for (i) boundary exchange, and (ii) the adjustment of a inner boundary to compensate for load imbalances. In other words mandatory message passing can be used to extend the local processor domain of more idle processors to ease the load of more busy processors. This strategy is implemented and evaluated in the present paper.

The general approach here is a two dimensional horizontal grid partitioning, where a rectangular subdomain is mapped on each processor node while neighbour relations can be maintained. This allows for the engagement of a considerably higher number of nodes than a slabwise, i. e. one dimensional distribution and, on the other hand, avoids the additional communication overhead of three dimensional partitioning. This overhead would be incurred to account for the various algorithmic dependency in modules with special vertical links, such as clouds, radiation,vertical transport and diffusion processes. Therefore, communication is only required for distributing the input data, the data exchange of the inner boundary to the nearest neighbours, and the recollection of output data.

The application of a fourth order advection scheme like the Bott algorithm [2] implies a boundary overlap of two grid points width.

Altogether four horizontal grid partitioning strategies with increasing level of sophistication have been implemented:

(i) Basic grid partitioning following the criterion of a static equal area partition (EAP). By physical processor allocation the two–dimensional local domains are selected as close as possible to a square stencil. The entire integration domain is divided by unbroken inner boundary lines. This is revealed in Fig. 2a with a 8x4 processor arrangement of which every processor occupies approximately the same number of grid points.

(ii) Static load balancing (SLB) relying upon the approximate knowledge of the mean distribution of computational work for each grid column, which may be obtained from a meteorologically similar or a preceeding model run and made available to the model initiation routine. Fig. 2b sketches an area of enhanced computational work by a shaded ellipsis. The common boundaries of the subdomains are adjusted in the way that the computational loads of processor lines and columns covering that area are levelled separately.

(iii) One–dimensional dynamic load balancing (DL1), which assesses the computa-

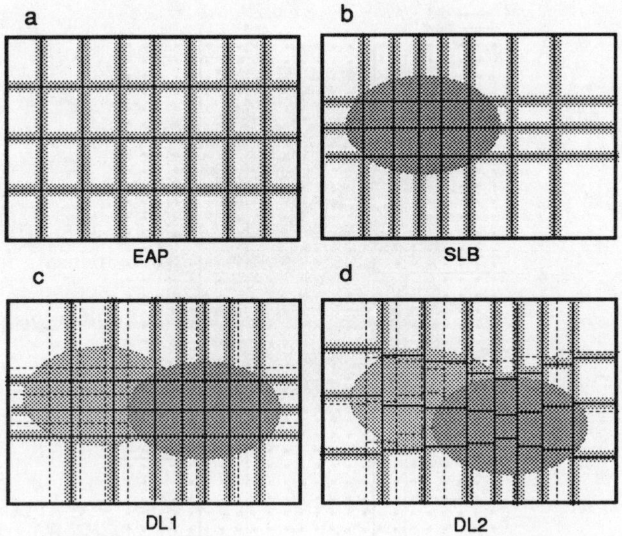

Figure 2: Grid partitioning strategies for balancing the computational load: a) trivial static equal–area partition (EAP), b) static load balancing with common boundary lines (SLB), c) dynamic load balancing with common boundary lines for 2× one–dimensional adjustment (DL1), and d) dynamic load balancing (DL2) with disjunct northern/southern boundaries for two–dimensional adjustment. The shaded areas denote domains with enhanced computational load at different instants where the variable light shading and dashed lines indicate temporal variances.

tional load for each processor column and processor row, as done in (ii), but updates the position of unbroken lines of north–south and east–west boundaries after each time step. This distribution strategy accounts for the temporal and spatial variability of patches with higher computational needs. In Fig. 2c two occurrences of these areas are sketched, distinguished by light and dark stippling to denote their transient appearance. The corresponding processor boundaries are marked by dashed and solid lines. The update of the new boundaries is based on two estimates: 1. The time measurements of the latest time step, i.e. a short term persistence of the work load distribution is assumed, refering slow temporal variances of the computational load. 2. For quickly varying computational loads induced by any external forcing a routine is provided which infers computational load portions by empirical parameters or recognition of regular imbalance patterns. This approach is applied to account for the solar radiation changes during dawn and dusk, when photochemistry calculations enforce small chemical time steps on limited areas. With the present extension of the integration domain, covering nearly entire Europe, the crossing of the terminator line takes about

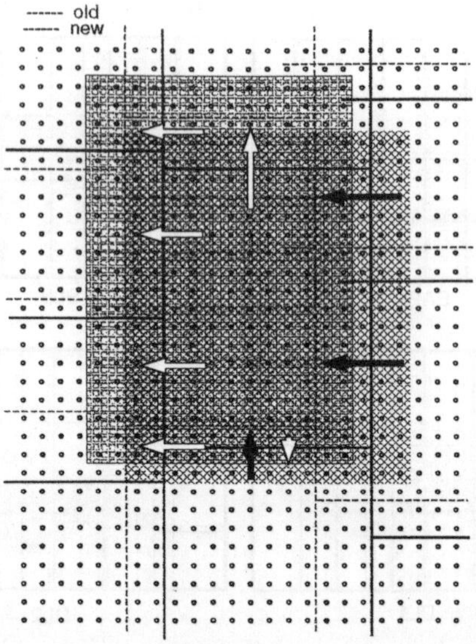

Figure 3: Local adjustment of the boundaries of a processor's subdomain with the two–dimensional dynamic load balancing scheme (DL2). The dashed lines indicate the old inner boundaries with the square–hashed rectangle denoting the associated subdomain including the overlap width of two gridpoints. The solid lines and the cross–hatched rectangle denote the new inner boundaries and extended subdomain after evaluation of the recent computational work load of each grid column. The solid and open arrows depict the required inbound and outbound width of communication volume respectively, where the length indicate the exchange thicknesses, based on an overlap area of two grid points width.

three hours. Neglecting this effect, the parallel performance would therefore be distorted during a quarter of the integration time, when simulating a full day or more.

(iv) Two–dimensional–dynamic load balancing (DL2), which is similar to (iii), but adjusts the local north–south processor boundaries separately in each processor column. By virtue of this strategy the new domain and thereby work load can be individually attributed to each processor. A consequence of this approach is that a processor can generally have more than one eastern or western neighbour. The rectangular shape of the subdomain is the only remaining restriction as depicted in Fig. 2d.

The movement of a processor's subdomain with the DL2 strategy is illustrated with the aid of Fig. 3. The black arrows indicate an inbound message flow opposite to the moving direction. The white arrows correspondingly denote outbound messages to the neighbour processors, which gather additional grid columns at their respective faces. In a balanced case a symmetric inbound and outbound data exchange would be executed as for the static balancing schemes.

4. Parallel performance

The performance runs were executed for each load balancing algorithm and for 2 to 138 processors. The sequential performance of a single processor is assessed by subtracting the parallel overhead and idle times of a two processor run, since the entire grid did not fit into the memory without paging memory. A spurious superlinear speed–up would then bias the results, favouring the parallel performance values. The error introduced by the procedure chosen here is supposed to be very limited and does not invalidate the comparison. To expose the parallel implementation to routine use conditions, the performance runs were applied to the simulation of 48 hours of a summer smog episode.

The parallel performance of the load balancing schemes applied are exhibited in Figs. 4 a, b, c, and for 1 to 128 processors used, where, recall, an individual processor's wallclock time is taken for reference as described above and I/O operation times are filtered out.

The four curves relate to the load balancing strategies outlined above. It is obvious that the static load balancing approach (SLB) is only slightly superior to the simple equal area partitition (EAP). This is already a strong indication that there is a significant temporal variability in load imbalances, which calls for a dynamic approach. As expected, the dynamic scheduling schemes perform better. In the case of the full adjustment scheme (DL2) the best results are obtained. Focussing attention to the 128 processor configurations where performance features are enhanced visible, a reduction of computing time from about 2100 second with the EAP scheme to 1310 seconds with DL2 can be seen in Fig. 4a. These values imply a speed–up of 52 and 94, respectively. Due to dynamic load balancing the efficiency increased from 36% to 69%, which is a reduction of idle times by a factor of about three. The excess percentages for boundary exchange with load balancing messages and idle times of the DL2 scheme is added in the right part of Fig. 1. Values for 20, 64, and 128 processor configurations are presented.

A mean allocation of processor boundaries for the DL2 scheme is shown in Fig. 5. The shading indicates a specific work intensity related to an average grid column for each processor.

It can be seen that most of computationally demanding areas, which are consequently covered by small subdomains, are concentrated over highly industrialized regions. Processor 0 in the south–western position is additionally occupied by administrative overhead.

Comparison of communication costs with the static schemes reveal that the addi-

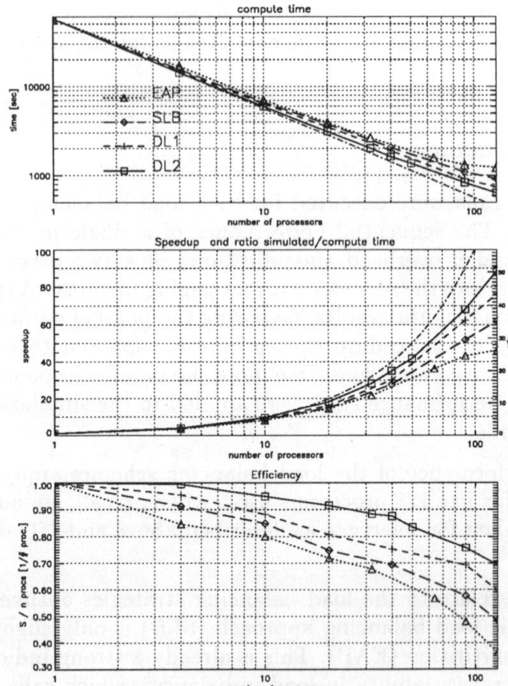

Figure 4: Parallel performances of the load balancing strategies as a function of the number of processors. I/O operation times are removed. The symbols indicate EAP: triangle, dotted line; SLB: diamond, long dashed line; DL1: cross, short dashed line; and DL2: squares, continuous line. The dash–dotted line indicates the position of perfect parallel performance. a) Top panel: model runtime, b) middle panel: speed–up (left ordinate) and ratio simulated–to–compute time (right ordinate), and c) bottom panel: efficiency

tional overhead due to the global decision base of work load and message passes are modest. The relative increase in time is 58% of stationary exchanges in the case of a 128 processor run.

The potential and limits of the implemented procedures to balance an irregular distribution of the computational load can be analysed in more detail. Formaly the work load $w(x_k, y_l, t_j)$ of the grid column at position (x_k, y_l) at time step t_j is decomposed $w(x_k, y_l, t_j) = \overline{w}(x_k, y_l) + w'(x_k, y_l, t_j)$, where the overbar denotes the stationary part during the simulated time span and the prime indicates the transient part.

The differences in coping with imbalances of the partition schemes and their de-

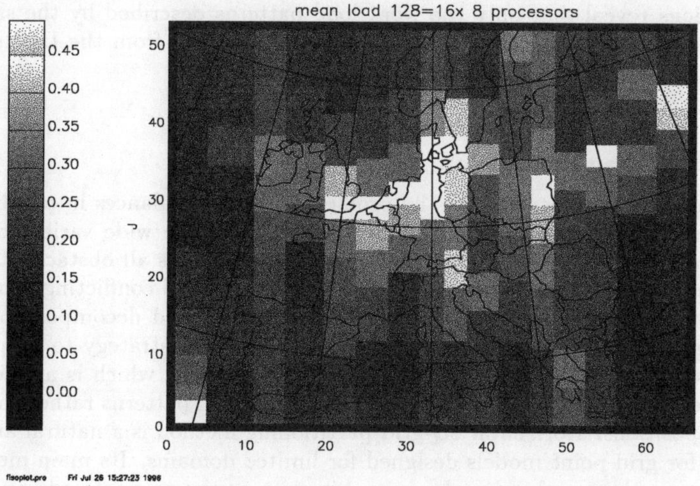

Figure 5: Integration domain and load balanced partitions of a horizontally decomposed model domain. The DL2 load balancing strategy is applied to a $16 \times 8 = 128$ processor configuration. The grey shading code indicates the specific computational load of each processor, normalized with respect to the mean value of the entire model domain. The individual processors' averages of the specific work loads are displayed to aid visibility of the interior boundaries.

pendence on the number of processors can be elucidated by Fourier analysis. An expansion of the work load distribution in terms of a Fourier series may read

$$
\begin{aligned}
w(x_k, y_l, t_j) \approx & \overline{W}_{00} + W'_{00}(t_j) + \sum_{\substack{m=-p_x/2 \\ m\neq 0}}^{p_x/2} (\overline{W}_{m0} + W'_{m0}(t_j)) \exp\left(2\pi i \frac{m\, x_k}{NX}\right) + \\
& \sum_{\substack{n=-p_y/2 \\ n\neq 0}}^{p_y/2} (\overline{W}_{0n} + W'_{0n}(t_j)) \exp\left(2\pi i \frac{n\, y_l}{NY}\right) + \\
& \sum_{\substack{m=-p_x/2 \\ m\neq 0}}^{p_x/2} \sum_{\substack{n=-p_y/2 \\ n\neq 0}}^{p_y/2} (\overline{W}_{mn} + W'_{mn}(t_j)) \exp\left(2\pi i \frac{m\, x_k}{NX}\right) \exp\left(2\pi i \frac{n\, y_l}{NY}\right)
\end{aligned} \qquad (3)
$$

where the truncation values $p_x/2$ and $p_y/2$ are limited by the number of processors p_x and p_y allocated in pseudo–zonal and pseudo–meridional direction, respectively. $W_{mn}(t_j)$ denotes the Fourier coefficient of wave number (m, n), to be decomposed into a stationary and a transient portion. Obviously, the primed terms admit only of dynamic scheduling schemes. For the SLB and DL1 type partitioning scheme, which are bounded to the constraint of mutually positioned processor boundaries, symmetry considerations reveal that only the work load patterns described by the single sum terms in (4) can be balanced. The specific benefits gained from the DL2 scheme is its ability to ibnclude the souble sum term.

5. Summary and discussion

This study mainly focussed on the problem of load imbalances in parallel implementations of chemistry transport models. In view of the wide variety of present parallel architectures, a load balancing scheme which avoids all obstacles to achieve full processor use, cannot be devised. A typical example of conflicting properties is the case of vector processing nodes, where fine grained grid decompositions favour cache usage but sacrifice vector performance. The general strategy to keep communication overhead and work load imbalances under control, which is applied in this study, relies upon the advantages of local communication patterns rather than global data transpositions. Horizontal 2D-grid partitioning method is a natural and robust candidate for grid point models designed for limited domains. Its main merits are a low communication overhead and minimal impact on the original code appearance, when porting to a parallel platform. There is only small interference with the basic modules so that independent developments are not obstructed.

Three major sources for temporal and spatial work load imbalances are identified: (i) variable time step lengths of the gas phase chemistry solver due to the diurnal cycle of emissions, (ii) variable time step length due to rapid changes of the photolysis rates during dawn and dusk, (iii) various processes simulated in clouds. Under these conditions, scheduling schemes are devised which are asserted to compensate for computational imbalances while retaining the advantages of horizontal grid partitioning. In order to avoid global data exchange operations with inevitably high administrative overhead and competition in network resources, strictly local domain adjustment techniques are applied.

The performance of one static and two dynamic load balancing strategies are compared with a grid partitioning approach based on equally partitioned integration areas. The prediction of individual computational loads of the processors for the subsequent time step takes the time measurements of the latest time step. A global decision base was selected to make amends for the performance via updating local boundaries of the processor domains. The data exchange with the adjacent neighbour processors for both exchanging the local boundaries and shifting of the boundaries to a new position with more promise for balanced loads, is executed by the same message passing operation. Thereby the overhead induced by the boundary adjustment is only moderate and the parallel implementation is shown to be efficient and scalable to a large number of processors. A 24 hours integration with 128 processors could be accelerated from 2100 seconds to 1310 seconds which gives an improvement of efficiency from 36% to 69%.

Although considerable improvements could be attained with the load balancing approach presented in this study, there is still a significant potential in reducing idle times. An obvious approach is the invention to a work distribution scheme acting already during the actual time step. However, in this case a frequent control data exchange on the relative work progress must then be managed between the processors. To level excess work which escaped the load balancing methods proposed in this study, a promising strategy could be the usage of the overlap regions as buffer area between immediate neighbours. This may then well serve as a fine grained balancing algorithm being complementary to our more coarse grained scheduling schemes.

Looking at some roots of load imbalances it may be possible that other ODE solvers than the QSSA scheme applied here, like e. g. multistep algorithms have a weaker impact on load imbalances due to a probably less rigorous adaption of the chemistry time step size. Nevertheless it must be expected that future developments in physics and chemistry of cloud modules as well as other operators, like soil models, will further imbalance the computational work severely. To name but a few, the modeling of the following processes are expected to have a major impact: activation, collection, delisquence and resuspension of aerosols, further sophistications of radiative scattering and absorpion parametrizations for clouds and haze, and solubility of aqueous phase chemicals. In the case of earth surface modeling, temporal variations of anthropogenic and biogenic emissions, deposition, and albedo effects may further impinge on the variable distribution of computational requirements.

For the future two major challenges to parallel load balancing management are expected to appear:

1. multiscale air quality modeling, where the computational load of an ensemble of nested grids with different resolutions, possibly generated during runtime, must be balanced, and

2. the coupling of a meteorological driver model, with its own inbalances, together with a CTM.

For both problems the dynamic scheduling schemes of this study are able to serve as a basis to further extensions.

Acknowledgements

The author is greatly indebted to Dr. Wolfgang Joppich, GMD St. Augustin, for fruitfull discussions. This work was supported by the German Federal Ministry of Research and Technology. We are grateful to the Central Institute for Applied Mathematics (ZAM) of the Research Centre Jülich (KFA) for giving computational support.

References

[1] Anthes, R.A., E.-Y. Hsie, Y.-H. Kuo (1987) Description of the Penn State/NCAR Mesoscale Model Version 4 (MM4). NCAR/TN/282+STR.

[2] Bott A (1989) A positive definite advection scheme obtained by nonlinear renormalization of advective fluxes. *Mon. Weather Rev.*, **117**, 1006–1015.

[3] Carmichael G R, Cohen D M, Cho S–Y, and Ogutzun M H (1989) Coupled transport–chemistry calculations on the massively parallel processor computer. *Comput. Ch. Eng.*, **13**, 1065–1073.

[4] Chang J S, Brost R A, Isaksen I S A, Madronich S, Middleton P, Stockwell W R, Walcek C J (1987) A three–dimensional Eulerian acid deposition model: physical concepts and formation. *J. Geophys. Res.*, **92**, 14681–14700.

[5] Dabdub D and Seinfeld S (1994) Air quality modeling on massively parallel computers. *Atmos. Env.*, **28**, 1679–1688.

[6] Lüling R, Monien B, and Ramme F (1991) A study of load balancing algorithms. Prodeedings of the third IEEE SPDP 91, 686–689.

[7] Shin W C and Carmichael G R (1992) Comprehensive air pollution modeling on a multiprocessor system. *Comput. Ch. Eng.*, **16**, 805–815..

[8] Zima H, Chapman B, Pantano M (1995) Compiler technology for scalable parallel architectures – a short overview. In: *Coming of age: Proceedings of the sixth ECMWF workshop on the use of parallel processors in meteorology.* Hoffmann G and Kreitz N, eds. World Scientific, Singapore, 240–249.

[9] Zlatev Z (1995) *Computer treatment of large air pollution models*, pp. 251–264. Kluwer Acacemic Plublishers, Dordrecht.

EUROPEAN NWP APPLICATIONS ON THE CRAY T3E

Deborah Salmond and Brian Bence
Silicon Graphics/Cray Research Ltd, 1530 Arlington Business Park, Theale, Reading, UK

1. Introduction

Massively parallel computer systems built around low cost microprocessors have been expected to dominate the high-end of technical computing since the early 1980's.

Like all new technologies it has taken much longer than many people predicted for the dream to become a reality. With a significant number of the European meteorological institutes acquiring operational MPP systems it is clear that this technology is no longer just a research tool.

This paper highlights the progress being made to use Cray MPP systems for operational weather forecasting in Europe with a brief outline of how the production codes were migrated to use a distributed memory architecture.

The effort required to port a shared memory application to use a distributed memory architecture can be significant, especially if good scaling is required. The amount of effort required to maintain good scaling is dependent on a number of underlying characteristics of the target system. The experience gained working with a number of Cray customers has shown that there are many features of the CRAY T3E architecture which simplify the porting effort and provide good scalability.

One of the main issues which has faced model developers is the choice of programming model for distributed memory systems. To date the production NWP models used on MPP architectures have been modified to use some form of data decomposition to distribute the computation and to use message passing for achieving the necessary data exchange. The effectiveness of this approach in the context of future hardware developments is examined.

2. Operational MPP systems in Europe

ECMWF and five national weather services (FMI, DNMI, SMHI, UKMO and DWD) in Europe have selected an MPP system for running their operational NWP suite. This followed careful benchmarking of MPP systems against the latest shared memory PVP systems using atmospheric NWP models. These choices have implied a significant amount of work to migrate the operational suite from a shared memory to a distributed memory platform.

3. Parallelisation strategies for NWP codes

For the unified model (UM) and Grid-Point HIRLAM the basic parallelisation method uses message passing with a 2-D horizontal domain decomposition with a halo (each processor has its own limited area model). This method gives rise to many synchronisations and short messages per time step and needs an MPP system with a low latency/high bandwidth interconnect between the processors. For good scalability it is essential to ensure that redundant calculations are minimised - for example, in the physics part no calculations are done in the halos. In UM and HIRLAM data is transposed so that FFT's, for filtering at the pole (UM) or the semi implicit part (HIRLAM), are done on a single processor using a distribution technique to ensure good load balance over all processors.

The HIRLAM countries will still run the data assimilation on their shared memory computer. The UKMO have parallelised their data assimilation using the same decomposition as the model and where this is not appropriate - for example, in the calculations of meteorological balance and in the recursive filtering, data has been redistributed and parallelised over levels.

Work on single processor optimisation has shown that for the CRAY T3E a 'vector like' structure for the FORTRAN code gives best performance so the look of the code will remain as it has been for the CRAY C90.

4. Main features of the CRAY T3E architecture

4.1 System interconnect

The CRAY T3E uses a fast low-latency interconnect topology where each processing element has six connections to its neighbours in a three dimensional torus; each link capable of sustaining a payload transfer rate of 480 Mbytes/sec. The latency for a remote memory access is between 2 and 3 microseconds including the software overhead.

The characteristics of this interconnect allow the model developer to parallelise the model in a straightforward manner without having to be too concerned about the amount or frequency with which data is exchanged.

4.2 I/O performance

The system is designed to provide I/O bandwidth to match the computational performance. Every eight processing elements can be connected to a one gigabyte per second I/O channel. These I/O channels are used to support all the different peripheral devices including disks which can operate at full device speed transferring data directly to or from the memory of any one or more of multiple processing elements. The operating system supports disk striping as well as RAID-3 and 5 for fibre channel disks. A transfer rate of 160 Mbytes/second is achievable to fibre channel disks for each controller connected to the system.

The potential of the I/O architecture has still to be fully exploited but the design offers a solution which scales with the system size and matches the computational speed.

4.3 Fast processor

The CRAY T3E was designed to support the microprocessor at a range of clock speeds. The product was initially introduced with a 300 Mhz microprocessor with a peak performance of 600 Mflops/second. The product has recently been enhanced by increasing the clock speed to 450 Mhz with the microprocessor now delivering a peak of 900 Mflops/second.

The higher performance processor is expected to provide at least a 25% performance improvement.

4.4 Single system image

A production NWP system has to offer high availability and maintainability. The CRAY T3E system has been designed to be operated and supported like the traditional Cray shared memory systems. The operating system provides a single system image provided by a single copy of the operating system, with one network address and one common file system. All these features make the system easier to support in an operational environment.

5. Exploiting future architectures

5.1 How we arrived at today's performance levels

In 1976 the CRAY-1 was introduced with a peak performance of 160 Mflops/second delivered by a single processor. Since that time the peak performance of a single processor has only advanced by around one order of magnitude - between 5 to 15 times depending whether it is a commodity microprocessor part or a specialised vector processor.

Today's MPP systems have been designed to support configurations with around 2000 processors giving three orders of magnitude improvement in the past 18 years. So it is clear that the main contribution to the increase in the peak performance of the high-end systems has been achieved by exploiting the parallelism available from multiple processors. This has taken time to achieve effectively as model developers have needed to invest significant effort in adapting codes to a distributed memory architecture.

There is no evidence concerning future microprocessor development to show that there will be a step change in the advancement of the single processor performance, indicating that future performance gains will have to be achieved by even higher levels of parallelism.

5.2 Application/architecture implications

In order to alleviate the single processor performance problem and to provide a more general purpose system solution, future high-end systems will combine multi-processor shared memory nodes connected by a CRAY T3E style interconnect.

This approach improves the general usability of the system for small to medium size jobs that run satisfactorily in the shared memory environment of a single node. It also maintains a programming model compatible with today's MPP systems for the high-end applications that use a distributed memory programming model with message passing.

Future system architectures should continue to provide a solution for those users who wish to advance by running higher resolution models. Users who wish to run problems at the same resolution in a significantly shorter elapsed time will find it harder to achieve their goal.

6. Conclusions

MPP systems have moved from the research laboratory to become production NWP systems. This has only been achieved after significant investment in adapting the production codes to a distributed memory model using message passing.

The characteristics of the CRAY T3E architecture have proved to simplify this task and to provide good scalability.

Future systems will require applications to exploit higher levels of parallelism as the improvement in single processor performance lags behind the continuing requirement for higher performance for NWP applications.

Those institutes that have invested in adapting their models to a distributed memory architecture are well positioned for the future. Despite the claims that easier solutions would be developed no alternative approach yet looks viable.

A DECADE OF PARALLEL METEOROLOGICAL COMPUTING ON THE FLOSOLVER

U N SINHA

Flosolver Unit, National Aerospace Laboratories, Bangalore 560017, India.

and

RAVI S NANJUNDIAH

Centre for Atmospheric & Oceanic Sciences, Indian Institute of Science, Bangalore 560012, India.

Abstract

Meteorological computations have been conducted on the Flosolver series of parallel computers since 1988. The models implemented range from a zonally symmetric model to a spectral weather-forecasting GCM. The parallel models are being used to study the various facets of monsoon variations. Our experience over this period indicates that parallel platforms provide a competitive, cost-effective method of carrying out simulations of global as well as regional atmospheric calculations. Also, when the same basic code is run on different platforms, the differences between outputs (which in isolated areas of the globe may not be entirely negligible) are less due to degrees of precision in computing than due to differences in standards specifying implementation of floating point computations, especially between microprocessor-based and high-power single processor systems.

1 Overview

Meteorological computations are being conducted on Flosolver, the parallel computers developed at the National Aerospace Laboratories, since 1988. An entire spectrum of atmospheric models, ranging from a simple zonally-symmetric climate model to a complex weather forecasting spectral model, have been successfully implemented. As early as 1988 [1], a parallel climate model implemented on the Flosolver was used to study the intraseasonal variations of the Indian Summer Monsoon. This, to the best of our knowledge, is the one of the first few applications of parallel computing to study a climate phenomenon.

The Flosolver project at NAL was started in 1986 with an explicit charter to "*design, develop and fabricate a suitable parallel processing system to solve fluid dynamical and aerodynamical problems*". However, the applications implemented on the Flosolver during the last decade encompass diverse fields e.g. from structural analysis to encryption algorithms. The emphasis has been on the development of *simple, cost-effective, moderately parallel computers* based on off-the-shelf hardware and software.

2 Climate Modelling with Flosolver Mk1 and Mk2

The earliest version of the Flosolver -Mk1, built in 1986, was based on the Intel 8086 microprocessor[2]. The processors were linked to each other by a communication bus - the Multibus I. All communication between processors was routed through a shared (global) memory.

Several parallelized Computational Fluid Dynamics (CFD) codes including the Transonic Small Perturbation equation, the Navier-Stokes equation and a 2-D Monsoon model were operational on this machine at 90-95 percent efficiency.

2.1 *The Flosolver Mk2*

The Flosolver Mk2 was developed in 1988 and was based on the Intel 80386 processor. This version could support upto 16 processing elements in clusters of 4 processors. This version supported two levels of communication: intra-cluster communication using global memory on the Multibus I and inter-cluster communication using a 16 bit parallel interface[3].

The software for implementing interprocessor communication and synchronisation for all versions of the Flosolver have been developed in-house at NAL.

2.2 *The Zonally Symmetric Model*

A zonally symmetric climate model was implemented on these versions of the Flosolver (in 1988)[1]. This was a primitive equation finite-difference model with complete non-linear physics including radiation, surface hydrological effects and cumulus convection. The method of domain decomposition was used for distributing the load amongst processors. Most of the communication was of the 'nearest neighbour' form. However computation of ice-boundaries needed global communication. This requirement prompted us to develop system calls for gather and broadcast. The speedups obtained using upto 16 processors was linear.

The simulations obtained from this parallel climate model were used to study intraseasonal variations of the Inter-Tropical Convergence Zone (ITCZ). The major features of the intraseasonal variations are:

1. Presence of two favoured regions of convection viz. over the heated continent and over the warm tropical ocean.
2. Poleward propagations of the convective cloud-bands with a time-interval of approximately forty days[4] (fig 1).

The existence of this parallel computer facilitated some of the simulations which were then not possible with the severely limited computational power then available inside the country. One such simulation was with this model at a higher resolution of 2°. This simulation established beyond doubt the capability of the model to simulate poleward propagations and showed that the structure of the simulated intraseasonal variations was insensitive to model resolution.

3 Atmospheric Modelling with Flosolver Mk3

3.1 *Flosolver Mk3*

The advent of RISC processors such as the Intel 860 saw a change in the architecture of the new version of Flosolver - the Mk3. While communication was still based on Multibus II, there no longer was a global memory. A dedicated message passing co-processor (linked to its counterpart through the system bus) was linked by a direct memory access (DMA) to the main CPU. The concept of clusters (and linkages between clusters) was also carried over to this version of Flosolver.

3.2 *The Global Climate Model*

The availability of larger computing power enabled us to use larger models. A finite-difference three-dimensional general circulation model (GCM) was implemented on this version of Flosolver [5]. This GCM had a resolution of 4° x 5°, with two levels in the vertical and used the staggered 'Arakawa B-grid' and a combination of predictor-corrector and leap-frog schemes for time integration. The computational load was distributed between processors using the method of domain-decomposition. Decomposition along latitude circles was preferred as the model used Fourier averaging near the poles (to prevent the timestep from becoming very small) and using this method of decomposition reduced communication overheads vis-a-vis decomposing along longitudes. Being a finite-difference model, the communication was mostly of the nearest neighbour form. However, to prevent occurrence of negative moisture due to excessive advection from a point, 'influence' technique was used in the sequential model. This required computations along a longitude circle, which has a two-fold affect:

1. Cause computations to be strictly sequential along a longitude
2. Communication would be required during the conduct of these computations along every longitude.

We re-ordered the loop to conduct these computations along latitudes circles which reduced these overheads considerably. Using four processors on the Flosolver Mk3 a one-day simulation approximately takes about 38 seconds for this model.

4 The Spectral GCM

The latest model to be implemented is the National Centre for Medium Range Weather Forecasts (NCMRWF) T-80 spectral model[6]. This has an approximate resolution of 1.4°x1.4° (128x256 gridpoints on a Gaussian grid). The radiation however is done on a coarser Gaussian grid of 81x162. The computational load has been distributed using the technique of domain decomposition. The domain has been decomposed along latitude circles. Inter-processor communication is required for the completion of summation of the Legendre transform (Fourier-latitude space to wavenumber space). Unlike in a finite-difference model this requires global communication. Presently the communication is conducted on a 'pipeline' pattern with the last processor broadcasting the global summation. A gather-scatter technique is used for the radiation computation and its extrapolation to the forecast grid. With eight processors (seven processors conducting the computations and a host handling dedicated I/O) a one day forecast takes about 69 minutes.

The latest version of the Flosolver - called the SuperSolver - is based on the MIPS-R8000 processor. Unlike previous versions, each processor runs the entire operating system. The communication between processors is through a high-speed switch (designed in-house) with a peak transfer rate of 10 MB/sec. The NCMRWF model has been successfully ported onto this machine. Using 4 processors a one-day forecast takes about 20 minutes.

4.1 *Scalability*

One of the major issues in the use of parallel computers is scalability. In table 1 we present the time taken for a time-step of integration forecast varying processor configuration. As can be observed, efficiency decreases as the number of processors is increased. This is due to the fact that the communication required for completing the summation of spectral co-efficients is global in nature (unlike finite difference techniques which generally require nearest-neighbour communication).

Procs	Time (sec)
1	200
2	109
4	62
6	48

Table 1: Time required (in seconds) for 1 time-step of forecast on Flosolver Mk3 with varying processor configuration

Additionally parts of the code such as horizontal diffusion, initializing of arrays etc are presently being done for the entire set of spectral co-efficients in all the processors. Though in the sequential mode, these parts of the code account for a small percentage of the total time, their share (as percentage of the total CPU time) increases as the number of processors is increased resulting in decreased parallel efficiency. The breakup of communication, parallel computation and sequential computation (i.e. computations for the entire domain on each of the processors) is shown in fig 1. As can be seen the sequential time remains constant but increases as a percentage of total computation time with increasing number of processors. Additionally, the communication time increases in actual terms (this being due to the nature of spectral techniques). The combined effect is to reduce the parallel efficiency.

Perhaps, some parallelization of computations in the spectral domain could increase the parallel efficiency in these parts of the codes. But this has to be seriously weighed against the increased communication overheads.

5 Validation

The process of validating this code on the Flosolver did provide us with certain interesting insights on the behaviour of the GCM on various platforms. Comparing the air-temperature forecasts between Flosolver and Cray (XMP216) we found that the results were in good agreement over most of the domain except at a few points over the Antarctica especially in the middle and upper troposphere.

Considering this to be occurring either due to an error in parallel implementation or due to loss of significance due to the usage of 32 bit arithmetic, we conducted an experiment on the Flosolver using the sequential version of the code but with 64-bit arithmetic (fig 3). To our surprise, the results were still quantitatively similar. The same code was also run on other platforms viz. the IBM-RS6000 (fig 4) and the MIPS-R8000 (fig 5) in the uni-processor mode. Notice that in all these figures, the maximum difference is near the Antarctic. The magnitudes of differences of each of these simulations with that of the Cray are also similar. Another region of significantly large differences is over the Tibetan plateau. One major commanlity between these two regions is that both are regions of high elevation. Comparing the vertical profile of Cray and the Flosolver (fig 6) we notice that the profile for Flosolver has a smoother variation vis-a-vis that of Cray[7].

The above comparisons clearly indicate that most these micro-processor based machines have a similar behaviour which is distinct from that of the Cray.

6 Conclusions and Future Plans

Meteorological computing on the Flosolver series of parallel computer began in 1988 and is still continuing. Parallel Computing has been a useful and cost-effective tool for the atmospheric scientists at CAOS and NAL to conduct studies which would not have been possible otherwise, given the constraints on their resources. Models ranging from a simple zonally symmetric model to a spectral GCM have been successfully implemented and extensively used.

Work is continuing on improving the efficiency of the implemented parallel NCM-RWF GCM. The idea of splitting the summation of the co-efficients and thus reducing the idling/sequential computing time on the processors is being considered. Alternate methods of decomposition such as decomposition in the vertical are also being considered. Using this technique of decomposition, the entire data required for transformations would either be within a processor (in the moderately parallel range of processors, up to 32 processors) or would be with the nearest neighbour. However, the tight coupling in the vertical due to computation of advection, diffusion cumulus convection and radiation needs to be carefully evaluated. Scalability aspects are also being seriously considered.

The NCMRWF forecast model has also been modified to facilitate seasonal and climate mode simulations. The climate version of the model is to be used to study the structure of intraseasonal and interannual variations of the Indian Summer Monsoon and the mechanisms governing them.

7 Acknowledgements

We thank Dr T S Prahlad, Director NAL, for his keen interest and continued support for the Flosolver project. We thank Prof Roddam Narasimha, without whose vision and deep interest in parallel computing and climate modelling, parallel computing at NAL/CAOS would not have been a reality.

We thank NCMRWF for its financial support to the Flosolver project for the parallel implementation of the NCMRWF T-80 GCM and also for their help and support over modelling issues at various stages of this project. We gratefully acknowledge the support and encouragement from our colleagues at CAOS and the Flosolver Unit.

8 References

1. R S Nanjundiah, *Simulation of Monsoon on Flosolver*, **PDAE 8813**,(National Aerospace Laboratories, Bangalore, 1988).

2. U N Sinha *et al.*, *Curr. Sci*, **57** (1988), 1277-1285.

3. U N Sinha *et al.*, *Super Computer*, **VI(4)** (1989), 37-42.

4. R S Nanjundiah *et al.*, *J of Jap. Met. Soc*, **70** (1992), 529-550.

5. R S Nanjundiah and U N Sinha, *Implementation of a Global Climate Model on Flosolver Parallel Computer*, **92AS 4** (Centre for Atmospheric Science, Indian Institute of Science, Bangalore, 1992).

6. U N Sinha *et al.*, *Curr. Sci*, **67** (1994), 178-184.

7. T N Venkatesh *et al.*, *Accuracy of Weather Computations on the Flosolver Mk3.* **FS 9606** *(National Aerospace Laboratories, Bangalore, 1996)*

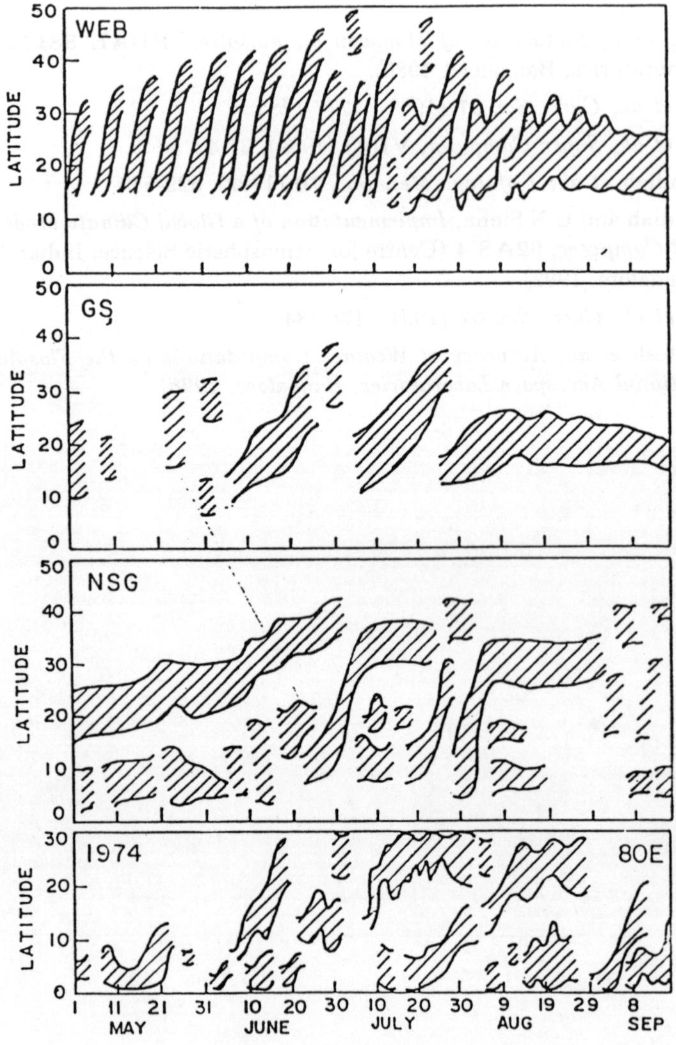

Figure 1: A typical output from the zonally symmetric model, showing the low-frequency oscillation of ITCZ and its comparision with observations[4]

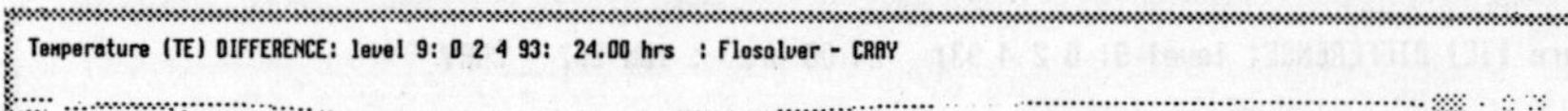

Figure 2: Differences between Flosolver and Cray XMP216 one day forecasts for temperature at level 9

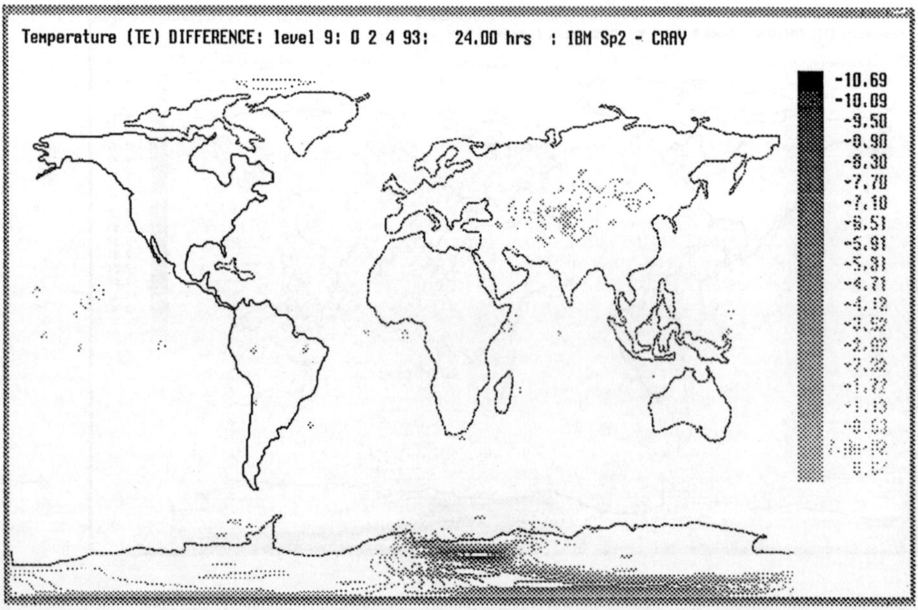

Figure 3: Differences between RS6000 and Cray XMP216 one day forecasts for temperature at level 9

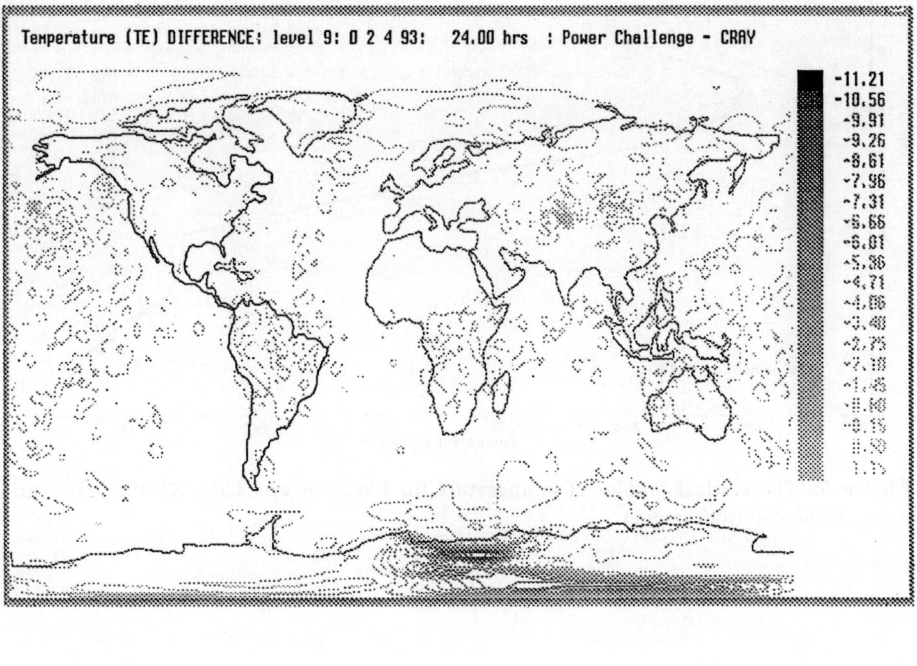

Figure 4: Differences between R8000 and Cray XMP216 one day forecasts for temperature at level 9

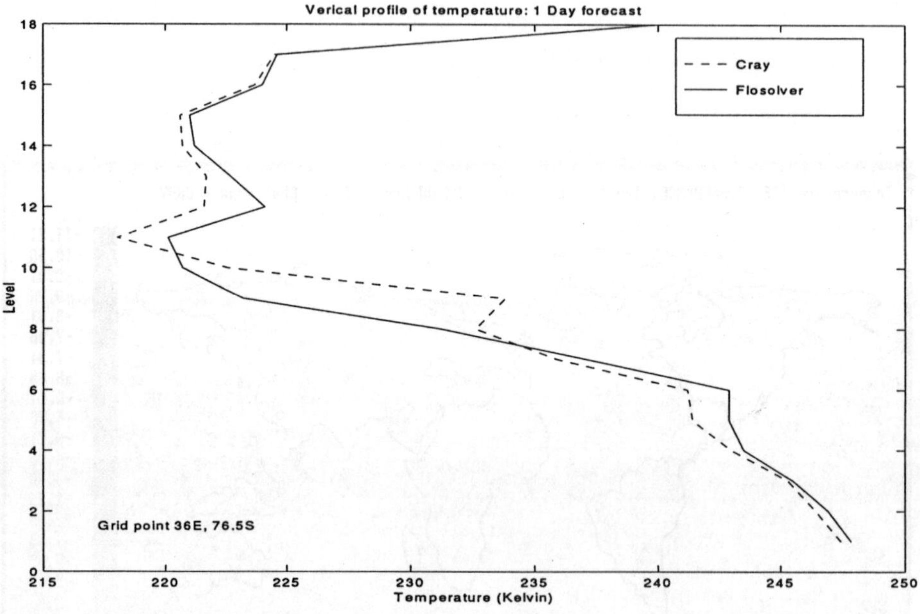

Figure 5: The vertical profiles of temperature for Flosolver and Cray XMP216 at a grid point in the Antarctic region.

Benchmarking of Medium Range Weather Forecasting Model on PARAM - A parallel machine

Akshara Kaginalkar and Sharad Purohit
Centre for Development of Advanced Computing (C-DAC)
Pune University Campus, Pune 4100 007, India.

Abstract

This paper describes the parallel processing efforts for medium range weather forecasting being pursued at the Centre for Development of Advanced Computing (C-DAC), India. The global spectral T80L18 model has been parallelised using latitude wise data decomposition on distributed memory parallel machine PARAM. The performance of parallel T80 for one day forecast on various platforms is discussed. The compute/communication optimisation issues are studied.

1 Introduction

Numerical weather prediction has emerged as one of the important discipline requiring increasing computing power. To have accurate timely forecasts, state-of-art computers are used all over the world. Climate and weather modelers were among the first users of parallel computers. In parallel processing different parts of application code are computed at the same time by different processors working in parallel. There are World wide efforts to port weather forecasting codes on parallel computers. Leading operational centres like ECMWF(European centre of Medium Range Weather Forecasting, UK) and NCEP(National Centre for Environmental Prediction, USA) are experimenting with parallel processing in numerical weather Prediction. This includes parallelisation of the IFS (Integrated Forecasting System) of ECMWF in collaboration with GMD(German National centre for computer Science)[1], NMC's parallel spectral model[2] and CHAMMP(Computer hardware, advanced mathematics and model physics) program[3] sponsored by DOE (Department of Energy, USA) on various supercomputers like CRAY T3D, Intel Paragon, IBM SP2, Meiko etc. In order to contribute in this new research endeavour in high performance computing, Department of Science and Technology, Government of India initiated the experiment of cost-effective computing solution to the weather forecasting models. A major step was taken to quantify the induction of parallel processing in the Indian context. Thus a project to port and parallelise the global prediction model T80L18 of NCMRWF (National Centre for Medium Range Weather Forecasting, New Delhi, India) was started with Centre for Development of Advanced Computing, Pune, India[4].

In this paper we describe the performance of medium range weather forecast code T80 on a distributed memory parallel machine - PARAM. The prime objective is to explore the parallelism for the spectral method in T80 code and to analyse its performance. The T80L18 model has triangular truncation and a spectral resolution of 80 waves. The parallel strategy involves data decomposition across the latitudes of the earth where independent computation for FFT, Legendre transform and grid calculations are done across the latitudes distributed over various processors.

In this paper, reproducibility of results, efficiency, compute/communication optimisation issues in parallelisation are discussed. It is demonstrated that the performance of parallel T80 for one day forecast outperforms the elapsed time for the same on CRAY XMP/216 at NCMRWF. Upgradation of Indian forecasting model to finer resolution codes like T213L31 or T126L28 on the C-DAC's parallel machine with the complete solution consisting of decoder, analysis, forecast and post-processing is proposed.

2 Global Spectral Model

Governing equations for global spectral weather model are derived from the conservation laws of mass, momentum, and energy. Vorticity, divergence, temperature, surface pressure and moisture equations are the main constituents of it[5]. Expansion of the global field is done using spherical harmonics. Finite difference operators are used to approximate the derivatives in the vertical direction and a semi-implicit time integration scheme is applied to the coupled equations of divergence, temperature and surface pressure. This is accomplished by applying a central difference scheme to the time derivative. For the non-linear terms calculation a transform method is used to convert the spectral quantities to the grid-point values in the physical space. They are transformed back to the spectral space after the computation[6,7].

The computations for spectral algorithms are performed in three discrete functional spaces the grid point domain, the Fourier domain, and the spectral domain. Linear terms like horizontal gradient are calculated in spectral space and non-linear terms are evaluated in the physical space. Physics calculation has radiation, surface physics gravity wave drag, kuo convection. Most of the computation is either in spectral or grid space and the variables are transformed between them. The time integration and the vertical integration take place in the spectral domain. Coefficients of expansion called the spectral coefficients are evaluated from the known values of the function in the physical space by Fourier and Legendre transforms.

2.1 The Algorithm

The T80 code has spectral resolution of 80 waves and 128 latitudes and 256 longitudes with 18 varying pressure levels in the vertical direction. Associated Legendre polynomials are evaluated at each latitude. Inner sum in the step 2 is calculated in the Legendre transform. Due to the symmetry of spectral transform, latitudes from northern and southern hemispheres are paired. At different stages of the algorithm the data dependencies are in different directions like FFT calculation is latitude dependent, whereas the Legendre transform depend on longitudes and in spectral space the calculation depends on wave number m.

Each time step of the algorithm consists of the following steps

- step 1: Input spectral coefficients $u_n^m(Z_k)$ for all m, n and k.
- step 2: Compute Fourier coefficients using inverse Legendre's transform.

$$u^m(\mu_j, Z_k) = \sum_{n=|m|}^{M} u_n^m(Z_k) P_n^m(\mu_j)$$

  for all m, j and k. P_n^m is the Legendre polynomial of degree n and order m.

- Step 3: Compute Gaussian grid point values $u(\lambda_l, \mu_j, Z_k)$ using the inverse Fourier transform

$$u(\lambda_l, \mu_j, Z_k) = \sum_{m=-M}^{M} u(\mu_j, Z_k) e^{im\lambda_l}$$

  for all l, j and k.

- step 4: Compute non-linear terms and physics in grid point domain on Gaussian grid. Update $u(\lambda_l, \mu_j, Z_k)$ for all i, j and k.
- step 5: Compute Fourier coefficients $u^m(\mu_j, Z_k)$ using the direct Fourier transform.

$$u^m(\mu_j, Z_k) = \frac{1}{K_2} \sum_{l=0}^{K_2-1} u(\lambda_l, \mu_j, Z_k) e^{-im\lambda_l}$$

- step 6: Compute spectral coefficients by direct Legendre's transform.

$$u_n^m(Z_k) = \sum_{j=1}^{K_1} \omega(\mu_j) u^m(\mu_j, Z_k) P_n^m(\mu_j)$$

  for all m, n and k, where $\omega(\mu_j)$ are the gaussian weights.

- step 7: perform calculations in spectral domain.

where

M	- truncation number and equals to 80
K_0	- number of vertical levels and equals to 18
K_1	- number of latitudes and equals to 128
K_2	- number of longitudes and equals to 256
$0 \leq m, n \leq M$	- spectral indices and their ranges
μ	- Gaussian latitude
$1 \leq j \leq K_1$	- latitude index and its range
λ	- longitude
$0 \leq l \leq K_2 - 1$	- longitude index and its range
Z	- vertical level
$1 \leq k \leq K_0$	- level index and its range

3 PARAM - A parallel machine

The parallel T80 code is ported successfully on various platforms of the PARAM series (Table 1), based on distributed memory parallel architecture. The initial sequential porting of the T80 code was done on single i860 processor with 64 Mbyte as main memory of the PARAM8600[8]. In addition to the compute processor i860, PARAM8600 has another microprocessor called the transputer for communication. Each transputer has links of 20 Mbits/sec speed. A 54 Mbyte /sec high speed bus called DRE(data restructuring engine) between a transputer and an i860 processor. For any i860 node the bandwidth to the external world is 320 Mbits/sec. It has 16 i860 processors and 64 transputers. The original CRAY code had asynchronous I/O related operations which were replaced by the PARAS special calls on PARAM8600. Later for faster execution these calls were modified to standard Fortran arrays using local memory.

Then the parallel T80 code was ported on next machine in the PARAM series the PARAM9000SS, the SUN SuperSparc based machine with 64 Mbyte main memory for each processor. The operating system is SOLARIS 2.5. The architecture is built around an interconnection network(CCP network) based on components complying to the IEEE1355 standard. PARAM9000/AA, has DEC ALPHA processor as compute node and networked by FDDI (Fibre distributed data interface) links. OSF is the operating environment on each of the nodes.

And very recently the parallel T80 on PARAM Openframe has been made available. The PARAM Openframe has SUN UltraSparc processor as main compute en-

Table 1: The PARAM series

Machine Name	Processor
PARAM8600	Intel i860
PARAM9000SS	SUN SuperSparc
PARAM9000AA	DEC ALPHA
PARAM Openframe	SUN UltraSparc

gine. The UltraSparc processor is based on SPARCV9 64-bit RISC architecture with SOLARIS 2.5.1. Fast ethernet with link speed of 100 Mbits/sec as well as Myrinet from Myricom(USA) with link speed of 1.2 Gbits/sec are the communication networks for the 8 Dual CPU SMP (Symmetric Multiprocessor) UltraSparcs. PARAM Openframe is a combination of shared memory and distributed memory architecture.

PARAS is the in-house developed message passing communication software which has a micro kernel running on all the processors of PARAM8600. The communication software for PARAM9000/SS, PARAM9000/AA, and PARAM Openframe is standard message passing interface MPI/PVM.

4 Parallelisation Strategy

The main task before deciding the parallelisation strategy of the code was to identify the nature of parallelism involved in the code and subsequent compute intensive parts and their data dependencies. There are many independent variables namely latitudes, longitudes, vertical levels and spectral indices for data decomposition. The most obvious and clear data independent compute intensive work is across the latitudes.

In this strategy of latitude wise data decomposition, data for latitudes is available with each processor. A pair of latitudes is placed on one processor. In spectral models major computation is done in Fourier and Legendre transform. The transforms are performed twice for physics and dynamics part of the code. In order to minimise the communication traffic, these transforms should be performed on each processor sequentially. With latitude distribution, FFT can be performed independently without any communication on each processor, but this leads to parallel Legendre transform (the most time consuming) where partial sums are performed on each processor. By this method the original algorithmic component of the sequential code remains the same.

Initially, an equal number of latitudes are distributed over the processors and accordingly data decomposition takes place. Each processor computes Fourier coefficients using the inverse Legendre transform and the gaussian grid point values using the inverse Fourier transform. Non-linear terms and physics in grid point domain calculations are done simultaneously on each processor. Then each processor computes Fourier coefficients only for those latitudes which are assigned to it. Partial sum of the Legendre transform is performed on each processor for given latitudes. Then these sums are circulated among all the processors to have global sum on each of the processor.

4.1 The Implementation

This parallelisation strategy was implemented for the T80L18 model. The original sequential code has many CRAY dependent routines written in CRAY assembly language (CAL). The CAL FFT was replaced by the Temperton's[9] Fortran algorithm. The important components of this code are the I/O, the forecast loop(GLOOPA), the radiation loop (GLOOPR) which is called once in twelve hours and the physics loop (GLOOPB), FFT and Legendre transform. GLOOPA and GLOOPB are called once in each time step of 15 minutes for the 24 hours forecast. The data distribution of parallel T80 is along the latitudes distributed over the processors. This is a master - workers model (Figure 1). The initial sigma and surface data is distributed along the latitudes to all the workers by the master. Since the original flow of the code has remained unchanged, similar computation is performed for given latitudes on all the processors. Each worker independently works for the radiation calculation and at the end of it master collects the radiation output from all the workers. Then master interpolates this data from radiation grid to a standard grid and is distributed back to all the workers. The dynamics and physics calculations are done simultaneously on all the workers. Global sum of the partial sums in Legendre transform for both dynamics and physics is the major communication for each time step for one day forecast. At the end the final surface and sigma forecast results are collected by the master from all the workers for zonal diagnostics and for the final output.

In this implementation the algorithmic component of the forecast code remains intact, only the loop indices for latitude computation are changed. Another feature of this implementation is that separate communication library for data movements. Depending on the architecture and the underlying communication software, appropriate communication primitives are to be attached. On PARAM8600, the communication library is PARAS and on PARAM9000/SS, PARAM9000/AA and PARAM Openframe, the communication library is written in generalised portable communication software MPI. Global reduction calls of MPI are used for the global sum calculation in the Legendre transform. Typical communication pattern is shown in (Figure 2). This full Fortran parallel T80 MPI version can be ported across any platform with distributed memory architecture.

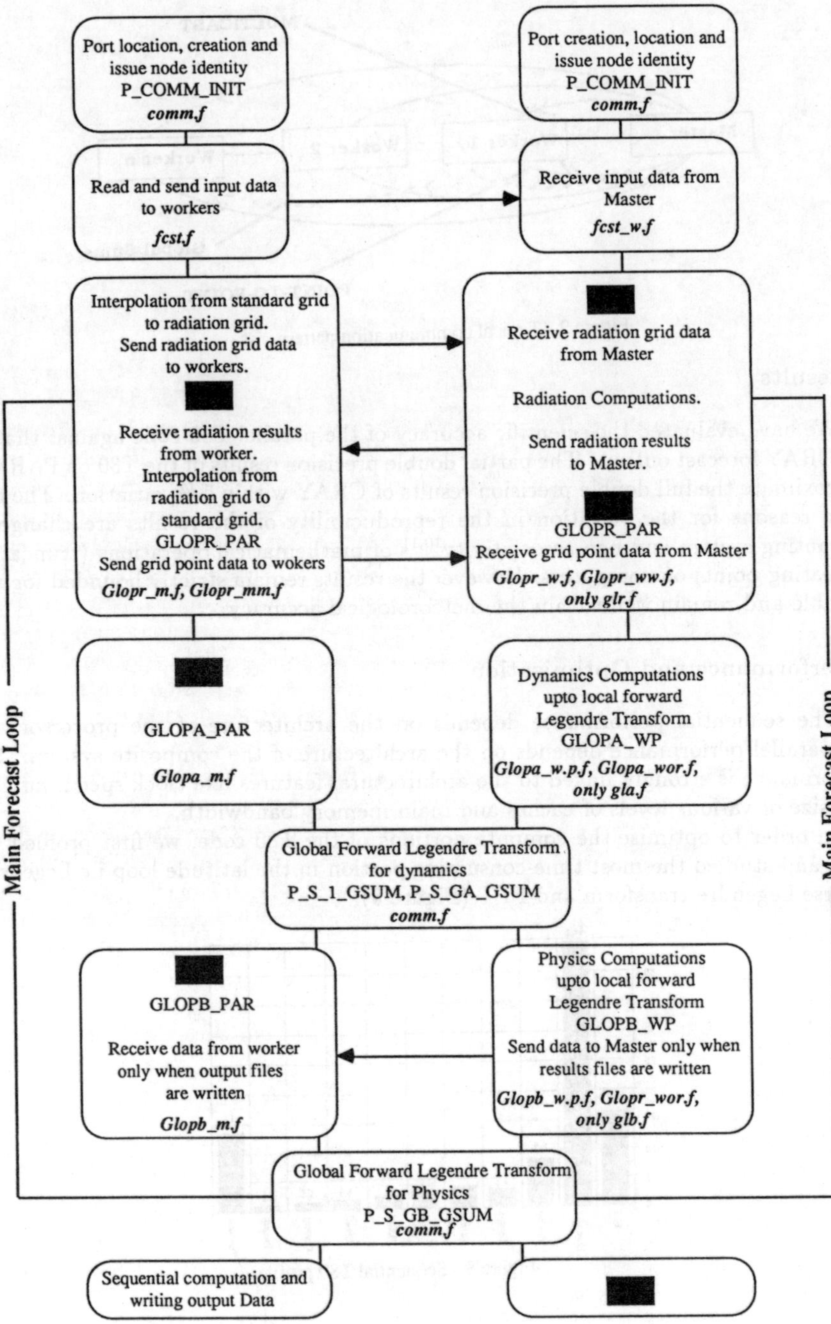

Figure 1 - Parallel T80 flowchart

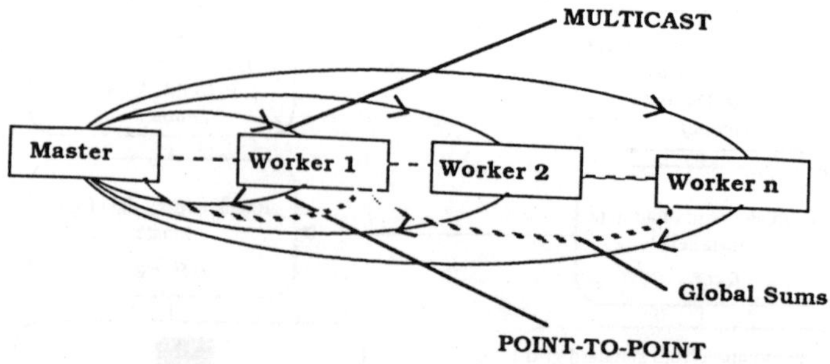

Figure 2 - Typical communication pattern

5 Results

We have evaluated the scientific accuracy of the parallel T80 code against that of the CRAY forecast output. The partial double precision results of the T80 on PARAM approximate the full double precision results of CRAY within 5 % variation. The two main reasons for the variation in the reproducibility of the results are change of computing system and non-associativity[10,11] of mathematical operations (truncation of floating point) on computers. However the results remain strictly bounded for any variable and remain well within the meteorological accuracy.

6 Performance and Optimisation

The sequential performance depends on the architecture of the processor and the parallel performance depends on the architecture of the composite system. The performance is strongly linked to the architectural features like clock speed, number and size of various levels of caches and main memory bandwidth.

In order to optimise the compute portions of the T80 code, we first profiled the code and studied the most time-consuming portion in the latitude loop i.e Legendre, inverse Legendre transform and FFT (Figure 3).

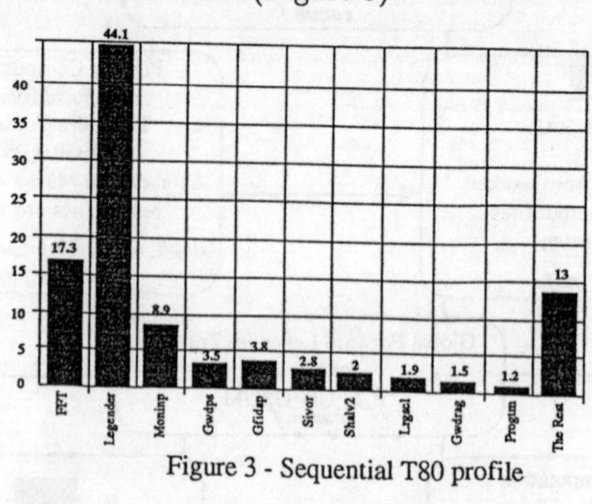

Figure 3 - Sequential T80 profile

Table 2: Sequential Performance of T80 for one day forecast on PARAM8600

Number of nodes	Optimisation	Time(minutes)
1	without optimisation	436
1	with cache utilisation for Legendre transform	280

We tuned the Legendre transform to use on-chip data cache of i860, where the vectors that are reused are retained. We replaced the core time consuming loop i.e matrix vector multiplication of Legendre transform by i860 assembly function 'zxpy'[12], which will use the on-chip cache registers. This is done without changing the Fortran loop structure and considerable time improvement was possible (Table 2).

Another improvement in the overall performance is by optimising the collective operations. Enhancing the conventional tree algorithm used in MPI to hypercube algorithm considering the topology requirement of the application with suitability of the architecture proved to be the best for 8 processor version of parallel T80. As a next level of optimisation specific to the 8 node dual-CPU SMP cluster, a modified hypercube algorithm[13] (Figure 4) is used with local processor on an SMP node communicating using shared memory and remote processes communicating via TCP/IP sockets.

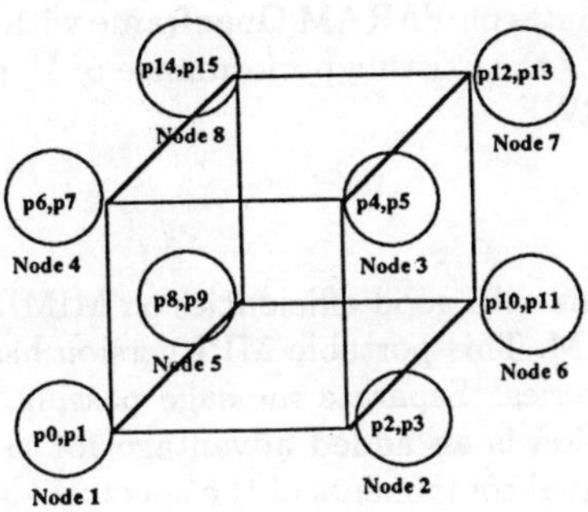

Figure 4 - Extended hypercube topology

We have experimented with different communication networks like fast ethernet, myrinet in the PARAM Openframe for the communication in parallel T80 code. Communication using Myrinet as network is faster than fast ethernet for same number of processors (Table 3).

Table 3: Performance of parallel T80 on various platforms of PARAM series

Platform	Nodes	Remarks	Time (minutes)
PARAM8600	1	Legendre transform optimisation	280
PARAM8600	16	Legendre transform optimisation	60
PARAM9000SS	8	CCP network	36
PARAM9000SS	8	Hypercube topology for communication	27
PARAM9000AA	6	FDDI network	20
PARAM Openframe	8	Fast ethernet	11
PARAM Openframe	8	Myrinet	9
PARAM Openframe	8	Myrinet + global communication optimisation	8
PARAM Openframe	16 (8 2-CPU SMP)	Myrinet + extended hypercube communication	7

Thus the performance of parallel T80 has varied from 280 minutes on single i860 for one day forecast to 7 minutes on PARAM Openframe with 8 dual CPU UltraSparc processors. This outperforms the existing performance of 15 minutes for the same on CRAY-XMP/216 at NCMRWF.

7 Conclusion

The parallel T80 performs with good efficiencies on MIMD type distributed memory supercomputer - PARAM. This portable MPI version has been tested on all the platforms of the PARAM series. Separate message passing library for the communication and data distribution is an added advantage for user friendly full Fortran parallel T80, where the original components of the spectral algorithm are intact. This model is scalable over number of processors.

The partial double precision results of parallel T80 code on PARAM scientifically approximate the full double-precision results of T80 on CRAY-XMP/216. Thus the parallel T80 code is easily portable with no specialised architectural dependent functionalities. Also it is easy for the future developments and maintenance from the software engineering point of view.

Further research is on to exploit fast global communication routines to reduce communication overheads as the number of processors added. Exploration of faster intrinsic functions and better utilisation cache of UltraSparc is in progress.

With the confidence gained by this porting, C-DAC along with NCMRWF has initiated to upgrade the existing weather forecasting codes with the better resolution models like T213L31 or T126L28. The activities to have complete parallel system for weather forecast with decoder, analysis and post-processing are in progress.

8 Acknowledgement

We wish to thank Dr.S.K.Mishra, Director General, National Centre for Medium Range Weather Forecasting(NCMRWF) and Dr. B.K.Basu, Director, NCMRWF, for useful suggestions and help during the successful implementation of this project at C-DAC. The interaction with our colleagues Dr. P.S.Narayanan and Dr. T.V. Singh was very fruitful and is hereby acknowledged.

9 References

1. S.R.M. Barros, et. al., *The IFS model : A parallel production weather code*, Parallel Computing, vol.21(1995), No.10, p. 1621.

2. J.G. Sela, *Weather forecasting on parallel architectures*, Parallel Computing, vol.21(1995), No.10, p. 1639.

3. J.Drake and I.Foster, *Design and performance of a scalable parallel community climate model*, Parallel Computing, vol.21(1995), No.10, p. 1571.

4. NCMRWF/DST Project, *Numerical Weather Prediction on PARAM*,1993

5. *Research version of Medium range forecast model (1988)*, Documentation - vol. 1, Hydrodynamics, physical parametrization and user's guide.

6. W. Bourke, et.al., *Global modelling of Atmospheric flow by spectral methods in computational physics*, vol. 7 : General Circulation modes of Atmosphere, ed. J. Chang. Academic Press, p 267-324.

7. S.C Purohit, P.S.Narayanan, T.V.Singh and A.Kaginalkar, *Development and implementation of parallel numerical weather prediction on PARAM*, progress report, (1994).

8. C-DAC technical reports,SSG-TR-1-93,(1993)

9. C.Temperton, *Selfsorting mixed radix Fast Fourier transform*, Journal of computational physics 52, 1-23,(1983).

10. D.H.Bailey, *RISC Microprocessors and Scientific computing*, Conference proceedings.

11. J.Drake, et. al., *PCCM2: A GCM Adapted for scalable parallel computers*, Math and Computer Sci. Division Argonne National Laboratory, (1993).

12. S.C. Purohit, et. al., *Global spectral medium range weather forecasting model on PARAM*, Supercomputer, Vol. XII, No. 3, (1996), p.27

13. C-DAC technical reports,SSG-TR-1-96, (1996).

Parallel implementation of the YH Limited Area Model on SP2

Jin Zhiyan
National Meteorological Center,
46 Baishiqiao Rd. ,Beijing 100081, P.R.China,
Zaphiris Christidis
IBM T. J. Watson Research Center
Yorktown Heights, NY 10598,U.S.A.

abstract

Parallel computer systems use high speed interconnect networks to link tens to hundreds of powerful RISC processors. The YH limited area weather model has been successfully parallelized and implemented on the IBM SP2 distributed memory scalable parallel system. The model employs explicit finite differencing techniques in time, and second order finite differencing schemes in space. It was parallelized using standard domain partitioning techniques in both horizontal directions. A parallel efficiency of 90% was realized by using 32 SP2 'thin' processors.

1. Introduction

The introduction of RISC processors in numerical weather prediction has resulted in significant research developments on the limited area numerical models[9]. As a direct consequence of the increased computational speed and large available computer memories, these models can cover larger areas with a much smaller grid resolution. The reduction on the grid size, often forces the execution of smaller time steps in order to achieve computational stability[3]. Furthermore, to increase computational accuracy, more sophisticated physical parameterization schemes and numerical techniques are often added. In addressing some of the most memory and computationally demanding weather applications, numerical weather modelers and computational scientists often resort to parallel processing.

In this paper we describe the results of implementing the YH limited area weather model on the SP2 distributed memory parallel system. In the second section we present the architecture of the SP2, while the YH model is described in section 3. Section 4 describes the parallel implementation technique, while results are presented in section 5. Finally we conclude with section 6 were we suggest areas of potential improvement and future research.

2. The SP2 system

The IBM SP2 used in this work is a distributed memory MIMD parallel computer and consists of 2 'wide' and 30 'thin' POWER2 RISC/6000 processor nodes which can communicate via a multistage interconnect network known as the High Performance Switch (HPS)[8]. Each node is packaged in 3 frames, and it has a peak performance of 266MFLOPS. Wide nodes have a larger cache and increased memory bandwidth compared to thin nodes. These nodes can communicate with each other by sending and receiving message packets through the HPS. The HPS provides a low-latency high-bandwidth communications fabric that can sustain point-to-point bandwidth of over 34 megabytes per second (MBS), while latency was measured close to 40 microseconds for a variety of applications. The interconnection network does not impose any particular topology consideration on the scientific application. A parallel program consists of a number of Unix tasks exchanging massages via the HPS. The messages can be encoded using the IBM Message Passing Library[6](MPL), which is a programming interface for FORTRAN, C and C++, and provides with a large number of subroutines supporting task-to-task message passing, as well as collective communications.

3. The YH Model

The YH model[4] is an one-way interactive, nested fine mesh regional numerical model. The model consists of two main parts: dynamics and physics. The dynamics part is based on a system of hydrodynamic equations describing the state of various meteorological parameters like horizontal winds u, v, temperature T, humidity q and the thickness layer π. These equations can be written in the coordinates x, y, σ as[1]:

$$\frac{\partial}{\partial t}\left(\frac{u}{m}\right)-q_v\frac{\pi v}{m}+\frac{\partial}{\partial x}\mathrm{K}+\dot{\sigma}\frac{\partial}{\partial \sigma}\left(\frac{u}{m}\right)=-\left(\frac{\partial \Phi}{\partial x}\right)_p+\frac{F_u}{m} \tag{1}$$

$$\frac{\partial}{\partial t}\left(\frac{v}{m}\right)-q_v\frac{\pi u}{m}+\frac{\partial}{\partial y}\mathrm{K}+\dot{\sigma}\frac{\partial}{\partial \sigma}\left(\frac{v}{m}\right)=-\left(\frac{\partial \Phi}{\partial y}\right)_p+\frac{F_v}{m} \tag{2}$$

$$\frac{\partial}{\partial t}\left(\frac{\pi}{m^2}\right)+\frac{\partial}{\partial x}\left(\frac{\pi u}{m}\right)+\frac{\partial}{\partial y}\left(\frac{\pi v}{m}\right)+\frac{\partial}{\partial \sigma}\left(\frac{\pi\dot{\sigma}}{m^2}\right)=0 \tag{3}$$

$$\frac{\partial}{\partial t}\left(\frac{\pi}{m^2}T\right)+\frac{\partial}{\partial x}\left(\frac{\pi u}{m}T\right)+\frac{\partial}{\partial y}\left(\frac{\pi v}{m}T\right)+\frac{\partial}{\partial \sigma}\left(\frac{\pi\dot{\sigma}}{m^2}T\right)=\frac{\pi Q}{C_p m^2}+\frac{\pi F_T}{m^2}+\frac{\pi\alpha\omega}{m^2 C_p} \tag{4}$$

$$\frac{\partial}{\partial t}\left(\frac{\pi}{m^2}q\right)+\frac{\partial}{\partial x}\left(\frac{\pi u}{m}q\right)+\frac{\partial}{\partial y}\left(\frac{\pi v}{m}q\right)+\frac{\partial}{\partial \sigma}\left(\frac{\pi\dot{\sigma}}{m^2}q\right)=\frac{\pi\overline{C}}{C_p m^2}-\frac{\partial}{\partial \sigma}\left(\frac{\overline{\omega' q'}}{m^2}\right)+\frac{\pi F_Q}{m^2} \tag{5}$$

Eq. (1) and Eq.(2) are the x and y momentum equations, Eq. (3) is the mass continuity equation, while Eq.(4) and Eq.(5) are the thermodynamic energy and the relative humidity equations respectively. The independent variable σ is a non-dimensional independent vertical coordinate, m is the map factor for the Lambert conformal projection in the forecasting region, ξ is the relative vorticity, ω is the vertical velocity in pressure coordinates, while $\dot{\sigma}$ is the vertical velocity in sigma coordinates. The parameters F_u, F_v, F_T and F_q are sinks for momentum, heat and humidity, while Q and $\overline{C}$ are relative heating and cooling rates respectively. The quantity K is proportional to the kinetic energy, Φ is the geopotential, while α is the specific volume. The above equations are solved on an Arakawa[7] C-type staggered grid, which has the property of conserving Potential Enstrophy and Energy.

The physics consist of physical parameterization schemes using horizontal and vertical eddy diffusion coefficients for the representation of turbulence in the Planetary Boundary Layer. Other parameterization schemes include dry and moist adiabatic adjustments, and atmospheric radiant transfer processes. The model includes a variety of ground physics and surface processes. Most of the physical parameterization schemes involve meteorological variables distributed in the vertical direction. The model employs second order numerical differences in space, while time integration is performed by using explicit time differencing (leap-frog), as well as the Euler backward differencing schemes.

4. Parallel Implementation

Since most of the model physical parameterizations occur in the vertical direction, the logical choice was to partition the computational domain in a way to eliminate data communication in the vertical direction, with the intent to keep the ratio of computation to communication high. Thus, a two-dimensional horizontal partitioning scheme was employed for parallelization. In this scheme, the computational domain was divided into as many sub-domains as to the number of available processors, in both east-west and north-south horizontal directions.

The large arrays representing all the independent variables of the problem were decomposed into local arrays, and assigned to each processor. Each local array was assigned an extra boundary row or column on each of the four sides and the data on them resident in the local memory of adjacent processors are supplied via message-passing among processors.

The explicit leap-frog time marching scheme was evenly partitioned among independent processors without undue concern for the size of the grid. All sub-domains were of the same size in order to minimize computational load imbalance.

The model was originally paralleled on transputers.[5] The same code was used with some modifications. These modifications included deleting some message-passing which can be avoided, combining of small message into a big one, changing the Express calls to message-passing library (MPL) data-exchange calls, in order to utilize the bi-directional data traffic capabilities of the HPS switch. No explicit synchronization calls were added to the parallelized code. Synchronization occurs indirectly as a result of the message passing. Whenever a processor gets ahead of its adjacent processors, it has to wait until those processors can compute and send their outer border values. No attempts were made to optimize the source code for the POWER2 architecture, other than the use of proper optimization flags during code compilation.

5. Results

At the beginning of each model simulation, One processor reads the initial and boundary conditions from local data files and distribute it to every processor. Each processor proceeds with the time stepping computations. The results obtained here, exclude timings required to read the data files. Timing is enabled after the initial and boundary conditions were processed. The area the model covered is $5000\times5000km^2$. Three kinds of resolution 80km, 40km and 20km are used. The numbers of grid points are 67×67, 131×131, 256×256 correspondingly. Comparing with the CRAY C90, an average of 26MFLOPS per node was realized during the model runs.

Performance curves for the parallel speedup and efficiency of the model execution on the SP2 are illustrated in Fig.1. and Fig. 2.

The speedup S_P in a parallel run can be defined as:

$$S_P = \frac{T_1}{T_p} \qquad (6)$$

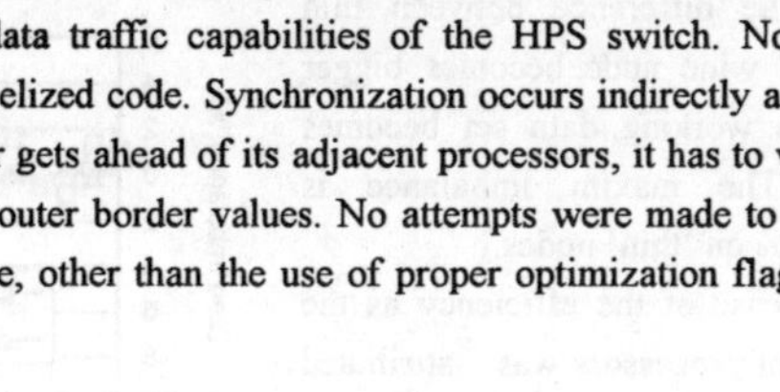

Figure 1: Speedup for the YH model

where T_1 and T_P are elapsed times from model runs on one and P processors respectively. In the ideal situation in which the run is completely parallelizable, the computation is perfectly divisible among the processors. Thus $S_P=P$. This behavior is represented by the ideal case in the Fig.1. In the actual parallel run, the obtained speedup rates deviated from the ideal.

Similarly, the parallel efficiency E_P is defined as:

$$E_p = \frac{T_1}{PT_p} \times 100\% = \frac{S_p}{P} \times 100\% \qquad (7)$$

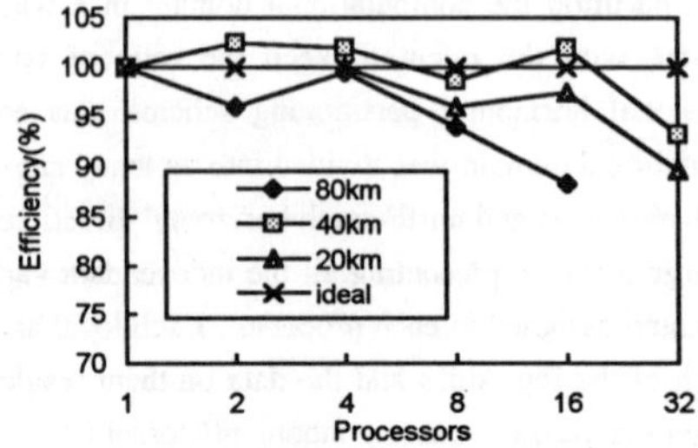

Figure 2: Efficiency of YH model

As it can be seen in Figure 2, efficiencies around 90% were achieved by using 32 processors, while a speedup of 29-30 was obtained for the same number of processors. It can be seen that at some point the parallel efficiency is above 100%, it is attributed to the decrease of the cash missing when the working data set becomes small. It covers some parts of the decrease of efficiency.

Figure 3 shows the load imbalance. The biggest imbalance is on processor 31 and 32, which are 'wide' nodes. The difference between thin node and wide node becomes bigger when the working data set becomes bigger. The maxim imbalance is around 6% on 'thin' nodes.

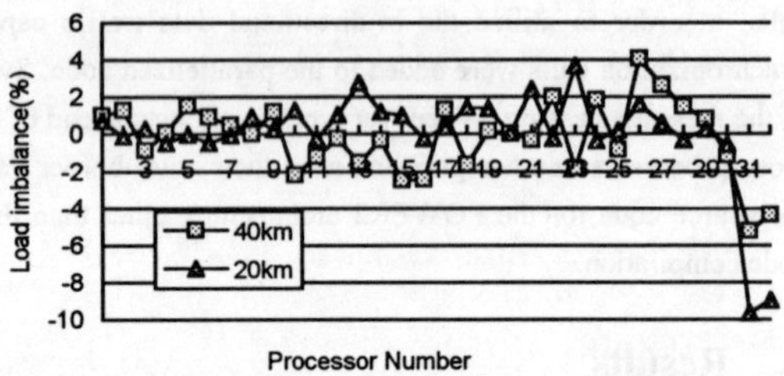

Figure 3: Load imbalance on 32 nodes

The decrease of the efficiency as the increase of processors was attributed to the necessary inter-processor communication for data swapping and the uneven work-load balance among the processors during the physics computations It is expected that the parallel efficiency will be increased if the number of vertical layers is further increased, as it would improve the ratio of computation to communication within each processor.

6. Conclusions

In this paper we have described the implementation of the YH model on the IBM SP2 parallel system. Though only 32 SP2 processors were used in this study, it is expected that the parallel efficiency will be improved significantly on a larger system if the relative granularity is increased by the addition of more vertical levels as well as by the addition of more complex physical processes. We also expect that with proper code tuning for the POWER2 architecture, we can further increase the current code performance on a single node.

7. Acknowledgments

The authors wish to wish acknowledge the support of the IBM China Limited. We would also like to thank Prof. Yan Hong for helpful discussions.

8. Reference

[1] Arakawa A. and Lamb V., Methods in Computational Physics,V, Vol 17, Academic Press, 173-265,1979.

[2] I. Foster and J. Michalakes, Massively Parallel Implementation of the Penn State/NCAR Mesoscale Model, in Proc. 9th Intl. Conf. on Interactive Information and Processing Systems for Meteorology, AMS, 1993.

[3] G. J. Haltiner, and R. T. Williams, Numerical Prediction and Dynamic Meteorology.V, Wiley, 477 pp,1979.

[4] Yan Hong, The Design of a Nested Fine-Mesh Model Over The Complex Topography, Parts One and Two, Plateau Meteorology. Vol 6, No 2, 1987.

[5] Hong Yan, zhiyan Jin , and Xiaoling Ding, Preliminary results of a Limited Area Model (YH Model) on Parallel computing environment, Proc. Fifth ECMWF Workshop on the Use of Parallel Processors in Meteorology: Parallel Supercomputing in Atmospheric Science, eds. G.-R. Hoffmann and T. Kauranne (World Scientific, London 1993).

[6] IBM Corp, IBM AIX Parallel Environment Parallel Programming Subroutine Reference, Second ed., June 1994. SH26-7228-01.

[7] F. Messinger and A. Arakawa, Numerical Methods used in Atmospheric Models", Vol 1 of GAPP Publication series 17. Geneva, Switzerland: WMO 1976.

[8] C. Stunkel, D. Shea, D. Grice, P. Hochschild and M. Tsao, The SP1 High Performance Switch, in Proc. Scalable High-Performance Computing Conf., (Knoxville, Tennessee), pp. 150-157, May 1994.

[9] G. Wightwick, L. Leslie and S. Wail, A Numerical Weather Prediction for the IBM SP2 Parallel Computer. IBM Technical Report IMS11223, 1994.

EXPERIENCE OF USING PARALLEL COMPUTERS IN METEOROLOGY

SUMMARY OF THE DISCUSSION

Norbert Kreitz, Pam Prior

European Centre for Medium-Range Weather Forecasts
Shinfield Park, Reading/Berks., RG2 9AX, United Kingdom

1. Introduction

After a long period of preparation - not least by this series of workshops - parallel machines in one form or another and especially parallel techniques have become rather common in the world meteorological community and are clearly making their mark. The final discussion session of this workshop concentrated on the following topics:

- Algorithms
- Hardware
- Input/Output
- Teraflop computing
- Future Workshops.

2. Algorithms

2.1 Type of Model

Previously, when there had been an enormous imbalance between computational power and available communications bandwidth, grid point models were thought to have the advantage over spectral models. Confirmed spectral modellers were not convinced that this would be the case in practice, and with bandwidth currently more in proportion to computing power, both types of models can run effectively on today's highly parallel machines. However, the growing importance of memory hierarchy may lead to spectral models gaining precedence, since the dynamic calculations in the spectral model are more amenable to memory hierarchy because of Legendre and FFTs, which lead to better space: volume ratios then straight grid point calculations.

Since the analysis is expected to become more expensive than the model, in terms of computational cost, and since increases in the number of observations used increase the time taken for the analysis, greater emphasis must be placed on the analysis in the future. There is a concern that proposed Kalman filter techniques might not be suitable for large number of processors and that new algorithmic solutions might need to be found.

2.2 Type of Grid

The type of grid used is not significant from the point of view of parallelisation but may be dictated by meteorological issues, provided that it has the following properties: it must be decomposable and it should be as homogeneous as possible, as both vector architectures and caches will benefit from unit stride access.

It was noted that grid point models entailed an additional communications overhead, due to interpolation between the various grids, for instance wind grids and temperature level grids; this could give a slight advantage to spectral methods. The problem of coupled ocean/global models using different grids for the ocean and the atmosphere should also be given consideration in the future.

2.3 Languages

There is little change with regard to programming languages being used in practise. HPF is still in the development phase but can definitely not be ruled out as a standard tool for the future. It is clear that simpler, especially regular, grids will ease the generation of more efficient programmes by HPF but the important advantage over Fortran/message passing is that the code will be easier to understand.

2.4 Parallelisation Strategies

The basic parallelisation strategies are, of course, by now well known and especially the transposition strategy is well established. Several years ago the emphasis was more on finding strategies which would provide good performance for the dynamical algorithms, whereas now with more and more physics to be computed the focus has shifted to solving the load imbalances in that area. People are increasingly looking into dynamical solutions of that problem by applying more sophisticated domain decompositions.

Another interesting development is the emergence of ensemble prediction systems which by their very nature can use parallel machines with the right hardware in a very efficient way, but one must not forget that the analysis has to be done as well.

2.5 Multiple Layers of Parallelism

All in all, with regard to the number of processors, one has to acknowledge that to a large extent over the years the interest and development has moved away from truly massive parallel processing (MPP) with many thousands of processors to a rather more modest parallel processing (mPP). Powerful vector units are as important as ever.

Relatively new in the field are the clusters of SMPs. On these architectures one has to run multiple layers of parallelism to exploit these architectures, the SMP paradigm for the node and, for instance, a messaging passing paradigm across the nodes. In principle, the scenario is not new and even a VPP700 can be seen as a machine with many nodes consisting of 9 processors each (1 scalar, plus 8 vector units). At the application programmer's levels one would like just to deal with the parallelism across the nodes only and leave the task to exploit parallelism within the nodes to compilers and operating software. For a small number of processors per node such techniques exist, but for future bigger and even more complex machines the outlook to hide the inner node parallelism is not optimistic. First a consensus on suitable programming models needs to be found and in general it is believed to be a very difficult problem. Nevertheless, SMP clusters are seen to be very useful especially in a heterogeneous user environment where the majority of users run programmes not exploiting the multiple layers of parallelism. However, some important applications like forecasting models do.

2.6 Accuracy

64 bit processing has become more and more the norm. 32 bit accuracy would usually be sufficient but in some areas like the variational analysis or within adjoint methods the higher accuracy is needed. Spectral models with ever higher triangular truncations fall into the same category. This means higher costs for memory and data transfer. As far as the computation itself is concerned considerable savings can be achieved by the use of half-precision intrinsics like SQRTH whenever possible.

3. Hardware

During this workshop Mr. Takamura stated that microprocessor development will naturally lead to vector processing. A time frame of 10-20 years is estimated for this development and convergence of cache based and vector architectures. In the meantime, it has to be realised that one has to program accordingly in order to use the hardware efficiently. Another expected trend is that with more processors and their increased distributed memory the cost of the memory might go up to 90% of the total cost.

The possible role of optical connections in future hardware was raised but obviously this area has not yet been investigated thoroughly enough to make predictions.

4. Input/Output

The neglect of I/O shows an unfortunate consistency in computer history. Parallel machines and parallel processing suffer from the same problem. Therefore, I/O techniques and especially the scalability of I/O should be an important topic for future workshops alongside all the other performance issues.

Naturally, the data volume to be shifted around when running meteorological applications on a parallel machine is substantial. At the moment this means that users have to have detailed and specific knowledge how to organise the I/O part of their application on their machine. Fortunately, it is expected that in future the data volume will not increase proportional to the MFLOPs used. The problem can be eased to some extent by being more selective as to what to store etc., but what is really needed are parallel I/O techniques and especially standards.

MPI-IO seems to be the emerging 'de facto' standard in this field. One distinct disadvantage of it is that it does not allow for data compression which makes it difficult to handle GRIB. Another problem is the expected size of the standard. Organisations and working groups can obviously restrict themselves to a manageable subset but vendors always have to implement the entire standard. This slows the process down and makes optimisation to a large extent impossible. In the interest of both sides, users and vendors, everybody should try to restrict the size of a standard.

5. Teraflop Computing

With teraflop machines on the horizon, if and how they could be used in meteorology needs to be considered. Whilst there is no doubt that teraflop computers will be useful other aspects are far less certain.

First of all, these machines must be affordable. This means costs have to stay more or less where they are now.

Algorithms will have to adapt to higher computing speeds, but no radical change is foreseen. There are ideas though to try new techniques for the parametrisation. So these topics will stay with us for future workshops. Hardwarewise evolution rather than revolution will lead to TFLOP machines. For the next order of magnitude, the petaflop machines, completely new hardware techniques will have to be found. With current architectures message passing is the generally accepted technique of communication. Other shorter and/or simpler ways would be appreciated on future architectures.

Batch has become the predominantly used processing environment with no change expected for the future. However, interactive processing remains important especially for debugging. On current machines, for instance a 128 processor T3D, Totalview works and is manageable, but will it work and will it be feasible on a machine with many more processors like a 1024 processor computer? Scheduling problems are envisaged for debugging and in general for bigger machines and remote batch processing in particular.

Another aspect of a future more powerful but presumably even more complex computing environment has to be made clear: research scientists want and need a simple environment in order to concentrate on their work. It is primarily those responsible for technical design who need the full flexibility and they should be the experts in areas like message passing or I/O. These more technically orientated users would appreciate to know which direction the hardware development will take. That way they could start work on appropriate interfaces for the scientific users.

6. Future Workshops

For future workshops there is a demand for both continuity and for change. The workshop is very much appreciated in its current form focussing on meteorological use of parallel computing. It provides a forum where one can get updated with the latest achievements of other people in this field. Vendor presentations are certainly an essential part as well. Newcomers can get an overview, trends in the development are exposed and as this workshop demonstrated, there are still many open issues. On the other hand, some change might be welcome. With teraflop computing in mind, one could run special sessions on suitable algorithms and parametrization, teraflop hardware, the infrastructure of a teraflop computer centre and the data handling in such an environment. Furthermore, the workshop could branch out into other directions like I/O or more abstract paradigms for programming, i.e. object oriented programming. For purely practical reasons, attendees from outside Europe would prefer one intense one-week workshop rather than several one or two-day workshops on specific issues.

The discussion about the scope and form of future workshops will certainly continue.

LIST OF PARTICIPANTS

Emanuel Anadranistakis	Hellenic National Meteorological Service, P.O.Box 73503, GR16 603 Hellinikon, Athens, Greece
Philippe d'Anfray	ONERA, 29 Av. de la div. Leclerc, B.P. 72, 92322 Chatillon Cedex, France
Mike Ashworth	DKRZ, Bundesstr. 55, 20146 Hamburg, Germany
Toine Beckers	NEC Deutschland GmbH, ESS, Emanuel-Leutze-Str. 4, D-40547 Düsseldorf, Germany
Brian Bence	SGI/Cray, 1530 Arlington Business Park, Theale, Reading/Berks., RG7 4SB, United Kingdom
Jean-Marie Bettems	Schweizerische Meteorologische Anstalt, Krähbühlstr. 58, CH-8044 Zürich, Switzerland
Dag Bjørge	Norwegian Meteorological Institute, P.O.Box 43, Blindern, 0313 Oslo, Norway
Graham Boydell	IBM UK Ltd., 1 New Square, Bedfont Lakes, Feltham/Middx. TW14 8HB, United Kingdom
Mark Bull	University of Manchester, Dept. of Computer Science, Oxford Road, Manchester, M13 9PL, United Kingdom
Paul Burton	Meteorological Office, London Road, Bracknell, Berks., RG12 2SZ, United Kingdom
Paul Caseau	Electricité de France, 2 Rue Louis Murat, 75384 Paris Cedex 08, France
Edmund Chadwick	Hitachi Dublin Laboratory, 0'Reilly Institute, Trinity College, Westland Row, Dublin 2, Ireland
Chin-lian Cheng	Institute For Information Industry, 9 FL., 106, Ho-Ping E. Rd. SEC2, Taipei, Taiwan, China

Zaphiris Christidis — IBM T.J. Watson Research Center, P.O.Box 218, 32-142, Yorktown Heights, N.Y. 10598, USA

Mike Clancy — Fleet Numerical Meteorology and Oceanography Center, 7 Grace Hopper Ave., Stop 1, Monterey, CA 93943, USA

François Courteille — NEC France, 56 Rue des Hautes Patures, 92924 Nanterre Cedex, France

Sushil Kumar Dash — Indian Institute of Technology, Centre for Atmosph. Sciences, Hauz Khas, New Delhi - 110 016, India

John Dennis — National Center for Atmospheric Research, 1850 Table Mesa Dr., Boulder, CO 80303, USA

Mike Desgagné — Environment Canada (AES), RPN, 2121 Trans-Canada Highway, 5th Fl., Dorval (Quebec), Canada H9P 1J3

Alan Dickinson — Meteorological Office, London Road, Bracknell, Berks., RG12 2SZ, United Kingdom

Ben Eddington — University of Reading, Dept. of Meteorology, 2 Earley Gate, Whiteknights, Reading RG6 6AU, United Kingdom

James Edwards — NOAA/FSL, R/E/FS1, 325 Broadway, Boulder, CO. 80303, USA

Kalle Eerola — Finnish Meteorological Institute, P.O.B. 503, 00101 Helsinki, Finland

Hendrik Elbern — Universität zu Köln, Institut f. Geophysik und Meteorologie, Aachener Str. 201-209, 50931 Cologne, Germany

Jean-François Estrade — METEO-FRANCE, SCEM/TTI/DEV/ST, 42 Av. G. Coriolis, 31057 Toulouse, France

Rupert Ford — University of Manchester, Dept. of Computer Science, Oxford Road, Manchester, M13 9PL, United Kingdom

Liliane Frappez	Institut Royal Meteorologique, Centre de Calcul, Av. Circulaire 3, B-1180 Brussels, Belgium
Kazuhisa Fujita	Research Organization for Information Science and Technology (RIST), 2-2-54, Nakameguro, Meguro-ku, Tokyo 153, Japan
Philippe Gire	NEC France, 56 Rue des Hautes Patures, 92924 Nanterre Cedex, France
Volker Gülzow	Deutsches Klimarechenzentrum, Bundesstr. 55, D-20146 Hamburg, Germany
William Hall	National Center for Atmospheric Research, P.O.Box 3000, Boulder, CO 80307-3000, USA
Leslie Hart	NOAA/FLS, R/E/FS5, 325 Broadway, Boulder, CO. 80303, USA
Pierre Hassid	Tera Computer Company - Europe, 5 Villa Alexandrine, 92100 Boulogne, France
Philippe Haye	Fujitsu Systems (Europe) Ltd., 16 Ave. Louis Bleriot, Wissous Cedex, France
Tom Heywood	Wolfson Electronic Business Centre, Dept. of Electronics & Comput. Science, University of Southampton, Highfield, Southampton SO17 1BJ, United Kingdom
Christopher Hill	Massachusetts Institute of Technology, 54-1511, Cambridge, MA 02139, USA
Kenichi Hori	NEC Corporation, 1-10, Nisshincho, Fuchu City, Tokyo 183, Japan
Zhiyan Jin	National Meteorological Center, 46 Baishiqiao Road, Beijing 100081, P.R. China
Jess Jørgensen	Danish Meteorological Institute, Lyngbyvej 100, DK-2100 Copenhagen, Denmark

Philip Jones — Los Alamos National Laboratory, T-3, MS B216, Los Alamos, NM 87545, USA

Wesley Jones — Silicon Graphics Computer Systems, Supercomputer Applications, 2011 N. Shoreline Bldg. MS 580, Mountain View, CA 94043, USA

Hong-wei Jyan — Central Weather Bureau, 64 Kung Yuan Rd. Taipei, Taiwan 100, China

Teruaki Kadokura — Fuji Research Institute Corp., 3-18-1, Kaigan, Minato-ku, Tokyo 108, Japan

Akshara Kaginalkar — Centre for Development of Advanced Computing, Pune University Campus, Ganeshkhind, Pune -411-007, India

Malvin Kalos — Cornell Theory Center, Cornell University, Engineering and Theory Center Bldg., Ithaca, New York 14853-3801, USA

Tuomo Kauranne — University of Joensun, P.O.Box 111, FIN-80101 Joensun, Finland

Nobuhiko Koike — NEC Europe Ltd., C&C Research Laboratories, Rathausallee 10, D-53757 St. Augustin, Germany

Takeshi Kondo — Fujitsu Limited, 9-3, Nakase 1-Chome, Mihama-ku, Chiba-city, Chiba 261, Japan

Elisabeth Krenzien — Deutscher Wetterdienst, Postfach 100465, 63004 Offenbach (Main), Germany

Martin Kücken — Potsdam Institut f'ür Klimafolgenforschung e.V., Postfach 601203, 14412 Potsdam, Germany

Christian Lantwin — NEC European Supercomputer Systems, Emanuel-Levefre-Str. 4, D-40547 Düsseldorf, Germany

Jay Larson — University of Madison, NASA GODDARD, JCESS/UMD, 7501 Forbes Rd., Seabrook, MD, USA

Pierre Leca	ONERA, 29 Av. de la div. Leclerc, B.P. 72, 92322 Chatillon Cedex, France
Magnus Lindskog	The Swedish Meteorological & Hydrological Institute, S-601 76 Norrköping, Sweden
Richard Loft	National Center for Atmospheric Research, 1850 Table Mesa Dr., Boulder, CO 80303, USA
Guy Lonsdale	NEC Europe Ltd., C&C Research Laboratories, Rathausallee 10, D-53757 St. Augustin, Germany
Peter Lyster	UMCP Joint Center for Earth System Science and NASA/GSFC Data Assimilation Office, 7501 Forbes Blvd, Seabrook, MD 20706, USA
Dimitris Maretis	Hofbreede 183, D-49078 Osnabrück, Germany
Michael McAtee	The Aerospace Corporation, DMSP Environmental Applications Center, P.O.B. 92957, Los Angeles, CA 90009, USA
Joe McCaffrey	Naval Research Lab., Code 7300, Stennis Space Center, MS 39529, USA
Ray McGrath	Irish Meteorological Institute, Glasnevin Hill, Dublin 9, Ireland
Michael Metcalf	CERN, 1211 Geneva 23, Switzerland
Nikolaos Missirlis	University of Athens, Department of Informatics, Panepistimiopolis, Athens 15710, Greece
Ajaya Mohanty	IBM, 522 South Rd. Poughkeepsie, N.Y. 12571, USA
Hitoshi Nagatani	High Performance Computing Center, General Purpose Computer Division, Hitachi Ltd., Hitachi Systemplaza Shinkawasaki, 890 Kashimada, Saiwai, Kawasaki Kanagawa 211, Japan
Ravi Nanjundiah	Centre for Atmos. & Oceanic Sciences (CAOS), Indian Institute of Science, Bangalore 560012, India

David Norton	Department of the Navy, 7 Grace Hopper Ave., Stop 2, Monterey, CA 93943-5502, USA
Bob Numrich	Silicon Graphics/Cray Research, 655E Lone Oak Dr., Eagan, MN 55121 USA
Jean Nunn-Price	EPSRC, Polaris House, North Star Avenue, Swindon SN2 1ET, United Kingdom
Enda O'Brien	Digital Equipment Corp., Ballybrit Business Park, Galway, Ireland
Matthew O'Keefe	Laboratory for Computational Science and Engineering, University of Minnesota, 4-174 EE/CSci Bldg., 200 Union St. SE, Minneapolis, MN 55455, USA
Denis Paradis	METEO-FRANCE, SCEM/TTI/DEV, 42 Av. G. Coriolis, 31057 Toulouse, France
Hans Plum	PALLAS GmbH, Hermülheimer Str. 10, D-50321 Brühl, Germany
Ben Ralston	IBM UK Ltd., 1 New Square, Bedfont Lakes, Feltham, Middx., United Kingdom
Jarmo Rantakokko	University of Uppsala, Dept. of Scientific Computing, P.O.Box 120, 75104 Uppsala, Sweden
Adriano Raspanti	Italian Met. Service, ITAV 2° Rep, P. le degli Archivi 34, 00144 Roma, Italy
Catherine Richmond	c/o CGAM, Dept. of Meteorology, University of Reading, P.O.B. 243, Earley Gate, Reading RG6 6BB, United Kingdom
Tom Rosmond	Department of the Navy, 7 Grace Hopper Ave., Stop 2, Monterey, CA 93943-5502, USA
Toshio Saito	FUJITSU Munich Liaison Office, c/o SNI, Otto-Hahn-Ring 6, D-81739 Munich, Germany

Deborah Salmond — SGI/Cray, 1530 Arlington Business Park, Theale, Reading/Berks., RG7 4SB, United Kingdom

Ulrich Schättler — Deutscher Wetterdienst, Postfach 100465, 63004 Offenbach (Main), Germany

Dieter Schiessl — World Meteorological Organization, 41 Ave. G. Motta, 1211 Geneva 2, Switzerland

Eric Schnepf — Siemens-Nixdorf, Scientific Computing, CC Supercomputers, Otto-Hahn-Ring 6, D-81739 München, Germany

Dieter Schröder — Bundesministerium für Verkehr, Postfach 20 01 00, D-53170 Bonn, Germany

Joseph Sela — National Center for Environmental Prediction, WWB, 5200 Auth Rd., Camp Springs, MD 20746, USA

Eric Sevault — METEO-FRANCE, SCEM/PREVI/ COMPAS, 42 Av. G. Coriolis, 31057 Toulouse, France

Peter Silva — Canadian Meteorological Centre, 2121 North Service Road, Trans-Canada-Highway, Suite 300, Dorval (Québec), H9P 1J3 Canada

Uday Narayan Sinha — National Aerospace Laboratories, Flosolver Unit, Post Bag No. 1779, Bangalore 560017, India

David Snelling — University of Manchester, Dept. of Computer Science, Oxford Road, Manchester M13 9PL, United Kingdom

Karl Solchenbach — PALLAS GmbH, Hermülheimer Str. 10, 50321 Brühl, Germany

Lois Steenman-Clark — University of Reading, Dept. of Meteorology, 2 Earley Gate, Whiteknights, Reading RG6 6AU, United Kingdom

Paul Swarztrauber — National Center for Atmospheric Research, 1850 Table Mesa Dr., Boulder, CO 80303, USA

Moriyuki Takamura	Fujitsu Ltd.,High Performance Computing Group, 1-1, Kamikodanaka 4 Chome, Nakahara-ku, Kawasaki 211-88, JAPAN
N. Tamura	Fujitsu Systems (Europe) Ltd., 2 Longwalk Road, Stockley Park, Uxbridge/Middx., UB11 1AB, United Kingdom
John Taylor	Quadrics Supercomputers World Ltd., 2440 The Quadrant, Aztec West, Bristol BS12 4QX, United Kingdom
Nicholas Taylor	AWE PLC, External Training, Bldg. H11.1, Aldermaston, Reading/Berks. RG7 4PR, United Kingdom
Peter Thomas	Fujitsu Systems (Europe) Ltd., 2 Longwalk Road, Stockley Park, Uxbridge/Middx., UB11 1AB, United Kingdom
Stephen Thomas	Recherche en Prévisions Numériques (RPN), Canadian Meteorological Centre, 2121 Trans Canada Highway, N. Service Road, 3rd Floor, Dorval, Québec, Canada H9P 1J3
Yasumasa Tomisawa	Fujitsu Systems (Europe) Ltd., 2 Longwalk Road, Stockley Park, Uxbridge, UB11 1AB, United Kingdom
Joseph-Pierre Toviessi	Canadian Meteorological Centre, 2121 Trans Canada Highway, N. Service Road, 3rd Floor, Dorval, Québec, Canada H9P 1J3
Peter Towers	Fujitsu Systems (Europe) Ltd., 2 Longwalk Road, Stockley Park, Uxbridge/Middx., UB11 1AB, United Kingdom
Yannick Trémolet	National Center for Environmental Prediction, W/NMC2, WWB, Washington DC 20233, USA
Jim Tuccillo	IBM STSS, 415 Loyd Rd., Peachtree City, GA 30269, USA

Keiichiro Uchida	Fujitsu European Centre For Information Technology, 2 Longwalk Road, Stockley Park, Uxbridge UB11 1AB, United Kingdom
Yasuhiko Ueno	Research Organization for Information Science and Technology (RIST), 2-2-54, Nakameguro, Meguro-ku, Tokyo 153, Japan
Christian Ulstad	Norwegian Meteorological Institute, P.O.Box 43, Blindern, 0313 Oslo, Norway
Roger Vanlierde	Institut Royal Météorologique, Avenue Circulaire 3, B-1180 Brussels, Belgium
Reiner Vogelsang	Cray Research GmbH, Bürohaus ATRIUM, Riesstr. 25, D-80992 München, Germany
Hon Gor Wai	Royal Observatory, 134A Nathan Road, Kowloon, Hong Kong
Martin Walker	Digital Equipment Corporation, 12 Ave. des Morgines, CH-1213 Petit Lancy 1, Switzerland
Peter Warn	NEC, c/o Roxel Sarl, 5 Ch. des Chalets, CH-1279 Chavannes-de-Bogis, Switzerland
Dietmar Wierzimok	IBM Deutschland GmbH, STSS Europe, Postfach 103068, 69020 Heidelberg, Germany
Fritz Wollenweber	German Military Geophysical Office, GMGO-WIII3, Mont Royal, 56841 Traben-Trarbach, Germany
Takuma Yamamoto	Fujitsu Ltd., 1-1, Kamikodanaka 4-Chome, Nakahara-ku, Kawasaki 211, Japan
Hong Yan	National Meteorological Center, 46 Baishiqiao Road, Beijing 100081, P.R. China
Hiroyuki Yoshida	Research Organization for Information Science and Technology (RIST), 2-28-8, Honkomagome, Bunkyo-ku, Tokyo 113, Japan

Jose Zero — NASA, Data Assimilation Office, Goddard Space Flight Center, Greenbelt, MD 20771, USA

Hans Zima — University of Vienna, Institute for Software Technology and Parallel Systems, Liechtensteinstraße 22, 1090 Vienna, Austria

ECMWF:

David Burridge	Director
Anton Beljaars	Head, Physical Aspects Section
Horst Böttger	Head, Meteorological Division
Massimo Capaldo	Head, Operations Department
Jens Daabeck	Head, Graphics Group
David Dent	Numerical Aspects Section
Rex Gibson	ERA Project Leader
Mats Hamrud	Numerical Aspects Section
John Hennessy	Head, Meteorol. Applications Section
Claus Hilberg	Head, Computer Operations Section
Geerd-R. Hoffmann	Head, Computer Division
Anthony Hollingsworth	Deputy Director & Head, Research Department
Mariano Hortal	Head, Numerical Aspects Section
Lars Isaksen	Consultant, Research Department
Norbert Kreitz	User Support Section
Andrew Lea	Head, User Support Section
Martin Miller	Head, Model Division
George Mozdzynski	Consultant, Research Department
Tim Palmer	Head, Diagnostic & Predictability Res. Section
Roger Saunders	Head, Satellite Data Section
Adrian Simmons	Head, Model Division
Neil Storer	Head, Systems Software Section
Bernard Strauss	Head, Meteorological Operations Section
Per Undén	Head, Data Assimilation Section
Walter Zwieflhofer	Head, Security, Internal Netw. & Workstat. Section